INTERNATIONAL CENTRE FOR MECHANICAL SCIENCES

COURSES AND LECTURES - No. 301

FINITE ELEMENT AND BOUNDARY ELEMENT TECHNIQUES FROM MATHEMATICAL AND ENGINEERING POINT OF VIEW

EDITED BY

E. STEIN
UNIVERSITÄT HANNOVER

W. WENDLAND
UNIVERSITÄT STUTTGART

Springer-Verlag Wien GmbH

Le spese di stampa di questo volume sono in parte coperte da contributi
del Consiglio Nazionale delle Ricerche.

This volume contains 99 illustrations.

ISBN 978-3-211-82103-9 ISBN 978-3-7091-2826-8 (eBook)
DOI 10.1007/978-3-7091-2826-8

Originally published by CISM, Udine in 1988.

PREFACE

The finite element methods "FEM" and the more recent boundary element methods "BEM" nowadays belong to the most popular numerical procedures in computational mechanics and in many engineering fields. Both methods have their merits and also their restrictions. Therefore, the combination of both methods becomes an improved numerical tool. The development of these methods is closely related to the fast development of modern computers. As a result one can see an impressive growth of FEM and BEM and also a rapid extension of the applicability to more and more complex problems. Nowadays, everywhere people are working on the improvement of the FEM and BEM methods which also requires detailed research of the mathematical and engineering background and of the properties of these methods. Currently, the refinement and formulation of algorithms and also the error analysis are the main fields of mathematical research in FEM and BEM in order to classify reliability and performance. Existing algorithms in engineering are extended and modified and also new engineering problems are tackled with FEM and BEM. Clearly this type of work requires the close interaction of engineers and mathematicians since sometimes even small improvements require rather deep knowledge of modern mathematical analysis and also the physical background as well as significant applications. The combination of mathematics, mechanics and numerical methods makes this field particularly attractive and interesting. However, there are not too many mathematicians and engineers who are willing to cooperate in this field. It was one of the goals of this course to encourage and initiate such a cooperation which requires much patience from both sides.

The aim of the course was to present significant basic formulations of FEM and BEM and to show their common practical and mathematical foundations, their differences as well as possibilities for their combination. These include:

1) Variational foundations in the general framework of complementary variational problems, which cover classical variational problems, variational inequalities as well as mixed and hybrid formulations.

2) A survey on non-linear finite element methods in continuum mechanics and on algorithms for the corresponding large systems of non-linear equations, furthermore the techniques of FEM in non-linear shell theory and in problems with elasto-plastic deformations and with constraints.

3) An introduction into the boundary element mehtods for classical potential theory, the Laplace equation, the treatment of time-dependent problems and the coupling of boundary elements and finite elements in the domain. (This lecture is not in these Notes. See:[Brebbia C.A., Telles J.C.F. and Wrobel L.C. (1984), Boundary Element Techniques, Springer-Verlag Berlin] Chapters 2 and 13 and [Brebbia C.A. and Nardini D. (1986), Solution of Parabolic and hyperbolic time dependent problems using boundary elements, Comp. and Maths. with Appls. 12B, pp. 1061-1072].

4) An introduction to the boundary element method in elastostatics and linear fracture mechanics in problems with body forces and thermal loading.

5) The numerical treatment of corner and crack singularities is still a problem in both methods, FEM and BEM. In particular, the pollution effect, graded mesh refinements, the singularity-subtraction method and the method of dual singularities and iteration procedures are presented. Some of these results are completely new providing significant improvements of FEM as well as BEM, in particular, for the efficient and accurate computation of stress intensity factors.

6) The application of BEM to plastic analysis combines the linear BEM technique with incremental and iteradtive methods involving the updating of field integrals. This method is based on variational BEM formulations leading to new algorithms.

7) The asymptotic error analysis of FEM as well as of BEM can both be based on the Galerkin method for variational problems. This allows also the analysis of some coupled FEM and BEM methods. Some current results are surveyed in this general framework, some of them are new.

The course was an exciting experience, all participants tried hard to develop mutual understanding and cooperation which culminated in a lively panel discussion on urgent open problems such as:
Nonlinear BEM,
Smoothing of singular boundaries,
Time depending problems,
Nonuniqueness of solutions,
Symmetrization in BEM,

Coupling of FEM and BEM,
Fundamental solutions for dynamic problems,
Nonhomogeneous bodies in the BEM,
Numerical integration in BEM.

After the course it took us almost two years to finish these Lecture Notes for two reasons: 1. The authors tried again to close the language gap between mathematics and engineering. 2. FEM and BEM develop so fast that the authors were tempted to include very current developments - a procedure without limits.

The more we are grateful to all the co-authors for their energetic engagement and the completion of these Lecture Notes. We also express our gratitude to all the participants in the course and to all the lecturers for the stimulating and encouraging discussions. We thank the International Center for Mechanical Sciences and its staff for the excellent organization of the course, the lovely surrounding, good working conditions and cordial care. We are grateful to the city of Udine, the "Stiftung Volkswagenwerk" and the "German Research Foundation" for their support and last not least to the Springer-Verlag for this splendid publication.

Erwin Stein and Wolfgang Wendland

CONTENTS

Page

Preface

Complementary Variational Principles
by W. Velte 1

Five Lectures on Nonlinear Finite Element Methods
by E. Stein, D. Bischoff, N. Müller-Hoeppe, W. Wagner, and P. Wriggers 33

Boundary Element Technique in Elastostatics and Linear Fracture Mechanics
by G. Kuhn 109

Numerical Treatment of Corner and Crack Singularities
by H. Blum 171

Plastic Analyis by Boundary Elements
by G. Maier, G. Novati, and U. Perego 213

On Asymptotic Error Estimates for Combined BEM and FEM
by W.L. Wendland 273

COMPLEMENTARY VARIATIONAL PRINCIPLES

W. Velte
Univesität Würzburg, Würzburg, F.R.G.

INTRODUCTION

As is well known, variational methods belong to the fundamental principles in mathematics and in mechanics. They are of interest not only from the theoretical but also from the numerical point of view. Indeed, when discretized the variational principles immediately provide numerical schemes for solving the underlying problem numerically. In particular, variational principles in various forms are videly used in order to establish Ritz - Galerkin schemes in finite element spaces.

In what follows, we will restrict ourselves to problems involving linear elliptic partial differential operators. There is now an abundance of papers on variational principles and finite element schemes, both, for classical boundary value problems and also for problems involving inequality constraints as, for instance, obstacle problems or problems involving unilateral boundary conditions.

In order to establish variational characterizations of the solution under consideration, various techniques have been introduced. It is the aim of these lectures to present, as fare as linear symmetric differential operators are concerned, a unified and straightforward approach to some of these variational characterizations and to exhibit the relations between them. In addition, emphasis is given to a posteriori error estimates in energy norm.

1 LINEAR BOUNDARY VALUE PROBLEMS

In this chapter, we will consider a certain class of linear boundary value problems for elliptic partial differential equations where the solution can be characterized by two different extremal principles, known as dual or complementary extremal principles.

In the applications we have in mind, the first of these two principles will always be the principle of minimal potential energy. Complementary principles, however, are not uniquely determined. As we will see, there is often some freedom to vary the complementary functional in a natural manner.

Pairs of dual or complementary extremal principles have been introduced long time ago: Dual to the principle of minimal potential energy is, for instance, Thomson's principle in electrostatics and Castigliano's principle of complementary work in linear elasticity. (See, for instance, Courant and Hilbert [1], vol.1.) The numerical methods of Ritz and of Trefftz rest upon dual extremal characterizations for the solution of the boundary value problem under consideration.

There are different techniques to establish pairs of dual extremal principles. Widely used are methods of convex analysis, in particular the theory of duality in convex optimization. This approach is very general, but it requires some knowledge in this field. (See, for instance, Ekeland and Temam [3] and the references given there.)

Another approach is due to Noble and Sewell. It applies to linear boundary value problems that can be written in the form $T^*Tu = f$, where T and T^* denote adjoint linear operators defined in appropriate function spaces. In Noble's and Sewell's approach, the problem is reformulated in terms of a Hamiltonian or a Lagrangian system of equations, respectively. (See, for instance, [4], [5], [6], [7] and the references given there.)

Here, we will present a simple and straightforward approach to dual extremal principles starting from the principle of minimal potential energy and the related variational equation (boundary value problem in weak form). This approach is very natural by two reasons: Firstly, the principle of minimal potential energy is fundamental for many problems in mechanics and in physics. Secondly, the related variational equation, which has to be satisfied by the solution, is the basis for a first important numerical method, namely the well known Ritz-Galerkin method. When appropriate finite element spaces are introduced, this approach yields so called conforming finite element methods.

From the variational equation we will construct in a straightforward manner dual functionals and dual extremal principles. Using both, the original and a dual principle, it will be possible to establish a posteriori error estimates for approximate solutions. (See also Velte [8].)

As we will see in chapter 2, the approach can be extended to a certain class of problems involving unilateral constraints, as obstacle problems or problems exhibiting unilateral boundary conditions. (In that case, hovewer, the solution has to satisfy a variational inequality. Nevertheless, there is a similar approach to dual extremal principles.) In this chapter , however, we will consider classical problems without unilateral constraints.

1.1 EXTREMAL FUNCTIONALS AND VARIATIONAL EQUATIONS

In this section, we will consider a certain class of linear boundary value problems for (systems of) elliptic partial differential equations where the solution can be characterized by the principle of minimal potential energy. However, in order to exhibit the essential features we will introduce the class of problems under consideration in the following manner:

By $H, (.,.), \|.\|$ we will denote a given real Hilbert space and by U a closed linear subspace of H. In our examples, H will be an appropriately chosen function space (Sobolev space) in which the problem is well defined. The subspace U of H will then consist of those functions which satisfy certain prescribed linear homogeneous boundary conditions. Since, by assumption, U is a closed linear subspace of H, it is itself a Hilbert space with respect to the scalar product $(.,.)$ and the norm $\|.\|$.

In the function space H we will consider firstly a real valued linear functional $\ell(.)$ and secondly a real valued symmetric bilinear form $a(.,.)$. We suppose that the following assumptions are satisfied: There are positive constants c_0, c_1 and c_2 such that

(A1) $\quad |\ell(u)| \leq c_0\|u\| \quad$ for $u \in H$

(A2) $\quad |a(u,v)| \leq c_1\|u\|\|v\| \quad$ for $u, v \in H$

(A3) $\quad a(u,u) \geq c_2\|u\|^2 \quad$ for $u \in U$

(A4) $\quad a(u,u) \geq 0 \quad$ for $u \in H$.

In other words, the linear functional $\ell(.)$ and the symmetric bilinear form $a(.,.)$ are by assumption bounded and hence continuous with respect to convergence in the norm $\|.\|$. Furtherly, the form $a(.,.)$ is supposed to be nonnegative on H and positive definite on the subspace U.

From (A2) - (A4) we see that the expression $|u|_a = a(u,u)^{1/2}$ represents a norm in U and a semi-norm in H. Finally we introduce the quadratic functional

$$J(u) = a(u,u) - 2\,\ell(u) \quad .$$

$J(.)$ and $|.|_a$ are known as energy functional and energy norm respectively, since in physical and mechanical applications $J(u)$ often represents (besides a factor 1/2) the potential energy of the system under consideration.

Now, we will formulate in terms of the functional $J(.)$ the principle of minimal potential energy. Here, however, the admissible functions are in general subjected to certain constraints. When linear inhomogeneous boundary conditions are prescribed, we may formulate the constraint as follows:

Let $u_o \in H$ denote any particular function satisfying the inhomogeneous boundary conditions. Let U denote the linear subspace of H which consists of all functions satisfying the corresponding homogeneous boundary conditions. Then the set U_{ad} of all admissible functions can be characterized as follows:

$$U_{ad} = u_o + U \quad .$$

In the special case of linear homogeneous boundary conditions we simply define $u_o = 0$ and $U_{ad} = U$.

Under the conditions (A1) - (A4) listed above, we now consider the following two problems:

<u>Problem (P)</u> (Extremal principle):

$$\text{Minimize } J(u) \quad , \quad u \in U_{ad} \quad .$$

<u>Problem (P)'</u> (Variational equation): Find $u \in U_{ad}$ where

$$a(u,\varphi) = \ell(\varphi) \qquad \text{for all } \varphi \in U \quad .$$

We shortly recall the following well known facts: Under the assumptions (A1) - (A4) Problem (P) has a uniquely determined solution. In addition, the two problems (P) and (P)' are equivalent: The minimizing function $\bar{u} \in U_{ad}$ is solution of (P)' and vice versa.

The equivalence is readily established by a standard argument in the calculus of variations: Considering for any fixed

Example 1.1.1

As function space H, $(.,.)$, $\|.\|$ we choose the Sobolev space $H^1(\Omega)$, $(.,.)_1$, $\|.\|_1$. For $u,v \in H^1(\Omega)$ and for a given function $f \in L^2(\Omega)$ we introduce

$$a(u,v) = \int_\Omega \operatorname{grad} u \cdot \operatorname{grad} v \, dx \quad , \quad \ell(u) = \int_\Omega f u \, dx \quad .$$

Clearly, (A1), (A2) and (A4) are satisfied.

Now we impose a boundary condition. Let Γ consist of two parts Γ_1 and Γ_2. On Γ_1 we impose the inhomogeneous boundary condition $u = g$. Here, we assume that g are the boundary values of an appropriate function $u_o \in H^1(\Omega)$.

As above we introduce $U_{ad} = u_o + U$, where

$$U = \{ u \in H^1(\Omega) \mid u = 0 \text{ on } \Gamma_1 \} \quad .$$

Clearly, the energy norm $|.|_a$ is only a semi-norm in $H^1(\Omega)$ since $|u|_a = 0$ for all constant functions. In U , however, $|.|_a$ is a norm and (A3) is satisfied. (This follows, for instance, from [9], Theorem 1.9. p. 20.)

Thus, the extremal problem (P) is well defined, and the equivalent variational equation (P)' is given by

$$u \in U_{ad} : \int_\Omega \operatorname{grad} u \cdot \operatorname{grad} \varphi \, dx = \int_\Omega f \varphi \, dx \quad \text{for all } \varphi \in U \, .$$

Integrating by parts we find, at least formally,

$$- \int_\Omega \Delta u \varphi \, dx + \int_{\Gamma_2} \frac{\partial u}{\partial n} \varphi \, ds = \int_\Omega f \varphi \, dx \quad \text{for all } \varphi \in U \, .$$

Here we have used $\varphi = 0$ on Γ_1. Δ denotes Laplace's operator. Since the equation must hold for any $\varphi \in U$, we conclude by a standard argument well known in the calculus of variations that

$$- \Delta u = f \text{ in } \Omega \, , \quad \frac{\partial u}{\partial n} = 0 \quad \text{on } \Gamma_2 \quad .$$

$\varphi \in U$ the family of admissible functions $\bar{u} + t\varphi$, $t \in \mathbb{R}$, we clearly have

$$J(\bar{u} + t\varphi) \geq J(\bar{u}) \qquad \text{for all} \quad t \in \mathbb{R} .$$

The minimum is attained for $t = 0$. It follows that $dJ(\bar{u} + t\varphi)/dt = 0$ at $t = 0$, which is equivalent with the variational equation (P)'.

1.2 EXAMPLES

In the examples given below, Ω will denote a bounded region in $\mathbb{R}^2$ or $\mathbb{R}^3$. By $H^k(\Omega)$ and $\overset{\circ}{H}{}^k(\Omega)$ we denote the usual (real) Sobolev spaces where, in our examples, $k = 1$ or $k = 2$. $H^1(\Omega)$, for instance, consists of all square integrable functions $u \in L^2(\Omega)$ having (at least in the weak sense) first derivatives in $L^2(\Omega)$. $H^1(\Omega)$ is a Hilbert space with respect to

$$(u,v)_1 = \int_\Omega \left(uv + \frac{\partial u}{\partial x_j}\frac{\partial v}{\partial x_j}\right) dx \quad , \quad \|u\|_1 = (u,u)_1^{1/2}$$

(summation over repeated subscripts).

The boundary Γ is supposed to be sufficiently smooth. That means in particular for functions $u \in H^1(\Omega)$, that boundary values of u at Γ are well defined, and that the following formula for integration by parts is valid:

$$\int_\Omega \operatorname{div} \underset{\sim}{v}\, \varphi \, dx = - \int_\Omega \underset{\sim}{v} \cdot \operatorname{grad} \varphi \, dx + \int_\Gamma (\underset{\sim}{v} \cdot \underset{\sim}{n})\, \varphi \, ds \quad .$$

Here, dx and ds denote volume and surface element, respectively. $\underset{\sim}{n}$ denotes the outward unit normal to the boundary Γ.

As is well known, $\overset{\circ}{H}{}^1(\Omega) = \{u \in H^1(\Omega) \mid u = 0 \text{ at } \Gamma\}$. The space $H^{1/2}(\Gamma)$ consists by definition of all functions $g \in L^2(\Gamma)$ that are boundary values of functions $u \in H^1(\Omega)$. For details we refer, for instance, to Nečas [9].

Hence, the variational equation (P)' is a weak form of the classical boundary value problem

$$-\Delta u = f \quad \text{in } \Omega , \quad u = g \quad \text{on } \Gamma_1 , \frac{\partial u}{\partial n} = 0 \quad \text{on } \Gamma_2 \quad .$$

Note, that the admissible functions $u \in U_{ad}$ have, by definition, only to satisfy the boundary condition on Γ_1 . The second boundary condition on Γ_2 is a consequence of the variational equation (P)'.

Generally speaking, boundary conditions which occur on the definition of U_{ad} are called essential boundary conditions. When it turns out that the solution of the variational equation (P)' satisfies automatically additional boundary conditions, these are called natural boundary conditions. (See als Courant-Hilbert [1] vol. 1.)

In the example above, $u = g$ on Γ_1 is the essential boundary condition and $\partial u/\partial n = 0$ on Γ_2 is the natural boundary condition.

Example 1.1.2

As function space we choose the Sobolev space $H^2(\Omega)$, $(.,.)_2$, $\|.\|_2$ where now in the norm $\|.\|_2$ the derivatives up to the second order occur. For $u,v \in H^2(\Omega)$ and for a given function $f \in L^2(\Omega)$ we introduce

$$a(u,v) = \int_\Omega \Delta u \, \Delta v \, dx \quad , \quad \ell(u) = \int_\Omega f u \, dx \quad .$$

Clearly, (A1), (A2) and (A4) are satisfied.

This time we impose two homogeneous boundary conditions, namely

$$u = 0 \quad , \quad \frac{\partial u}{\partial n} = 0 \quad \text{on } \Gamma \quad .$$

The set U_{ad} of admissible functions is given by $U_{ad} = U$ where

$$U = \overset{\circ}{H}{}^2 = \{ u \in H^2(\Omega) \mid u = \frac{\partial u}{\partial n} = 0 \quad \text{on} \quad \Gamma \} .$$

As is well known, the norm $\|\cdot\|_2$ and the energy norm $|\cdot|_a$ are equivalent norms in $\overset{\circ}{H}{}^2(\Omega)$ (see, for instance, [9]). Hence assumption (A3) is satisfied.

Thus the extremal problem (P) is well defined, and the equivalent variational equation (P)' is given by

$$\int_\Omega \Delta u \, \Delta\varphi \, dx = \int_\Omega f\varphi \, dx \quad \text{for all} \quad \varphi \in \overset{\circ}{H}{}^2(\Omega) .$$

Integrating by parts we find, at least formally,

$$\int_\Omega \Delta\Delta u \varphi \, dx = \int_\Omega f\varphi \, dx \quad \text{for all} \quad \varphi \in \overset{\circ}{H}{}^2(\Omega) ,$$

whence $\Delta\Delta u = f$ in Ω without additional boundary conditions. Hence the variational equation (P)' is a weak form of the classical boundary value problem

$$\Delta\Delta u = f \quad \text{in} \quad \Omega \quad , \quad u = \frac{\partial u}{\partial n} = 0 \quad \text{on} \quad \Gamma .$$

In this example both boundary conditions are essential conditions. There is no natural boundary condition.

1.2 COMPLEMENTARY EXTREMAL PRINCIPLES

In what follows we will consider linear boundary value problems which are related to an extremal principle (P) satisfying the assumptions (A1) - (A4). We will start from the variational formulation (P)' of the problem under consideration, and we will show how in a simple and straightforward manner different extremal principles can be obtained which are in a certain sense complementary to the original extremal principle (P).

As we will see, those pairs of complementary extremal principles will also provide error estimates for approximate solutions.

Adopting the notations of section 1.1, let us recall the formulation of problem (P) and of problem (P)':

Problem (P): Minimize $J(u) = a(u,u) - 2\,\ell(u)$, $u \in U_{ad}$.

Problem (P)': Find $u \in U_{ad} = u_o + U$ such that

$$a(u,\varphi) = \ell(\varphi) \qquad \text{for all} \quad \varphi \in U \; .$$

Now we drop in problem (P)' the condition $u \in U_{ad}$ and consider all solutions $v \in H$. We define

$$V_{ad} = \{\, v \in H \mid a(v,\varphi) = \ell(\varphi) \quad \text{for all} \quad \varphi \in U \,\} \; .$$

Clearly, the solution $\bar{u}$ of (P) and (P)' is contained in V_{ad}. More precisely, we have $U_{ad} \cap V_{ad} = \{\bar{u}\}$.

In terms of the classical formulation of the boundary value problems under consideration, the functions $u \in U_{ad}$ have to satisfy the essential boundary conditions, whereas the functions $v \in V_{ad}$ have to satisfy the differential equation together with the natural boundary conditions (if there are any).

We now define a complementary energy functional $J_c(.)$ by

$$J_c(v) = -a(v,v) + 2a(v,u_o) - 2\,\ell(u_o) \; , \qquad v \in H \; .$$

In the case of homogeneous boundary conditions $(u_o = 0)$ the functional reduces to $J_c(v) = -a(v,v)$.

<u>Theorem 1.2.1.</u> The functionals $J(.)$ and $J_c(.)$ are complementary in the following sense:

$$J_c(v) \le J(\bar{u}) \le J(u) \qquad \text{for all} \quad u \in U_{ad} \, , \; v \in V_{ad} \; .$$

Equality holds for $v = \bar{u}$ and for $u = \bar{u}$, respectively. Furtherly, when u and v are considered as approximations to $\bar{u}$, then the following a posteriori error estimate holds:

$$|u-\bar{u}|_a^2 + |v-\bar{u}|_a^2 = |u-v|_a^2 = J(u) - J_c(v) .$$

Proof. By (A4) and by the definition of U_{ad} and V_{ad} we find for $\varphi = u - u_o$

$$0 \leq |u-v|_a^2 = |u|_a^2 + |v|_a^2 - 2a(v,u-u_o) - 2a(v,u_o)$$

$$= J(u) - J_c(v) .$$

On the other hand, introducing $\varphi = u - \bar{u}$ in the variational equality, we find $a(v-\bar{u},u-\bar{u}) = 0$ and hence

$$|v-u|_a^2 = |(v-\bar{u})-(u-\bar{u})|_a^2 = |v-\bar{u}|_a^2 + |u-\bar{u}|_a^2 .$$

Theorem 1.2.1 applies also to problems where linear constraints in terms of differential equations occur. We give an example:

Example 1.2.1

In a bounded region $\Omega \subset \mathbb{R}^3$ we consider vector fields $\underset{\sim}{u} = (u_1,u_2,u_3)$ where $u_j \in H^1(\Omega)$. Then, for a given function $\underset{\sim}{f} = (f_1,f_2,f_3)$ where $f_j \in L^2(\Omega)$ we define

$$a(\underset{\sim}{u},\underset{\sim}{v}) = \int_\Omega u_{j,k}\, v_{j,k}\, dx \quad , \quad \ell(\underset{\sim}{u}) = \int_\Omega f_j u_j\, dx$$

($u_{j,k} = \partial u_j/\partial x_k$, summation over repeated subscripts).

Furtherly we define

$$U_{ad} = U = \{ \underset{\sim}{u} \in \overset{o}{H}{}^1(\Omega)^3 \mid \operatorname{div} \underset{\sim}{u} = 0 \text{ in } \Omega \} .$$

Then the variational equation (P)' is a weak form of the STOKES boundary value problem for velocity $\underset{\sim}{u}$ and pressure p

$$-\Delta \underset{\sim}{u} + \operatorname{grad} p = \underset{\sim}{f} \quad \text{in } \Omega$$

$$\operatorname{div} \underset{\sim}{u} = 0 \quad \text{in } \Omega \quad , \quad \underset{\sim}{u} = 0 \text{ on } \Gamma .$$

Now, we drop the condition $u \in U_{ad}$. The resulting variational equation for $\underset{\sim}{v}$ with components in $H^1(\Omega)$ is a weak form of

$$-\Delta \underset{\sim}{v} + \operatorname{grad} p = \underset{\sim}{f} \qquad \text{in } \Omega \quad .$$

For any pair $\underset{\sim}{v}, p$ satisfying the differential equation and any $\underset{\sim}{u} \in U_{ad}$ follows

$$-|\underset{\sim}{v}|_a^2 \le J(\bar{\underset{\sim}{u}}) \le J(\underset{\sim}{u}) \quad .$$

Equality holds for $\underset{\sim}{v} = \underset{\sim}{u} = \bar{\underset{\sim}{u}}$.

GENERALIZATION

In general it is possible to take advantage of the special structure of the bilinear form $a(.,.)$ and to construct complementary extremal functionals that differ from the functional $J_c(.)$ considered in Theorem 1.2.1 .

If, for example, the bilinear form $a(.,.)$ is given by

$$a(u,v) = \int_\Omega \operatorname{grad} u \cdot \operatorname{grad} v \, dx \qquad , \qquad u,v \in H^1(\Omega)$$

or by

$$a(u,v) = \int_\Omega \Delta u \, \Delta v \, dx \qquad , \qquad u,v \in H^2(\Omega) \, ,$$

then we may also write

$$a(u,v) = \int_\Omega Tu \; Tv \, dx$$

where $Tu = \operatorname{grad} u$ and $Tu = \Delta u$, respectively.

More general, let be given besides of H, $(.,.)$, $\|.\|$ a second Hilbert space E, $(.,.)_E$, $|.|_E$ together with a linear operator $T : H \to E$ such that

(A5) $$a(u,v) = (Tu,Tv)_E \qquad \text{for all } u, v \in H \, .$$

Then we define a complementary energy functional $F(.)$ and a set $W_{ad} \subset E$ as follows:

$$F(v) = -(v,v)_E + 2(v,Tu_o)_E - 2\ell(u_o)\ , \qquad v \in E$$

$$W_{ad} = \{\ v \in E \mid (v,T\varphi) = \ell(\varphi) \ \text{ for all } \varphi \in U\ \}.$$

Note, that $F(Tu) = J_c(u)$ and that $T\bar{u} \in W_{ad}$.

Theorem 1.2.2. The functionals $J(.)$ and $F(.)$ are complementary in the following sense:

$$F(v) \leq J(\bar{u}) \leq J(u) \qquad \text{for all } \ u \in U_{ad}\ ,\ v \in W_{ad}\ .$$

Equality holds for $v = T\bar{u}$ and for $u = \bar{u}$ only. Furtherly, when v is considered as approximation to $T\bar{u}$ and u as approximation to $\bar{u}$, then the following a posteriori error estimate in energy norm holds:

$$|u - \bar{u}|_a^2 + |v - T\bar{u}|_E^2 = |Tu - v|_E^2 = J(u) - F(v) \ .$$

Proof. The proof is very similar to the proof of Theorem 1.2.1.

Remark. We want to emphasize that the minimizer of $J(u)$ is $\bar{u}$, whereas the maximizer of $F(v)$ is given by $T\bar{u}$. Thus, the latter extremal principle yields in the examples given above approximations for the derivatives of $\bar{u}$ (or combinations of the derivatives of $\bar{u}$) and not for $\bar{u}$ itself. This is an important point in the applications, for instance in linear elasticity, where one is not only interested in the displacements $\underset{\sim}{u} = (u_1,u_2,u_3)$ but also in the stresses which are functions of the derivatives of the displacements. For given body forces $\underset{\sim}{f} = (f_1,f_2,f_3)$, the functional

$$J(\underset{\sim}{u}) = \int_\Omega \frac{1}{4} c_{jkmn}(u_{j,k} + u_{k,j})(u_{m,n} + u_{n,m})dx - 2\int_\Omega f_j u_j\, dx$$

is (besides a factor 1/2) the potential energy, where c_{jkmn} are the coefficients of a symmetric, positive definite quadratic form. Introducing the strains $\underset{\sim}{e}$ and the stresses $\underset{\sim}{s}$, we may define the linear Operator T by $T\underset{\sim}{u} = \underset{\sim}{s}$, where

$$e_{jk} = \frac{1}{2}(u_{j,k} + u_{k,j}) \quad , \quad s_{jk} = c_{jkmn}\, e_{mn} \quad \text{(Hooke's law)} \ .$$

1.3 VARIATIONAL PRINCIPLES AND LAGRANGIAN MULTIPLIERS

As we have seen, the admissible functions in the pairs of complementary extremal problems have to satisfy certain constraints: The functions $u \in U_{ad}$ in problem (P) have to satisfy the essential boundary conditions, whereas the admissible functions in the complementary problem have to satisfy (at least in a weak form) a differential equation and eventually natural boundary conditions.

It is possible to remove the constraints by the technique of Lagrangian multipliers. This technique can be applied to the problem (P), but also to complementary extremal problems. We will explain the main idea by means of simple model problems.

1.3.1 PROBLEM (P) AND LAGRANGIAN MULTIPLIERS

For given $f \in L^2(\Omega)$ and $g \in H^{1/2}(\Gamma)$ we consider in $H^1(\Omega)$

$$J(u) = \int_\Omega \operatorname{grad} u \cdot \operatorname{grad} u \, dx - 2 \int_\Omega f u \, dx \quad ,$$

$$U_{ad} = \{ u \in H^1(\Omega) \mid u = g \text{ on } \Gamma \} , \quad U = \overset{o}{H}{}^1(\Omega) .$$

We want to remove the constraint $u = g$ on Γ which can be put in variational form

$$\int_\Gamma \psi (u - g) ds = 0 \qquad \text{for all } \psi \in L^2(\Gamma) .$$

Introducing the multiplier λ we define the Lagrangian

$$L(u,\lambda) = J(u) + 2 \int_\Gamma \lambda (u - g) ds \quad ,$$

where $u \in H^1(\Omega)$ and $\lambda \in L^2(\Gamma)$. By definition, the functional $L(.,.)$ is stationary for a pair of functions u,λ if and only if u,λ satisfies the following two variational equations:

(i) $\int_\Gamma \psi(u - g)ds = 0$ for all $\psi \in L^2(\Gamma)$

(ii) $a(u,\varphi) + \int_\Gamma \lambda\,\varphi\, ds = \ell(\varphi)$ for all $\varphi \in H^1(\Omega)$.

(i) is equivalent with the boundary condition $u = g$ on Γ, and (ii) is a weak form of $-\Delta u = f$ in Ω together with $\lambda = -\partial u/\partial n$ on Γ.

Theorem 1.3.1. The system (i), (ii) has a unique solution $\bar{u}$, $\bar{\lambda}$. Furtherly

$$L(v,\lambda) \leq J(\bar{u}) \leq L(u,\mu)$$

for any pair u,μ satisfying (i) and for any pair v,λ satisfying (ii).

Proof. Uniqueness: From (i) follows $\bar{u} = g$ on Γ or $\bar{u} \in U_{ad}$. (ii) holds in particular for all $\varphi \in \overset{o}{H}{}^1(\Omega)$. Thus, $\bar{u}$ is also solution of the variational equality (P)' and hence uniquely determined. The same is true for $\bar{\lambda}$. Existence: As is easily seen, $\bar{u}$ and $\bar{\lambda} = -\partial\bar{u}/\partial n$ is a solution of (i), (ii).

Let u,μ satisfy (i) , i.e. $u = g$ on Γ . Then

$$L(u,\mu) = J(u) \quad .$$

Let v,λ satisfy (ii). For $\varphi = v$ we find

$$L(v,\lambda) = -|v|_a^2 - 2\int_\Gamma \lambda\, g\, ds \quad .$$

On the other hand, introducing a function $u_o \in H^1(\Omega)$, $u_o = g$ on Γ , we find from (ii) for $\varphi = u_o$

$$-\int_\Gamma \lambda\, g\, ds = a(v,u_o) - \ell(u_o)$$

and hence $L(v,\lambda) \doteq J_c(v)$. Combining the results we find

$$L(v,\lambda) = J_c(v) \leq J(\bar{u}) \leq J(u) = L(u,\mu) \quad .$$

Remark. Theorem 1.3.1 shows that $\bar{u},\bar{\lambda}$ can be characterized as saddle point of the functional $L(.,.)$. As we have seen, the functional $L(.,.)$ is closely related to the functionls $J(.)$ and $J_c(.)$. The essential point is that the characterization by means of $L(.,)$ and the system (i), (ii) yields, when disretized, a different numerical scheme for approximating the solution $\bar{u}$. Those schemes resting upon saddle point characterizations of the solution are commonly used in finite elements.

1.3.2 COMPLEMENTARY PROBLEMS AND LAGRANGIAN MULTIPLIERS

Let us again consider the same boundary value problem as in section 1.3.1 where, however, we restrict ourselves for simplicity to the homogeneous boundary condition $u = 0$ on Γ . Then, adopting the notations of section 1.2 (and in particular of theorem 1.2.2), the functional $F(.)$ is given by

$$F(\underset{\sim}{v}) = - \int_\Omega v_j v_j \, dx \quad ,$$

and the admissible functions $v \in V_{ad}$ are the solution of the variational equation

$$\int_\Omega \underset{\sim}{v} \cdot \text{grad}\,\varphi \, dx = \int_\Omega f \varphi \, dx \qquad \text{for all} \quad \varphi \in \overset{o}{H}{}^1(\Omega)$$

which is a weak form of $-\text{div}\,\underset{\sim}{v} = f$ without boundary conditions. We want to remove this condition which occurs as constraint in the complementary extremal principle for the functional $F(.)$.

Introducing the Lagrangian multiplier $\lambda \in \overset{o}{H}{}^1(\Omega)$, we define

$$L(\underset{\sim}{v},\lambda) = F(\underset{\sim}{v}) + 2 \int_\Omega (\underset{\sim}{v}\,\text{grad}\,\lambda - f\lambda)\,dx \quad .$$

By definition, the functional is stationary for a pair $\underset{\sim}{v},\lambda$ if the first variations vanish. This is equivalent with the two equations

$$\text{(i)} \quad \int_\Omega \underset{\sim}{v} \cdot \underset{\sim}{\psi}\, dx = \int_\Omega \operatorname{grad}\lambda \cdot \underset{\sim}{\psi}\, dx \qquad \text{for all } \underset{\sim}{\psi} \in E$$

$$\text{(ii)} \quad \int_\Omega \underset{\sim}{v} \cdot \operatorname{grad}\varphi\, dx = \int_\Omega f\varphi\, dx \qquad \text{for all } \varphi \in \overset{o}{H}{}^1(\Omega) .$$

(i) is equivalent with $\underset{\sim}{v} = \operatorname{grad}\lambda$, and (ii) is a weak form of $-\operatorname{div}\underset{\sim}{v} = f$ without boundary conditions. (i) and (ii) both together yield $-\Delta\lambda = f$. Hence, $\lambda = \bar{u}$ and $v = \operatorname{grad}\bar{u}$ where $\bar{u}$ denotes the solution of the underlying boundary value problem.

<u>Theorem 1.3.2.</u> The system (i), (ii) has a unique solution, namely $\underset{\sim}{v} = \operatorname{grad}\bar{u}$, $\lambda = \bar{u}$. Furtherly

$$L(\underset{\sim}{v},\mu) \le F(\operatorname{grad}\bar{u}) \le L(\underset{\sim}{w},\lambda)$$

for any pair $\underset{\sim}{w},\lambda$ satisfying (i) and any pair $\underset{\sim}{v},\mu$ satisfying (ii).

<u>Proof.</u> The first statement is already proved. Let $\underset{\sim}{w},\lambda$ satisfy (i). Then $\underset{\sim}{v} = \operatorname{grad}\lambda$, whence

$$L(\underset{\sim}{w},\lambda) = J(\lambda) \ge J(\bar{u}) .$$

Note that $\lambda \in \overset{o}{H}{}^1(\Omega)$. Now, let $\underset{\sim}{v},\mu$ satisfy (ii). Then

$$L(\underset{\sim}{v},\mu) = F(\underset{\sim}{v}) \le F(\operatorname{grad}\bar{u}) = J(\bar{u}) .$$

This finishes the proof.

<u>Remark.</u> Supposing $\operatorname{div}\underset{\sim}{v} \in L^2(\Omega)$ and $\operatorname{div}\underset{\sim}{\psi} \in L^2(\Omega)$, we get from (i), (ii) integrating by parts

$$\text{(i)}' \quad \int_\Omega (\lambda \operatorname{div}\underset{\sim}{\psi} + \underset{\sim}{v}\cdot\underset{\sim}{\psi})dx = 0 \qquad \text{for all } \underset{\sim}{\psi} \in E$$

$$\text{(ii)}' \quad \int_\Omega (\operatorname{div}\underset{\sim}{v} + f)\,\varphi\, dx = 0 \qquad \text{for all } \varphi \in \overset{o}{H}{}^1(\Omega) .$$

The variational principle in the form (i)' , (ii)' was studied by Raviart and Thomas [16] in the context of a mixed method for finite elements. See also Brezzi [17].

2 PROBLEMS INVOLVING INEQUALITY CONSTRAINTS

Now we turn to problems involving inequality constraints. Those problems play an important role in various fields of applications. In elasticity, well known examples are obstacle problems (deformation of an elastic body under given forces in the presence of a fixed rigid obstacle) or problems of Signorini type (linear differential equations together with boundary conditions in terms of inequality constraints). See, for instance, Duvaut and Lions [10], Kinderlehrer and Stampacchia [11], Fichera [12] and the references given there.

In this chapter we will establish complementary extremal principles for a certain class of problems involving inequality constraints. In particular, we will recover a number of results distributed in the literatur, especially results concerning plate problems obtained by Haslinger, Hlaváček, Brezzi and others. In addition, we will present error estimates in energy norm.

2.1 EXTREMAL PROBLEMS AND VARIATIONAL INEQUALITIES

As in chapter 1, we will consider bilinear forms $a(.,.)$ and linear functionals $\ell(.)$ where, by assumption, (A1)-(A4) is satisfied. Again, we will consider quadratic energy functionals $J(u) = a(u,u) - 2\,\ell(u)$.

In chapter 1, the set of admissible functions was given in the form $U_{ad} = u_o + U$ where $u_o \in H$ was a given element and U a closed linear subspace in which (A3) holds. This time we consider a class of problems where the set U_{ad}

is given by

$$U_{ad} = u_o + K .$$

Here, K denotes a closed convex set contained in U . In the particular case K = U we fall back to the definition of U_{ad} given in chapter 1. Now, let us consider for $U_{ad} = u_o + K$ the following problems.

Problem (P) (Extremal principle):

$$\text{Minimize} \quad J(u) \quad , \quad u \in U_{ad} .$$

Problem (P)' (Variational inequality): Find $u \in U_{ad}$ satisfying

$$a(u, v - u) - \ell(v - u) \geq 0 \quad \text{for all} \quad v \in U_{ad} .$$

As is well known (see, for instance, Lions and Stampacchia [13]), problem (P) and problem (P)' are equivalent, and they admit a unique solution $\bar{u} \in U_{ad}$.

The extremal problem (P) is a minimization problem for a convex functional defined on a closed convex set. Existence of a unique solution $\bar{u}$ follows by standard arguments of convex analysis. Equivalence of (P) and (P)' is readily establisehd by the following argument:

For arbitrary $u, v \in U_{ad}$ and $t \in [0,1]$ the convex combination

$$v_t = (1-t)u + tv = u + t(v-u)$$

belongs to U_{ad} , too. Now let u minimize the energy functional J . Then $J(u) \leq J(v_t)$ for $t \in [0,1]$. The minimum is attained for t = 0 , whence

$$\frac{d}{dt} J(v_t) \geq 0 \quad \text{at} \quad t = 0 .$$

On the other hand,

$$J(v_t) = J(u) + 2t[a(u,v-u) - \hat{\ell}(v-u)] + t^2|v-u|_a^2$$

for any $v \in U_{ad}$. Thus the solution u of (P) is also solution of (P)' and vice versa.

Examples of convex sets

Consider the real function space $H = H^1(\Omega)$. We introduce the set U_{ad} of all functions $u \in H^1(\Omega)$ satisfying $u = 0$ on Γ and $u \geq u_o$ in Ω, where u_o is a given obstacle function satisfying itself $u_o = 0$ on Γ . Equivalently we may write

$$U_{ad} = u_o + K \quad \text{where} \quad K = \{u \in \overset{o}{H}{}^1(\Omega) \mid u \geq 0 \text{ in } \Omega\} .$$

Consider in the real function space $H = H^1(\Omega)$ the functions satisfying the Signorini type constraint $u \geq 0$ on Γ . Here, we may write

$$U_{ad} = K \quad \text{where} \quad K = \{u \in H^1(\Omega) \mid u \geq 0 \text{ on } \Gamma\} .$$

In both examples, K is a closed convex set generated by one-sided constraints. It turns out that in both examples K has the following special structure: K is a cone with vertex at zero which means that

(i) $\quad u + v \in K \quad$ for all $u, v \in K$

(ii) $\quad \alpha u \in K \quad$ for all $u \in K$, $\alpha \geq 0$.

There are, however, also important examples in which the convex set K is not a cone. This happens, for instance, when K is generated by two-sided constraints of Signorini type like

$$U_{ad} = K = \{u \in H^1(\Omega) \mid 0 \leq u \leq 1 \text{ on } \Gamma\} .$$

2.2 COMPLEMENTARY EXTREMAL PRINCIPLES (GENERAL CASE)

Let us firstly consider extremal problems (P) for $U_{ad} = u_o + K$, $K \subset U$, where K is a closed convex set of general type. We recall the equivalent variational inequality (P)':

$$u \in U_{ad} : \quad a(u,v-u) - \ell(v-u) \geq 0 \quad \text{for all } v \in U_{ad} .$$

Now, similarly as in section 1.2, we drop the constraint $u \in U_{ad}$ in (P)' and consider all functions $w \in H$ satisfying the variational inequality

$$w \in H : \quad a(w,v-w) - \ell(v-w) \geq 0 \quad \text{for all } v \in U_{ad} .$$

The set of all solutions w will be denoted by W_{ad}. Note that $\bar{u} \in W_{ad}$ where $\bar{u}$ is the solution of (P) and of (P)'.

<u>Theorem 2.2.</u> For any $w \in W_{ad}$ and any $u \in U_{ad}$ follows

$$J(w) \leq J(\bar{u}) \leq J(u) .$$

Equality holds for $w = \bar{u}$ and for $u = \bar{u}$ only. Furtherly, when u and w are considered as approximations to $\bar{u}$ then the following a posteriorie estimates in energy norm hold:

$$|u - \bar{u}|_a^2 + |w - \bar{u}|_a^2 \leq J(u) - J(w) ,$$

$$4 \left| \frac{u+w}{2} - \bar{u} \right|_a^2 \leq 2[J(u) - J(w)] - |u - w|_a^2 .$$

<u>Proof.</u> Using (A4) together with the variational inequality for $w \in W_{ad}$ we get

$$0 \leq |u - w|_a^2 = |u|_a^2 - 2a(w,u-w) - |w|_a^2 \leq J(u) - J(w).$$

In particular, we get

$$|u - \bar{u}|_a^2 \leq J(u) - J(\bar{u}) \quad \text{and} \quad |\bar{u} - w|_a^2 \leq J(\bar{u}) - J(w) .$$

Adding these two inequalities we get the first a posteriori estimate. The second estimate follows then easely from

$$|(u-\bar{u}) + (w-\bar{u})|_a^2 + |(u-\bar{u}) - (w-\bar{u})|_a^2 = 2|u-\bar{u}|_a^2 + 2|w-\bar{u}|_a^2 .$$

2.3 COMPLEMENTARY EXTREMAL PRINCIPLES (SPECIAL CASE)

Let us now consider the special case where the closed convex set K is a cone with vertex at zero. Again we start from the variational inequality (P)'

$$u \in U_{ad} : \quad a(u,v-u) - \ell(v-u) \geq 0 \quad \text{for all } v \in U_{ad} .$$

Since K is a cone, (P)' is equivalent with the two relations

$$\text{(i)} \quad a(u,u_o-u) - \ell(u_o-u) = 0$$

$$\text{(ii)} \quad a(u,\varphi) - \ell(\varphi) \geq 0 \quad \text{for all } \varphi \in K ,$$

Indeed, since $v = u_o \in U_{ad}$, we get from (P)'

$$a(u,u_o-u) - \ell(u_o-u) \geq 0 .$$

On the other hand, since K is a cone with vertex at zero, $v = u + (u - u_o) \in U_{ad}$. Hence we get from (P)'

$$a(u,u-u_o) - \ell(u-u_o) \geq 0 .$$

Thus we get (i). Finally, for $u \in U_{ad}$, we have $v = u + \varphi \in U_{ad}$ for all $\varphi \in K$. Introducing $\varphi = v \quad u$ in (P)' we get (ii). Vice versa, when (i) and (ii) hold we come back to (P)'.

Now we drop the condition $u \in U_{ad}$ and also the condition (i) and consider the variational inequality (ii) in the form

$$w \in H : \quad a(w,\varphi) - \ell(\varphi) \geq 0 \quad \text{for all } \varphi \in K .$$

The set of all solutions w will be denoted by W_{ad}. Note that $\bar{u} \in W_{ad}$ where $\bar{u}$ denotes the unique solution of (P) and of (P)'.

Furtherly as in section 1.2 we introduce the complementary energy functional

$$J_c(w) = -a(w,w) + 2a(w,u_o) - 2\ell(u_o) \quad , \quad w \in H .$$

Theorem 2.3.1. The functionals J(.) and $J_c(.)$ are complementary in the following sense:

$$J_c(w) \leq J(\bar{u}) \leq J(u)$$

for all $u \in U_{ad}$ and all $w \in W_{ad}$. Equality holds for $u = \bar{u}$ and $w = \bar{u}$. Furtherly, when u and w are considered as approximations to $\bar{u}$, the following a posteriori estimates in energy norm hold:

$$|u-\bar{u}|_a^2 + |w-\bar{u}|_a^2 \leq J(u) - J_c(w) \quad ,$$

$$4\left|\frac{u-w}{2} - \bar{u}\right|_a^2 \leq 2[J(u) - J_c(w)] - |u-w|_a^2 \quad .$$

Proof. From (A4) together with the variational inequality (ii) satisfied by $w \in W_{ad}$ we get

$$\begin{aligned} 0 \leq |u-w|_a^2 &= |u|_a^2 - 2a(w,u-u_o) - 2a(w,u_o) + |w|_a^2 \\ &\leq |u|_a^2 - 2\ell(u-u_o) - 2a(w,u_o) + |w|_a^2 \\ &= J(u) - J_c(w) \quad . \end{aligned}$$

This inequality holds in particular for $u = \bar{u}$ and for $w = \bar{u}$, respectively. Noting that by (i)

$$J_c(\bar{u}) = -|\bar{u}|_a^2 + 2a(\bar{u},\bar{u}) - 2\ell(\bar{u}) = J(\bar{u}) \quad ,$$

we find $J_c(w) \leq J(\bar{u}) \leq J(u)$ where equality holds for $u = \bar{u}$ and for $w = \bar{u}$, respectively. The proof of the a posteriori inequalities is very similar to the corresponding inequalities in theorem 2.2.

Similar as in section 1.2 it is possible to extend the considerations to the case where a second Hilbert space E, $(\cdot,\cdot)_E$, $|\cdot|_E$ and a linear operator $T: H \to E$ are given such that

$$a(u,v) = (Tu,Tv)_E \quad \text{for all} \quad u, v \in H .$$

Then we introduce the variational inequality

$$w \in E : \quad (w,T\varphi)_E - \ell(\varphi) \geq 0 \qquad \text{for all} \quad \varphi \in K .$$

The set of all solutions will be denoted by W_{ad}^E . Furtherly, as in section 1.2 we introduce instead of the functional $J_c(\cdot)$ now the functional $F(\cdot)$ given by

$$F(w) = -|w|_E^2 + 2(w,Tu_o)_E - 2\,\ell(u_o) \quad , \quad w \in W_{ad}^E .$$

<u>Theorem 2.3.2.</u> The functionals $J(\cdot)$ and $F(\cdot)$ are complementary in the following sense:

$$F(w) \leq J(\bar{u}) \leq J(u)$$

for all $u \in U_{ad}$ and all $w \in W_{ad}^E$. Equality holds for $u = \bar{u}$ and for $w = T\bar{u}$ only. Furtherly, when u and w are considered as approximations to $\bar{u}$ and to $T\bar{u}$, respectively, then the following a posteriori estimates in energy norm hold:

$$|u-\bar{u}|_a^2 + |w-T\bar{u}|_E^2 \leq J(u) - F(w) \quad ,$$

$$4\left|\frac{Tu+w}{2} - T\bar{u}\right|_E^2 \leq 2[J(u) - F(w)] - |Tu-w|_E^2 \quad .$$

<u>Proof.</u> The proof of theorem 2.3.2 is very similar to the proof of theorem 2.3.1 and will be omitted here.

2.4 EXAMPLES

Example 2.4.1: A problem with one-sided obstacle.

Let be given an obstacle function $u_o \in \overset{o}{H}{}^1(\Omega)$ and a function $f \in L^2(\Omega)$. Let U_{ad} consist of all functions $u \in \overset{o}{H}{}^1(\Omega)$ satisfying $u \geq u_o$ in Ω. Then we consider the extremal problem:

$$\text{Minimize} \quad J(u) = \int_\Omega |\operatorname{grad} u|^2 dx - 2 \int_\Omega f u \, dx \ .$$

We may write $U_{ad} = u_o + K$, where K denotes the closed convex cone given by

$$K = \{ u \in \overset{o}{H}{}^1(\Omega) \mid u \geq 0 \ \text{ in } \Omega \} \ .$$

Defining $H = H^1(\Omega)$, $U = \overset{o}{H}{}^1(\Omega)$, the assumptions (A1) - (A4) are satisfied. Furtherly, we have $K \subset U$, and K is a cone with vertex at zero. Thus we may apply either theorem 2.3.1 or theorem 2.3.2.

In order to apply theorem 2.3.1, we have to introduce the set W_{ad} consisting of all solutions of the variational inequality

$$w \in H^1(\Omega) : \int_\Omega \operatorname{grad} w \cdot \operatorname{grad} \varphi \, dx - \int_\Omega f \varphi \, dx \geq 0 \quad \text{for all} \quad \varphi \in K$$

which is a weak form of $-\Delta w \geq f$ in Ω without boundary conditions. $J_c(w)$ has to be defined as in theorem 3.2.1.

In order to apply theorem 2.3.2 for $\Omega \subset \mathbb{R}^n$, we introduce $E = [L^2(\Omega)]^n$ together with $Tu = \operatorname{grad} u$. Then the set W^E_{ad} consists of all solutions $\underset{\sim}{w} = (w_1, \dots, w_n)$ of the variational inequality

$$\underset{\sim}{w} \in E: \int_\Omega \underset{\sim}{w} \cdot \operatorname{grad} \varphi \, dx - \int_\Omega f \varphi \, dx \geq 0 \quad \text{for all} \ \varphi \in K$$

which is a weak form of $-\operatorname{div} \underset{\sim}{w} \geq f$ in Ω without boundary conditions. The complementary energy functional $F(.)$ in theorem 2.3.2 is given by

$$F(\underset{\sim}{w}) = -\int_\Omega \underset{\sim}{w}\cdot\underset{\sim}{w}\,dx + 2\int_\Omega \underset{\sim}{w}\cdot \operatorname{grad} u_o\,dx - 2\int_\Omega f\,u_o\,dx \quad .$$

Essentially the same complementary functional was used by Haslinger [14] in the context of finite element approximations. Haslinger derived it establishing first a saddle point problem via Lagrangian multipliers. His complementary functional is given by

$$G(\underset{\sim}{w}) = -\int_\Omega \underset{\sim}{w}\cdot\underset{\sim}{w}\,dx + 2\int_\Omega \underset{\sim}{w}\cdot \operatorname{grad} u_o\,dx$$

which differs from $F(\underset{\sim}{w})$ by a constant term only. Thus, the extremal principles based on $F(.)$ and on $G(.)$ are equivalent. Note, however, that theorem 2.3.2 includes also an a posteriori error estimate in terms of $F(.)$.

<u>Example 2.4.2:</u> A problem with one-sided boundary condition.

Let us consider the following extremal problem:

$$\text{Minimize}\quad J(u) = \int_\Omega (\,|\operatorname{grad} u|^2 + u^2)dx - 2\int_\Omega f\,u\,dx$$

where $f \in L^2(\Omega)$ and where the set of admissible functions is given by

$$U_{ad} = \{\, u \in H^1(\Omega) \mid u \geq 0 \text{ on } \Gamma \,\} .$$

Here, $U_{ad} = K$ is a closed convex cone with vertex at zero. When we drop the constraint $u \in U_{ad}$ in the variational inequality (P)', we get the following variational inequality defining W_{ad} in theorem 2.3.1:

$$w \in H^1(\Omega) : \int_\Omega (\operatorname{grad} w \cdot \operatorname{grad} \varphi + w\varphi)\,dx - \int_\Omega f\varphi\,dx \geq 0$$

for all $\varphi \in K$, which is a weak form of

$$-\Delta w + w = f \text{ in } \Omega ,\quad \frac{\partial w}{\partial n} \geq 0 \text{ on } \Gamma \quad .$$

Similarly, theorem 2.3.2 can be applied as follows: We introduce the Hilbert space $E = (L^2(\Omega))^{n+1}$ consisting of vector functions $\underset{\sim}{w} = (w, w_1, \ldots, w_n)$, and we define the linear operator $T : H \to E$ by

$$Tu = (u, \frac{\partial u}{\partial x_1}, \ldots, \frac{\partial u}{\partial x_n}) .$$

Then the space W^E_{ad} consists of all solutions $(w, w_1, \ldots, w_n)$ of the variational inequality

$$\int_\Omega \left(\sum_{j=1}^n w_j \frac{\partial \varphi}{\partial x_j} + w\varphi \right) dx - \int_\Omega f\varphi \, dx \geq 0 \quad \text{for all } \varphi \in K$$

which is a weak form of

$$- \sum_{j=1}^n \frac{\partial w_j}{\partial x_j} + w = f \text{ in } \Omega , \sum_{j=1}^n w_j \nu_j \geq 0 \quad \text{on } \Gamma ,$$

$\underset{\sim}{n} = (\nu_1, \ldots, \nu_n)$ denoting the outward normal to the boundary Γ. Clearly, it is easier to find admissible functions $\underset{\sim}{w} \in W^E_{ad}$ than functions $w \in W_{ad}$.

FINAL REMARKS

It was the aim of these lectures to present the main ideas in terms of simple model problems. Clearly, it is possible to extend them to a wider class of problems.

For instance, the energy functional $J(.)$ may also contain boundary integrals so that the natural boundary conditions become inhomogeneous.

Furtherly, it is possible to incorporate the technique of so called Trefftz functionals. Here the energy functional $J(.)$ is replaced by an other quadratic functional $\mathfrak{J}(.)$, where one or more of the natural conditions related to $J(.)$ now occur as essential conditions for the new functional $\mathfrak{J}(.)$, and where $J(u) = \mathfrak{J}(u)$ for all functions satisfying these natural conditions which are essential ones for the Trefftz functional $\mathfrak{J}(.)$. Then it is possible to construct extremal functionals complementary to $\mathfrak{J}(.)$ in the same way as described in these lectures for the energy functional $J(.)$.

Finally, it is possible to extend the technique of Lagrangian multipliers as presented in section 1.3 also to problems involving unilateral constraints. In that case, the multipliers have to be submitted to unilateral constraints.

Summarizing we want to emphasize that a variety of variational principles so fare considered in papers on finite elements and established by different approaches can be derived and understood in the frame of the simple ideas presented in these lectures.

Concerning the references given below, we want to emphasize that the list of references is fare from being complete.

References

[1] COURANT, R. and HILBERT, D.:
Methoden der mathematischen Physik. Springer-Verlag. Heidelberger Taschenbücher vol. 30 und 31.

[2] TREFFTZ, E.:
Ein Gegenstück zum Ritzschen Verfahren. Verhandl. d. 2. Internat. Kongreß f. techn. Mechanik. Zürich 1926, 131-138.

[3] EKELAND, I. and TEMAM, R.:
Convex analysis and variational problems. North-Holland Publ. Company, Amsterdam 1976.

[4] SEWELL, M.J.:
Dual approximation principles and optimization in continuum mechanics. Phil. Trans. Roy. Soc. (London) A 265 (1969) 319-351.

[5] NOBLE, B., SEWELL, M.J.:
On dual extremum principles in Applied Mathematics. J. Inst. Math. Appl. 9 (1972) 123-193.

[6] ROBINSON, P.D.:
Complementary variational principles. In: Rall. L.B. (Ed.): Nonlinear functional analysis and applications. Academic Press 1971, pp. 509-576.

[7] ODEN, J.T. and REDDY, J.N.:
Variational methods in theoretical mechanics. Springer-Verlag Berlin, Heidelberg, New York 1976.

[8] VELTE, W.:
Direkte Methoden der Variationsrechnung. Teubner Studienbücher Mathematik. Teubner Verlag Stuttgart 1976.

[9] NEČAS, J.:
Les méthodes directes en théorie des équations élliptiques. Masson et Cie., Paris, Academia, Prag 1967.

[10] DUVAUT, G. and LIONS, J.L.:
Les inéquations an mécanique et en physique. Dunaud, Paris 1972.

[11] KINDERLEHRER, D., STAMPACCHIA, G.:
An introduction into variational inequalities and their applications. Academic Press 1980.

[12] FICHERA, G.:
Boundary value problems of elasticity with unilateral constraints. S.392-424 in: Flügge, S. (Herausg.): Handbuch der Physik vol. VI a/2. Springer-Verlag Berlin, Heidelberg, New York.

[13] LIONS, J.L., STAMPACCHIA, G.:
Variational inequalities. Comm. Pure Appl. Math. 20 (1969) 493-519.

[14] HASLINGER, J.:
Dual finite element analysis for an inequality of the 2nd order. Aplikace Matematiky 24 (1979), 118-132.

[15] BREZZI, F., HAGER, W.W., RAVIART, P.A.:
Error estimates for the finite element solution of variational inequalities. Part I. Primal theory. Part II. Mixed methods. Numer.Math. 28 (1977) 431-443. 31 (1978) 1-16.

[16] RAVIART, P.A., THOMAS, J.M.:
A mixed finite element method for 2nd oder elliptic problems. In: Mathematical aspects of the FEM (Rom, December 1975). Springer Lecture Notes in Math. vol. 606.

[17] BREZZI, F.:
On the existence, uniqueness and approximations of saddle-point problems arising from Lagrangian multipliers. R.A.I.R.O. Anal. Numér. 11 (1977) 209-216.

Additional references

ALLEN, G.:
Variational inequalities, complementary problems and duality theorems.
J. f. Math. Anal. and Appl. 58 (1977) 1-10.

ARTHURS, A.M.:
Complementary variational principles.
Oxford Math. Monographs, Oxford 1970.

COTTLE, R.W., GIANESSI, F., LIONS, J.-L.:
Variational Inequalities and Complementary Problems. Theory and Applications.
(Internat. School of Math.,19.-30. Juni 1978 in Erice Sicily) John Wiley and Sons 1980 .

FRAEIJS de VEUBEKE, B.:
Displacement and equilibrium models in the finite element method.
S. 145-197 in: O.C. Zienkiewicz, G.S. Holister (Eds.) Stress Analysis, John Wiley 1965.

FRAEIJS de VEUBEEKE, B.:
The duality principles of elastodynamics.
Finite elements applications.
S. 357-377 in: Lectures in finite element methods in continuum mechanics.
Ed. by J.T. Oden, E.R. De Arantes e Oliveira
The Univ. of Alabama Press, Huntsville 1973.

GWINNER, J.:
Nichtlineare Variationsgleichungen mit Anwendungen.
Verlag Haag u. Herchen, Frankfurt/M. 1978

GWINNER, J.:
Bibliography on non-differentiable optimization and
non-smooth analysis. [Vierhundert Referenzen.]
J. of Computational and Applied Math. 7 (1981) 277-285.

PRAGER, W., SYNGE, J.L.:
Approximations in elasticity based on the concept of
function space.
Quart. Appl. Math. 5 (1947) 241-259.

TONTI, E.:
Variational principles in elastostatics.
Mechanica 2 (1967) 201-208.

WERNER, B.:
Complementary variational principles and non-conforming
Trefftz elements.
Internat. Series Numer.Math. (ISNM) vol. 56 (1981) 180-192.

BABUŠKA, I.:
The finite element method with Lagrangian multipliers.
Numer. Math. 20 (1973) 179-192.

BRAMBLE, J.H.:
The Lagrange multiplier method for Dirichlet's problem.
Math. of Computation 155 (1981) 1-12.

BREZZI, F., RAVIART, P.A.:
Mixed finite element methods for 4th order elliptic
equations.
In: J.H. Miller (ed.):
Proc. Royal Irish Academy Conference on Numerical Analysis.
Topics in Numerical Analysis III, 1976. Academic Press
1977, 33-56.

BUFLER, H.:
Variationsgleichungen und finite Elemente. Bayer. Akad.
d. Wiss. Math.-Nat. Klasse Bd. 1975, 155-187.

CIARLET, P.G., RAVIART, P.A.:
A mixed finite element method for the biharmonic equation.
In: C. de Boor (Ed.): Mathematical aspectes of finite
elements in partial differential equations.
Academic Press 1874, pp. 125-145.

HASLINGER, J.:
Dual finite element analysis for an inequality of the
2nd order. Aplikace Matematiky 24 (1979)118-132.

HLAVÁČEK, I.:
Dual finite element analysis for unilateral boundary value
problems. Aplikace Matematiky 22 (1977) 14-51.

HLAVÁČEK, I.:
Convergence of dual finite element approximations for unilateral boundary value problems. Aplikace matematiky 25 (1980) 375-386.

QUARTERONI, A.:
On mixed methods for fourth-order problems. Computer methods in applied mechanics and engineering 24 (1980) 13-34.

RANNACHER, R.:
On nonconforming and mixed finite element methods for plate bending problems. The linear case.
RAIRO Numer. Anal. 13 (1979) 369-387.

SCAPOLA, T.:
A mixed finite element method for the biharmonic problem.
R.A.I.R.O. Anal. Numér. 14 (1980) 55-79.

SCHOLZ, R.:
Approximation von Sattelpunkten mit finiten Elementen.
Bonner Math. Schriften 89 (1976) 53-66.

SCHOLZ, R.:
A mixed method for 4th order problems using linear finite elements. R.A.I.R.O. Anal. Numer. 12 (1978) 85-90.

WUNDERLICH, W.:
Mixed models for plates and shells: Principles - Elements - Examples. S. 215 - 241 in: S.N.Atluri, R.H.Gallagher, O.C. Zienkiewicz (ed.): Hybrid and finite element methods.
John Wiley 1983.

FIVE LECTURES ON
NONLINEAR FINITE ELEMENT METHODS*

E. Stein, D. Bischoff, N. Müller-Hoeppe, W. Wagner and P. Wriggers
Universität Hannover, Hannover, F.R.G.

Summary and Scope

This course will give a modern concept of finite-element-method in nonlinear solid mechanics using material (Lagrangian) coordinates. Elastic post-buckling analysis of shells is treated as an essential example for the geometrically nonlinear behaviour of structures, and elastic-plastic deformations are introduced in the context of ultimate load analysis of thin-walled steel structures. Lastly, problems with unilateral constraints, such as incompressible elastic deformations and contact problems of several deformable bodies are treated. A main feature of this course is the derivation of consistent linearizations of the weak forms of equilibrium within the same order of magnitude, taking also into account the material laws and - if present unilateral constraints, in order to get Newton-type iterative algorithms with quadratic convergence.

The lectures are intended to introduce into effective discretizations and algorithms based on a well founded mechanical and mathematical analysis.

*Dedicated to Prof. Dr. Drs. h.c. J.H. Argyris on the occasion of his 70th birthday

1 Survey of Nonlinear Continuum Mechanics

In this chapter a short summary of the continuum mechanics of solids is given. Since only the basic ideas of continnum mechanics can be considered here we restrict ourselves to a theory with small strains but large rotations which will be the basis of the finite element formulation in chapter 2. Applications of this class of problems are e.g. buckling of shells or elasto–plastic limit analysis of folded plates. Problems which exhibit large strains like recasting processes or the analysis of rubber bearings are not considered here. The restriction to small strains effects the constitutive theory, the reader who wishes to get a deeper insight may consult standard textbooks on continuum mechanics (Malvern 1969, Truesdell and Noll 1965, Marsden and Hughes 1983).

1.1 Kinematics

The motion of a body B can be described by introducing an undeformed configuration B_o (the subcript $_o$ denotes reference to the undeformed configuration in the following) and the deformed configuration B as shown in Figure 1.1.

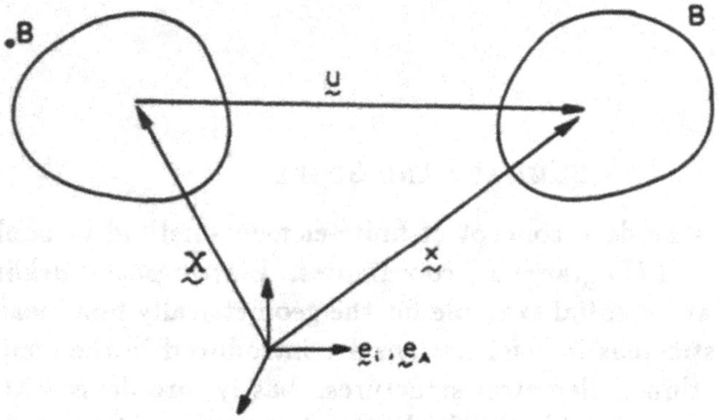

Fig. 1.1 Motion of a body B .

A particle X in the undeformed configuration is decribed by the position vector $\mathbf{X} = X_A\,\mathbf{e}_A$. At time t the position occupied by the particle X is given by

$$\mathbf{x} = \hat{\mathbf{x}}\,(\mathbf{X}, t\,) = x_i\,\mathbf{e}_i. \tag{1.1}$$

If the base vectors $\mathbf{e}_A$ and $\mathbf{e}_i$ are identical (see Figure 1.1) then the shifter between both coordinate systems reduces to the Kronecker symbol $\mathbf{e}_i \cdot \mathbf{e}_A = \delta_{iA}$. Furthermore, we can introduce the displacement vector of the particle X (see Figure 1.1)

$$\mathbf{u} = \mathbf{x} - \mathbf{X} = (\,\delta_{iA}\,x_i - X_A\,)\,\mathbf{e}_A. \tag{1.2}$$

The deformation gradient $\mathbf{F}$ is defined as the tensor which associates an infinitesimal vector $\mathbf{dX}$ at $\mathbf{X}$ with a vector $\mathbf{dx}$ at $\mathbf{x}$: $\mathbf{dx} = \mathbf{F}\mathbf{dX}$. Therefore, the components of $\mathbf{F}$ are the partial derivatives $\frac{\partial x_i}{\partial X_A} = x_{i,A}$. With (1.1) and (1.2) we obtain

$$\mathbf{F} = F_{iA}\,\mathbf{e}_i \otimes \mathbf{e}_A = Grad\,\mathbf{x} = \mathbf{1} + grad\mathbf{u},$$

$$F_{iA} = x_{i,A} = \delta_{iA} (\delta_{AB} + u_{A,B}). \tag{1.3}$$

The equations of continuum mechanics may be formulated with respect to the deformed or the undeformed configuration of the body. Throughout this lecture we refer all quantities to the undeformed configuration B_o (material description). However, in section 1.5 the transformation of the principle of virtual work from B_o to the deformed configuration B is shown.

Knowing the deformation gradient $\mathbf{F}$ we can introduce the Green-Lagrangian strain tensor $\mathbf{E}$ which is defined by

$$\mathbf{E} = E_{AB} \mathbf{e}_A \otimes \mathbf{e}_B = \frac{1}{2} (\mathbf{F}^T \mathbf{F} - \mathbf{1}),$$

$$E_{AB} = \frac{1}{2} (F_{iA} \, F_{iB} - \delta_{AB}) = \frac{1}{2} (u_{A,B} + u_{B,A} + u_{C,A} u_{C,A} - \delta_{AB}) \tag{1.4}$$

1.2 Balance Laws

In this section we will summarize differential equations expressing locally the conservation of mass and momentum. These equations are associated with the fundamental postulates of continuum mechanics. A detailed derivation may be found e.g. in Malvern (1969,Ch. 5). Within the material description the conservation of mass is given by $\rho_o = det\, \mathbf{F} \rho$ where ρ is the density. From this equation we obtain a relation between the volume elements in the deformed and undeformed configuration

$$dv = \frac{\rho_o}{\rho} dV = det\, \mathbf{F} dV. \tag{1.5}$$

The conservation of momentum or the local form of the equation of motion is given by

$$Div \mathbf{P} + \rho_o \mathbf{b} = \rho_o \ddot{\mathbf{x}},$$

$$P_{Ai,A} + \rho_o b_i = \rho_o \ddot{x}_i. \tag{1.6}$$

Here $\mathbf{P}$ denotes the unsymmetric first Piola-Kirchhoff stress tensor which associates the stress vector $\mathbf{t}$ with a normal vector $\mathbf{n}$ by $\mathbf{t} = \mathbf{P} \mathbf{n}$, see e.g. Malvern (1969). This relation is well known as Cauchy theorem. $\rho_o \mathbf{b}$ is the body force (e.g. gravity force) and $\rho_o \ddot{\mathbf{x}}$ is the inertial force which will be neglected in the following. The balance of moment of momentum leads to the following condition for the 1. Piola-Kirchhoff stress tensor

$$\mathbf{P} \mathbf{F}^T = \mathbf{F} \mathbf{P}^T \tag{1.7}$$

which shows the unsymmetry. Obviously it is more convenient to work with symmetric stress tensors. Therefore we may introduce other stress tensors. Since a given stress vector does not change its magnitude when refered to the undeformed or deformed configuration we may write

$$\int_{\partial B_o} \mathbf{t} \, dA = \int_{\partial B_o} \mathbf{P} \, \mathbf{n}_o \, dA = \int_{\partial B} \frac{1}{det \mathbf{F}} \mathbf{P} \, \mathbf{F}^T \mathbf{n} \, da = \int_{\partial B} \boldsymbol{\sigma} \, \mathbf{n} \, da \tag{1.8}$$

which defines the symmetric Cauchy stress tensor $\boldsymbol{\sigma}$ refered to the deformed configuration

$$\boldsymbol{\sigma} = \frac{1}{det\mathbf{F}}\mathbf{P}\,\mathbf{F}^T$$

$$\sigma_{ik} = \frac{1}{det(x_{i,A})} F_{iA}\, P_{Ak}. \tag{1.9}$$

The symmetric second Piola-Kirchhoff stress tensor is refered to B_o and defined by

$$\mathbf{S} = \mathbf{F}^{-1}\,\mathbf{P},$$

$$S_{AB} = F_{Ai}\, P_{Bi}. \tag{1.10}$$

S itself does not represent a physically meaningful stress but plays an important role in constitutive theory because it is the conjugated stress measure to the Green-Lagrangian strain tensor (1.4) when the expression of the internal virtual work is refered to B_o.

1.3 Constitutive Law

The constitutive theory describes the macroscopic behaviour of a material. By the constitutive equations strains are related to the stresses within a body. Since real materials behave in very complex ways we have to approximate physical observations of the real materials response. This approximations can be done seperately for different material responses (e.g. elastic, plastic or viscoplastic behaviour).

In this section we restrict ourselves to pure elastic material behaviour and small strains. The constitutive equation for the elastic response of the second Piola-Kirchhoff stress tensor may be derived from the hyperelastic potential ψ by

$$\mathbf{S} = \rho_o \frac{\partial \psi(\mathbf{E})}{\partial \mathbf{E}}; \qquad S_{AB} = \rho_o \frac{\partial \psi(E_{AB})}{\partial E_{AB}}. \tag{1.11}$$

The restriction to small strains does not limit the magnitude of the rotations of material line elements $\mathbf{dx}$. The polar decomposition theorem for $\mathbf{F}$: $\mathbf{F} = \mathbf{R}\,\mathbf{U}$ renders a possibility for an estimate of the magnitudes of the rotations $\mathbf{R}$ and the stretches $\mathbf{U}$. Inserting the polar decomposition into (1.4) the Green strain is $\mathbf{E} = 1/2(\mathbf{U}^2 - \mathbf{1})$. We see that the rotations do not play any role for the magnitude of the strains. Small stretches may be defined by the following relation

$$\|\mathbf{U} - \mathbf{1}\| = \epsilon = o(0.01) \tag{1.12a}$$

The linearization of the right stretch tensor $\mathbf{U}$ yields

$$\mathbf{U} = \mathbf{1} + \frac{1}{2}\,(\,Grad\mathbf{u} + Grad\mathbf{u}^T\,). \tag{1.12b}$$

Then for small values of *Grad*u the right stretch tensor **U** fulfills relation (1.12a) and the deformation gradient reduces to $\mathbf{F} = \mathbf{R}$. The incorporation of these relations in the constitutive theory leads to a linear relation between **S** and **E** often refered to as a St. Venant material (Ciarlet 1982)

$$\mathbf{S} = \lambda\, tr(\mathbf{E})\, \mathbf{1} + 2\,\mu\, \mathbf{E}$$

$$S_{AB} = \lambda\, E_{LL}\, \delta_{AB} + 2\,\mu\, E_{AB} = C_{ABCD}\, E_{CD} \tag{1.13}$$

where λ, μ are the Lamé -constants.

1.4 Boundary conditions

To obtain a well-defined mathematical problem boundary conditions must be added to the formulation given above. Boundary conditions can be prescibed for the displacement field **u** on∂B_u and for the surface tractions **t** on ∂B_σ in the form

$$\hat{\mathbf{u}} = \mathbf{u} \quad on \quad \partial B_u$$

$$\hat{\mathbf{t}} = \mathbf{P}\, \mathbf{n}_o \quad on \quad \partial B_\sigma \tag{1.14}$$

1.5 Variational principles

The variational or weak form of the equations derived in the previous sections are very useful for the following reasons. They are helpful in the study of existence and uniqueness and furthermore basis for many numerical methods e.g. the finite element method. The energy functional for elasticity problems is given for a displacement formulation by

$$\Pi(\mathbf{u}) = \int_{B_o} \psi(\mathbf{E})\, dV - \int_{B_o} \rho \mathbf{b} \cdot \mathbf{u}\, dV - \int_{\partial B_{\sigma}} \hat{\mathbf{t}} \cdot \mathbf{u}\, dA \quad \Rightarrow \quad Min. \tag{1.15}$$

where $\psi(\mathbf{E})$ denotes the stored energy function. $\Pi(\mathbf{u})$ must be a minimum for **u** to be a solution of the associated boundary value problem. The derivative of Π at **u** in the direction $\boldsymbol{\eta}$ is given by

$$D\,\Pi(\mathbf{u}) \cdot \boldsymbol{\eta} = \frac{d}{d\alpha} \Pi(\mathbf{u} + \alpha\, \boldsymbol{\eta}) \Big|_{\alpha=0}$$

and often called first variation of Π in the literature (or Gateaux – derivative)

$$g(\mathbf{u}, \boldsymbol{\eta}) = D\Pi \cdot \boldsymbol{\eta} = \int_{B_o} \frac{\partial \psi}{\partial \mathbf{E}} \cdot (D\mathbf{E} \cdot \boldsymbol{\eta})\, dV - \int_{B_o} \rho \mathbf{b} \cdot \boldsymbol{\eta}\, dV - \int_{\partial B_{\sigma}} \hat{\mathbf{t}} \cdot \boldsymbol{\eta}\, dA = 0 \tag{1.16}$$

The directional derivative $g(\mathbf{u},\boldsymbol{\eta})$ must vanish for Π to be a minimum at $\mathbf{u}$. The variation of $\mathbf{E}$ yields

$$\delta\mathbf{E} = D\mathbf{E}\cdot\boldsymbol{\eta} = \frac{1}{2}\left(Grad\boldsymbol{\eta} + Grad^T\boldsymbol{\eta} + Grad^T\boldsymbol{\eta}\, Grad\mathbf{u} + Grad^T\mathbf{u}\, Grad\boldsymbol{\eta} \right),$$

$$\delta E_{AB} = \frac{1}{2}\left(\eta_{A,B} + \eta_{B,A} + \eta_{C,A} u_{C,B} + u_{C,A}\eta_{C,B} \right). \tag{1.17}$$

With (1.17) and (1.11) equation (1.16) may be rewritten

$$g(\mathbf{u},\boldsymbol{\eta}) = \int_{B_o} \mathbf{S}\cdot\delta\mathbf{E}\, dV - \int_{B_o} \rho\mathbf{b}\cdot\boldsymbol{\eta}\, dV - \int_{\partial B_{\sigma o}} \hat{\mathbf{t}}\cdot\boldsymbol{\eta}\, dA = 0 \tag{1.18}$$

The first term in (1.18) is often refered to as stress divergence term. The last two terms denote the virtual work of the applied loads. Formulation (1.18) is also called principle of virtual work or weak form of the boundary value problem of elastostatics. The principle of virtual work, however, may be derived from equation (1.6) without using the constitutive relation (1.11). Then we can not use algorithms which need the potential $\psi(\mathbf{E})$ as a basis. On the other side the principle of virtual work can be applied to problems for which no potential exist like frictional sliding or non-associated plasticity.

Since (1.18) is nonlinear in $\mathbf{u}$ often iterative numerical methods are employed to solve (1.18). For this purpose — especially for Newton-type methods — the linearization $D\,g$ of (1.18) at $\bar{\mathbf{u}}$ may be necessary

$$g(\bar{\mathbf{u}},\boldsymbol{\eta},\Delta\mathbf{u}) = g(\bar{\mathbf{u}},\boldsymbol{\eta}) + D\,g(\bar{\mathbf{u}},\boldsymbol{\eta})\cdot\Delta\mathbf{u} = 0. \tag{1.19}$$

(1.19) leads to the Newton-Raphson procedure for the solution of the nonlinear boundary value problem (1.18). The linearization of g yields

$$D\,g(\bar{\mathbf{u}},\boldsymbol{\eta})\cdot\Delta\mathbf{u} == \int_{B_o} \{[\frac{\partial\mathbf{S}}{\partial\mathbf{E}}\cdot(D\mathbf{E}\cdot\Delta\mathbf{u})]\cdot\delta\mathbf{E} + \bar{\mathbf{S}}\cdot D(\delta\,\mathbf{E})\cdot\Delta\mathbf{u}\}\, dV. \tag{1.20}$$

With

$$\mathcal{C} = \frac{\partial\mathbf{S}}{\partial\mathbf{E}} \tag{1.21}$$

$$\Delta\mathbf{E} = D\mathbf{E}\cdot\Delta\mathbf{u} = \frac{1}{2}\left(Grad\Delta\mathbf{u} + Grad^T\Delta\mathbf{u} + Grad^T\Delta\mathbf{u}\, Grad\,\bar{\mathbf{u}} + Grad^T\bar{\mathbf{u}}\, Grad\Delta\mathbf{u} \right),$$

$$\Delta E_{AB} = \frac{1}{2}\left(\Delta u_{A,B} + \Delta u_{B,A} + \Delta u_{C,A}\,\bar{u}_{C,B} + \bar{u}_{C,A}\,\Delta u_{C,B} \right). \tag{1.22}$$

$$D(\delta\mathbf{E})\cdot\Delta\mathbf{u} = \frac{1}{2}\left(Grad^T\Delta\mathbf{u}\, Grad\boldsymbol{\eta} + Grad^T\boldsymbol{\eta}\, Grad\Delta\mathbf{u} \right),$$

$$\Delta\delta\, E_{AB} = \frac{1}{2}(\,\Delta u_{C,A}\,\eta_{C,B} + \eta_{C,A}\,\Delta u_{C,B}\,). \tag{1.23}$$

the linearization of (1.18) can be expressed by

$$\begin{aligned} D\,g\,(\bar{\mathbf{u}},\boldsymbol{\eta}\,)\cdot\Delta\mathbf{u} &= \int_{B_o}\{\delta\,\mathbf{E}\cdot\mathcal{C}[\Delta\mathbf{E}\,] + \bar{\mathbf{S}}\cdot\Delta\delta\,\mathbf{E}\,\}\,dV \\ &= \int_{B_o}\{\delta\,E_{AB}\,C_{ABCD}\,\Delta E_{CD} + \bar{S}_{AB}\,\Delta\delta E_{AB}\}\,dV \end{aligned} \tag{1.24}$$

REMARK: Some numerical algorithms are based on the so called updated Lagrangian formulation (Bathe, Ramm and Wilson 1976) which is equivalent to the linearization of (1.18) with reference to the current configuration B of the body.

The principle of virtual work in the current configuration and its linearization may be obtained from (1.18) and (1.24) using the following transformations:

$$\begin{aligned} \mathbf{c}\,[\,\mathbf{a}\,] &= \frac{2}{det\,\mathbf{F}}\,\mathbf{F}\,\mathcal{C}\,[\,\mathbf{F}^T\,\mathbf{a}\,\mathbf{F}\,]\,\mathbf{F}^T \\ c_{iklm} &= \frac{2}{det\,\mathbf{F}}\,F_{iA}\,F_{kB}\,F_{lC}\,F_{mD}\,C_{ABCD} \end{aligned} \tag{1.25}$$

and

$$\begin{aligned} grad\,(\ldots) &= Grad\,(\ldots)\,\mathbf{F}^{-1} \\ (\ldots)_{i,k} &= \frac{\partial(\ldots)_i}{\partial x_k} = \frac{\partial(\ldots)_i}{\partial X_A}\,\frac{\partial X_A}{\partial x_k} = \delta_{iB}\,(\ldots)_{B,A}\,x_{A;k}. \end{aligned} \tag{1.26}$$

together with (1.9) and (1.10) we arrive at the weak form

$$g\,(\mathbf{u},\boldsymbol{\eta}\,) = \int_B \boldsymbol{\sigma}\cdot grad\boldsymbol{\eta}\,dv - \int_B \rho\mathbf{b}\cdot\boldsymbol{\eta}\,dv - \int_{\partial B_\sigma}\hat{\mathbf{t}}\cdot\boldsymbol{\eta}\,da = 0 \tag{1.27}$$

and its linearization

$$\begin{aligned} D\,g\,(\bar{\mathbf{u}},\boldsymbol{\eta}\,)\cdot\Delta\mathbf{u} &= \int_B\{grad\boldsymbol{\eta}\cdot(\,\mathbf{c}\,[\,\frac{1}{2}\,(\,grad\Delta\mathbf{u} + grad^T\Delta\mathbf{u}\,)] + grad\Delta\mathbf{u}\,\bar{\boldsymbol{\sigma}}\,)\,\}\,dv \\ &= \int_B\{\eta_{i,k}\,(\frac{1}{2}c_{iklm}\,(\Delta u_{l,m} + \Delta u_{m,l}\,) + \bar{\sigma}_{ik}\,\Delta u_{i,m})\}\,dv \end{aligned} \tag{1.28}$$

1.6 Matrix formulation

For computational purposes it is convenient to express vectors and tensors in matrix form. Using the following definitions

$$\mathbf{u} = \begin{Bmatrix} u_1 \\ u_2 \\ u_3 \end{Bmatrix}; \quad \boldsymbol{\eta} = \begin{Bmatrix} \eta_1 \\ \eta_2 \\ \eta_3 \end{Bmatrix}; \quad \boldsymbol{\epsilon} = \begin{Bmatrix} E_{11} \\ E_{22} \\ E_{33} \\ 2\,E_{12} \\ 2\,E_{23} \\ 2\,E_{13} \end{Bmatrix}; \quad \mathbf{S} = \begin{Bmatrix} S_{11} \\ S_{22} \\ S_{33} \\ S_{12} \\ S_{23} \\ S_{13} \end{Bmatrix} \tag{1.29}$$

and analogous definitions for $\mathbf{b}$, $\mathbf{t}$, $\delta\mathbf{E}$, $\Delta\mathbf{E}$ the weak form (1.18) and its linearization (1.24) yield

$$g(\mathbf{u},\boldsymbol{\eta}) = \int_{B_o} \delta\boldsymbol{\epsilon}^T\,\mathbf{S}\,dV - \int_{B_o} \boldsymbol{\eta}^T \rho\mathbf{b}\,dV - \int_{\partial B_{\sigma_o}} \boldsymbol{\eta}^T\,\hat{\mathbf{t}}\,dA = 0 \tag{1.30}$$

and

$$D\,g(\bar{\mathbf{u}},\boldsymbol{\eta})^T\,\Delta\mathbf{u} = \int_{B_o} \{\delta\,\boldsymbol{\epsilon}^T\,\mathbf{C}\,\Delta\boldsymbol{\epsilon} + \Delta\delta\,\boldsymbol{\epsilon}^T\,\bar{\mathbf{S}}\,\}\,dV. \tag{1.31}$$

Remark: If we restrict ourselves to two-dimensional problems like plane stress or plane strain applications the definitions (1.29) used in (1.30) and (1.31) reduce to

$$\mathbf{u} = \begin{Bmatrix} u_1 \\ u_2 \end{Bmatrix}; \quad \boldsymbol{\eta} = \begin{Bmatrix} \eta_1 \\ \eta_2 \end{Bmatrix}; \quad \boldsymbol{\epsilon} = \begin{Bmatrix} E_{11} \\ E_{22} \\ 2\,E_{12} \end{Bmatrix}; \quad \mathbf{S} = \begin{Bmatrix} S_{11} \\ S_{22} \\ S_{12} \end{Bmatrix} \tag{1.32}$$

Furthermore, the constitutive matrix $\mathbf{C}$ is given for plane strain by

$$\mathbf{C} = \begin{bmatrix} \lambda + 2\,\mu & 0 & 0 \\ 0 & \lambda + 2\,\mu & 0 \\ 0 & 0 & \mu \end{bmatrix}. \tag{1.33}$$

2 Finite element formulation for solid problems

In this chapter the finite element formulation for the boundary value problem of elastostatics stated in chapter 1 will be considered. For this purpose we introduce a standard finite element discretization

$$B^h = \bigcup_{e=1}^{n_e} \Omega_e$$

with n_e elements and the volume of an element denoted by Ω_e.

For two-and three-dimensional solid applications isoparametric finite elements are widely used. Their advantage can be seen in the good modeling of complicated geometries. The isoparametric concept means that the geometry $\mathbf{X}$ and the displacement field $\mathbf{u}$ are interpolated within an element Ω_e by the same functions (see Figure 2.1)

$$\mathbf{x}_e^h = \sum_{i=1}^{n} N_i \mathbf{x}_i \,,$$

$$\mathbf{u}_e^h = \sum_{i=1}^{n} N_i \mathbf{v}_i \tag{2.1}$$

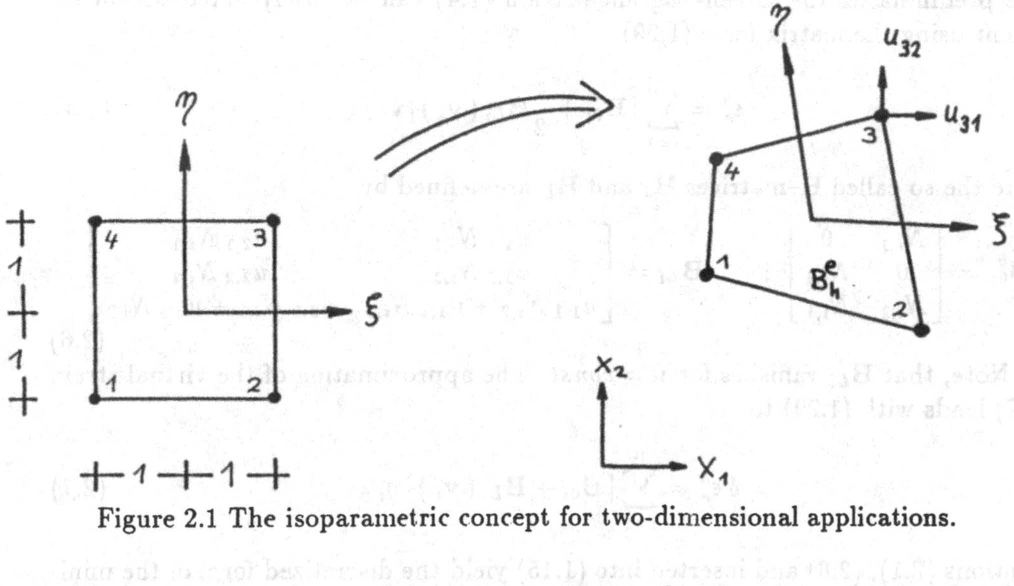

Figure 2.1 The isoparametric concept for two-dimensional applications.

The functions N_i are called shape functions. One basic requirement for the choice of N_i — the C^o - continuity — stems from (1.15). Furthermore N_i should be a complete polynominal in x_1 , x_2 and x_3 (see Zienkiewics (1979), Ch. 7.). There are several ways for the generation of shape functions. Here we restrict ourselves to the Lagrangian family. The reader may consult standard textbooks on finite elements (e.g. Zienkiewics(1979) or Bathe(1982)). An example for the choice of shape functions within the Lagrangian family is given for an 8-node brick element by

$$N_i = \frac{1}{2}(1+\xi_i\xi)\frac{1}{2}(1+\eta_i\eta)\frac{1}{2}(1+\phi_i\phi). \tag{2.2}$$

The last term in (2.2) vanishes in two-dimensional applications. The shape functions are defined in a local coordinate system ξ, η, ϕ therefore a transformation to the global coordinates x_1, x_2, x_3 is necessary (see Figure 2). Since the following formulation is the same for two- and three- dimensional problems we will restrict ourselves to two-dimensional applications for the ease of understanding. In the weak form (1.18) the derivatives of the displacement field are used, within the isoparametric concept we obtain

$$\frac{\partial \mathbf{u}_e^h}{\partial X_A} = \sum_{i=1}^{n} \frac{\partial N_i(\xi,\eta,\phi)}{\partial X_A}\mathbf{v}_i \tag{2.3}$$

where the chain rule has to be used for the evaluation of the partial derivative of N_i with respect to X_A (see e.g. Zienkiewicz (1979), Ch. 8)

$$\begin{Bmatrix} N_{i,1} \\ N_{i,2} \end{Bmatrix} \frac{1}{det\,\mathbf{J}} \begin{bmatrix} x_{2,\eta} & -x_{2,\xi} \\ -x_{1,\eta} & x_{1,\xi} \end{bmatrix} \begin{Bmatrix} N_{i,\xi} \\ N_{i,\eta} \end{Bmatrix} \quad with \quad \mathbf{J} = \begin{bmatrix} x_{1,\xi} & -x_{2,\xi} \\ -x_{1,\eta} & x_{2,\eta} \end{bmatrix} \tag{2.4}$$

where $det\,\mathbf{J}$ is the *Jacobian* of the transformation between $d\xi$, $d\eta$ and dx_1, dx_2. With these preliminaries the Green-Lagrange strain (1.4) can be interpolated within an element using the matrix form (1.29)

$$\boldsymbol{\epsilon}_e^h = \sum_{i=1}^{n} [\mathbf{B}_{0i} + \frac{1}{2}\mathbf{B}_{Li}(\mathbf{v}_i)]\,\mathbf{v}_i \tag{2.5}$$

where the so called B–matrices $\mathbf{B}_0$ and $\mathbf{B}_L$ are defined by

$$\mathbf{B}_{0i} = \begin{bmatrix} N_{i,1} & 0 \\ 0 & N_{i,2} \\ N_{i,2} & N_{i,1} \end{bmatrix} ; \qquad \mathbf{B}_{Li} = \begin{bmatrix} u_{1,1}\,N_{i,1} & u_{2,1}\,N_{i,1} \\ u_{1,2}\,N_{i,2} & u_{2,2}\,N_{i,2} \\ u_{1,1}\,N_{i,2} + u_{1,2}\,N_{i,1} & u_{2,2}\,N_{i,1} + u_{2,1}\,N_{i,2} \end{bmatrix} \tag{2.6}$$

Note, that $\mathbf{B}_{Li}$ vanishes for $\mathbf{u} = const.$ The approximation of the virtual strain (1.17) leads with (1.29) to

$$\delta\boldsymbol{\epsilon}_e^h = \sum_{i=1}^{n} [\mathbf{B}_{0i} + \mathbf{B}_{Li}(\mathbf{v}_i)]\,\boldsymbol{\eta}_i \tag{2.7}$$

Equations (2.1), (2.6) and inserted into (1.15) yield the discretized form of the minimum principle for an element Ω_e

$$\Pi_e^h = \sum_{i=1}^{n}\sum_{k=1}^{n}\frac{1}{2}\int_{\Omega_e^h} [(\mathbf{B}_{0i} + \frac{1}{2}\mathbf{B}_{Li})\mathbf{v}_i]^T\,\mathbf{C}\,[(\mathbf{B}_{0k} + \frac{1}{2}\mathbf{B}_{Lk})\mathbf{v}_k]\,d\Omega - \text{"}load\ terms\text{"} \tag{2.8}$$

Furthermore the weak form (1.18) or (1.30) is obtained with the aid of (2.7) for one element Ω_e

$$g_e^h(\mathbf{v},\boldsymbol{\eta}) = \sum_{i=1}^{n} \boldsymbol{\eta}_i^T \{ \int_{\Omega_e} (\mathbf{B}_{0i} + \mathbf{B}_{Li})^T \mathbf{S}_i \, d\Omega - \int_{\Omega_e} \mathbf{N}_i^T \rho \mathbf{b}_i \, d\Omega - \int_{\partial\Omega_e} \mathbf{N}_i^T \hat{\mathbf{t}}_i \, d(\partial\Omega) \}. \tag{2.9}$$

The assembly process of all element contributions to the global algebraic system of equations for the problem at hand shall be denoted by the operator which incorporates boundary and transition conditions for the displacements into the global structure of the system of equations. With this notation the so called residual $\mathbf{r}$ in the global form is given by

$$\boldsymbol{\eta}^T \mathbf{r}(\mathbf{v}) = \bigcup_{e=1}^{n_e} g_e^h; \qquad \forall \boldsymbol{\eta} \epsilon \mathcal{R}^K \tag{2.10}$$

where K denotes the total number of degrees of freedom. Equation (2.10) yields for arbitrary $\boldsymbol{\eta}$

$$\mathbf{r}(\mathbf{v}) = \mathbf{f}(\mathbf{v}) - \mathbf{p} = 0. \tag{2.11}$$

The linearization (1.24) leads to the definition of the tangential stiffness matrix. With (1.22), (1.23), (1.29), (1.31), (2.1), (2.6) and (2.7) the following form of the tangent matrix is obtained

$$D g_e^h(\bar{\mathbf{v}},\boldsymbol{\eta}_i)\,\Delta\mathbf{v}_k = \sum_{i=1}^{n}\sum_{k=1}^{n} \boldsymbol{\eta}_i^T \int_{\Omega_e^h} [(\mathbf{B}_{0i} + \mathbf{B}_{Li})^T \mathbf{C} (\mathbf{B}_{0k} + \mathbf{B}_{Lk}) + \mathbf{I} G_{ik}] \, d\Omega \, \Delta\mathbf{v}_k \tag{2.12}$$

where for two-dimensional problems G_{ik} is defined by

$$G_{ik} = [N_{i,1} \quad N_{i,2}] \begin{bmatrix} S_{11} & S_{12} \\ S_{21} & S_{22} \end{bmatrix} \begin{Bmatrix} N_{k,1} \\ N_{k,2} \end{Bmatrix}.$$

The assembly procedure leads to the global tangential stiffness matrix $\mathbf{K}_T$

$$\boldsymbol{\eta}^T \mathbf{K}_T \Delta\mathbf{v} = \bigcup_{e=1}^{n_e} D g_e^h \Delta\mathbf{v}_e. \tag{2.13}$$

In (2.10) and (2.12) the evaluation of the integrals is performed according to the Gaussian quadrature formulas (see e.g. Zienkiewicz (1979), Ch.8). Finally the discretized form of the Newton-Raphson procedure (1.19) for the solution of the nonlinear boundary value problem may be stated

$$g(\bar{\mathbf{v}},\boldsymbol{\eta})\,\Delta\mathbf{v} = \boldsymbol{\eta}^T (\mathbf{r}(\mathbf{v}) + \mathbf{K}_T \Delta\mathbf{v}) = 0. \tag{2.14}$$

3 Algorithms

After discretizing the continuous problem by means of finite elements we get the following finite-dimensional problem: We have to look for a vector $\mathbf{v}_0 \in \mathcal{R}^K$ (e. g. nodal displacements) such that

$$\Pi(\mathbf{v}_0) = \min_{\mathbf{v}\in\mathcal{R}^K} \Pi(\mathbf{v}) \tag{3.1}$$

with

$$\Pi(\mathbf{v}) := W(\mathbf{v}) - \mathbf{p}^T\mathbf{v} \tag{3.2}$$

or, by discretizing the principle of virtual work

$$\mathbf{f}(\mathbf{v}_0) = \mathbf{p} \qquad \text{i. e.} \quad \mathbf{r}(\mathbf{v}_0) = 0 \tag{3.3}$$

(with $\mathbf{f} = \nabla W$ if $\mathbf{f} : \mathcal{R}^K \to \mathcal{R}^K$ is a potential function).

To solve these (nonlinear) problems there come into question essentially three types of algorithms: Descent methods, path following methods or smoothing methods. Their basic ideas and their applicability for discretized problems in structural mechanics are the subject of the following chapter.

3.1 Descent methods

3.1.1 Basic idea

There is a fundamental underlying structure for almost all descent algorithms. One starts at an initial point; determines, according to a fixed rule, a direction of movement; and then moves in that direction to a (relative) minimum of the energy on that line. At the new point a new direction is determined and the process is repeated. The primary difference between algorithms rest with the rule by which successive directions of movement are selected. Once the selection is made, all algorithms call for movement to the minimum point on the corresponding line. (This process of determining the minimum point on a given line is called "line search").

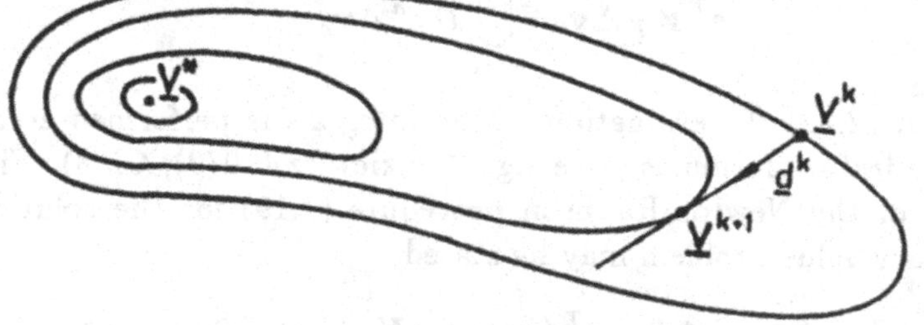

Fig. 3.1 Descent step

$\mathbf{v}_0$: solution of the discretized problem,
$\mathbf{d}^k$: descent direction,
$\mathbf{v}^{k+1} := \mathbf{v}^k + \alpha_k \mathbf{d}^k$,
α : line search parameter.

3.1.2 Examples for descent directions

a) Steepest descent:

$$\mathbf{d}^k := -\nabla\Pi(\mathbf{v}^k) = -\mathbf{r}(\mathbf{v}^k) \qquad \text{(residual force)} \tag{3.4}$$

Convergence rate: linear

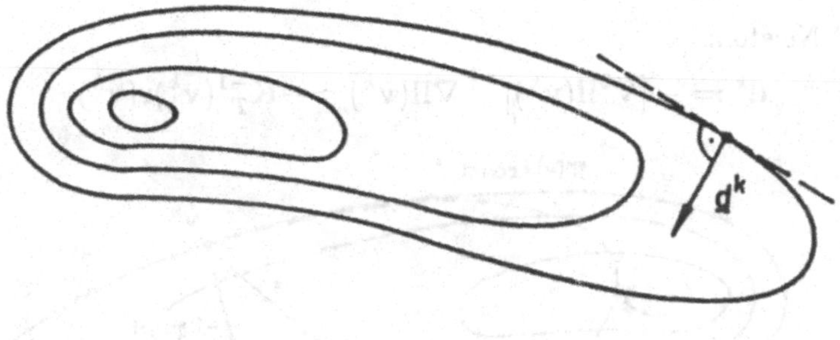

Fig. 3.2 Gradient step

Stepwise convergence is governed by cond$\nabla^2\Pi(\mathbf{v}_0)$, which essentially is given by the ratio of the largest to the smallest eigenvalue of $\nabla^2\Pi(\mathbf{v}_0)$. There is no prediction for the length of the step , thus an extensive use of the line search is necessary.

b) Conjugate Gradients (CG):

$$\mathbf{d}^0 := -\nabla\Pi(\mathbf{v}^0) = -\mathbf{r}(\mathbf{v}^0),$$
$$\mathbf{d}^k := -\nabla\Pi(\mathbf{v}^k) + \beta_k \mathbf{d}^{k-1}, \qquad k \geq 1, \tag{3.5}$$

where

$$\beta_k := \frac{\|\nabla\Pi(\mathbf{v}^k)\|^2}{\|\nabla\Pi(\mathbf{v}^{k-1}\|^2} \qquad \text{(Fletcher/Reeves).} \tag{3.6}$$

The motivation for these formulas is given by the quadratic termination property, since for problems of the form:

$$Q(\mathbf{v}) := \frac{1}{2}\mathbf{v}^T\mathbf{K}\mathbf{v} - \mathbf{p}^T\mathbf{v} \tag{3.7}$$

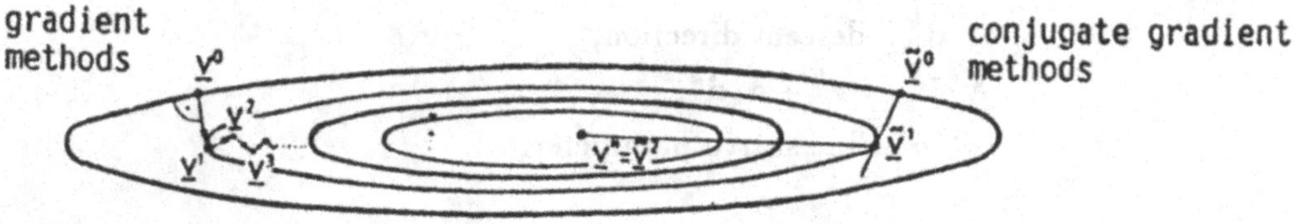

Fig. 3.3 Comparison of gradient and conjugate gradient methods

one gets

$$\mathbf{d}^{iT}\mathbf{K}\mathbf{d}^{j} = 0 \qquad \text{for } i \neq j. \tag{3.8}$$

A geometric comparison of conjugate gradients and gradient methods is given in Fig. 3.3.

c) (Damped) Newton:

$$\mathbf{d}^k := -\left[\nabla^2\Pi(\mathbf{v}^k)\right]^{-1}\nabla\Pi(\mathbf{v}^k) = -\mathbf{K}_T^{-1}(\mathbf{v}^k)\mathbf{r}(\mathbf{v}^k) \tag{3.9}$$

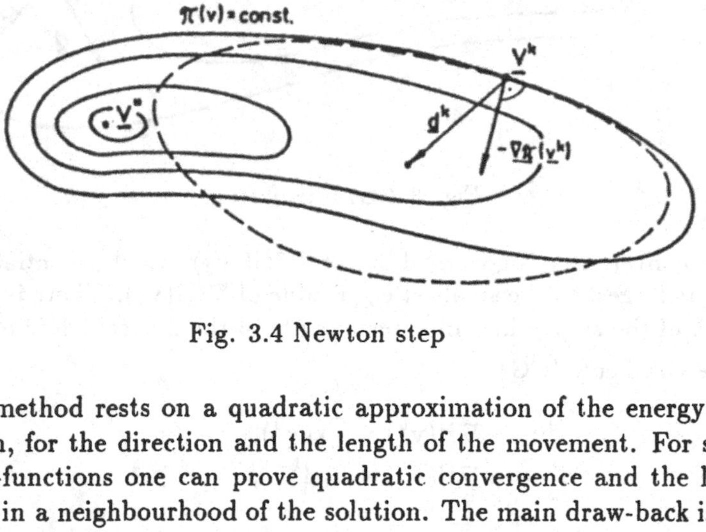

Fig. 3.4 Newton step

Newton's method rests on a quadratic approximation of the energy and gives a proposal both, for the direction and the length of the movement. For sufficiently smooth energy-functions one can prove quadratic convergence and the line search can be omitted in a neighbourhood of the solution. The main draw-back is the large effort in order to compute and to factorize the Hessian (tangential stiffness matrix) $\nabla^2\Pi$ for large scale problems.

d) Quasi-Newton:

$$\mathbf{d}^0 := -\mathbf{H}_0\nabla\Pi(\mathbf{v}^0) = -\mathbf{H}_0\mathbf{r}(\mathbf{v}^0) \tag{3.10}$$

(where $\mathbf{H}_0 := \left[\nabla^2\Pi(\mathbf{v}^0)\right]^{-1}$ or a good approximation of the Hessian)

$$\mathbf{d}^k := -\mathbf{H}_k\nabla\Pi(\mathbf{v}^k) \tag{3.11}$$

where $\mathbf{H}_k^{-1}$ should be positive definite, symmetric and subsequent iterationpoints should fulfil the quasi-Newton equation:

$$\mathbf{H}_k(\nabla\Pi(\mathbf{v}^{k+1}) - \nabla\Pi(\mathbf{v}^k)) = \mathbf{v}^{k+1} - \mathbf{v}^k \qquad \text{(secant equation).} \tag{3.12}$$

The most popular and perhaps one of the most clever schemes for constructing the inverse Hessian was proposed by Broyden, Fletcher, Goldfarb and Shanno (BFGS-method). It has very nice properties concerning the numerical stability, which is superior to most other update formulas, creates conjugate directions for quadratic problems and the update formula could be given in a product form which avoids explicit updating of the inverse Hessian

$$\mathbf{H}_{k+1} := (\mathbf{I} - \mathbf{p}^k\mathbf{q}^{kT})\mathbf{H}_k(\mathbf{I} - \mathbf{q}^k\mathbf{p}^{kT}) \tag{3.13}$$

(which could be used in a recursive way) where

$$\mathbf{p}^k := \left(\frac{\boldsymbol{\delta}^{kT}\boldsymbol{\gamma}^k}{\boldsymbol{\delta}^{kT}\mathbf{H}_{k-1}\boldsymbol{\delta}^k}\right)^{-\frac{1}{2}}\mathbf{H}_{k-1}\boldsymbol{\delta}^k - \boldsymbol{\gamma}^k, \tag{3.14}$$

$$\mathbf{q}^k := \frac{1}{\boldsymbol{\delta}^{kT}\boldsymbol{\gamma}^k}\boldsymbol{\delta}^k, \tag{3.15}$$

$$\boldsymbol{\delta}^k := \mathbf{v}^k - \mathbf{v}^{k+1}, \tag{3.16}$$

$$\boldsymbol{\gamma}^k := \nabla\Pi(\mathbf{v}^k) - \nabla\Pi(\mathbf{v}^{k-1}). \tag{3.17}$$

The last point is the main reason for its applicability for finite element methods, since explicit updating would destroy the sparse structure of the Hessian and would lead to an enormous amount of storage place. The method has superlinear convergence properties, needs additional storage place only for the updating vectors $\mathbf{p}^k$, $\mathbf{q}^k$ and demands for a line search in order to preserve the positive definiteness of $\mathbf{H}_k$ (Matthies, Strang, 1979).

3.1.3 Preconditioning

Since it is the ratio of eigenvalues that influences convergence properties of the algorithms one can hope to find (by a transformation of variable) a problem with nicer convergence properties.

With $\mathbf{w} := \tilde{\mathbf{K}}^{-1/2}\mathbf{v}$ one transforms the quadratic problem $Q(\mathbf{v})$ to $\hat{Q}(\mathbf{w})$, where

$$\begin{aligned}\hat{Q}(\mathbf{w}) : &= \frac{1}{2}\mathbf{w}^T\tilde{\mathbf{K}}^{-1/2T}\mathbf{K}\tilde{\mathbf{K}}^{-1/2}\mathbf{w} - \mathbf{p}^T\tilde{\mathbf{K}}^{-1/2}\mathbf{w} \\ &= \frac{1}{2}\mathbf{w}^T\hat{\mathbf{K}}\mathbf{w} - \hat{\mathbf{p}}^T\mathbf{w}\end{aligned}$$

with

$$\hat{\mathbf{K}} := \tilde{\mathbf{K}}^{-1/2T}\mathbf{K}\tilde{\mathbf{K}}^{-1/2},$$
$$\hat{\mathbf{p}}^T := \mathbf{p}^T\tilde{\mathbf{K}}^{-1/2}.$$

The optimal choice for $\tilde{\mathbf{K}}$ would be $\tilde{\mathbf{K}} = \mathbf{K}$, i. e. $\hat{\mathbf{K}} = \mathbf{I}$. If this is not possible it should be desireable to choose $\tilde{\mathbf{K}}$ in such a way, that $\tilde{\mathbf{K}}$ is a good approximation of $\mathbf{K}$ (in the sense, that the ratio of eigenvalues of $\hat{\mathbf{K}}$ is small) and that the evaluation and factorization of $\tilde{\mathbf{K}}$ is not too expensive. The effect of such a transforamtion on the descent direction would be for the gradient method:

$$\mathbf{d}^k := -\tilde{\mathbf{K}}^{-1}\nabla\Pi(\mathbf{v}^k) \tag{3.18}$$

and for the preconditioned conjugate gradients:

$$\mathbf{d}^k := -\tilde{\mathbf{K}}^{-1}\nabla\Pi(\mathbf{v}^k) + \tilde{\beta}_k\mathbf{d}^{k-1} \tag{3.19}$$

with

$$\tilde{\beta}_k := \frac{\nabla\Pi(\mathbf{v}^k)\tilde{\mathbf{K}}^{-1}\nabla\Pi(\mathbf{v}^k)}{\nabla\Pi(\mathbf{v}^{k-1})\tilde{\mathbf{K}}^{-1}\nabla\Pi(\mathbf{v}^{k-1})}. \tag{3.20}$$

In the following we will give an example for the construction of a preconditioning matrix by hierarchical shape-functions.

Let $\{\bar{N}_i\}_{,i=1,\ldots,n}$ be the hierarchical basis (HB) belonging to a given mesh, constructed by m uniform refinements of a coarse mesh (n_k: Number of basisfunctions belonging to the k-th mesh) and $\{N_i\}_{i=1,\ldots,n}$ the corresponding nodal basis (NB). Let $\mathbf{T}$ be the matrix which transforms the displacement vector with respect to the hierarchical basis into the dispacement vector with respect to the nodal basis.

One gets:

(i) The Hessian with respect to the (HB) is given by $\bar{\mathbf{K}} = \mathbf{TKT}^T$ which has the property: cond($\bar{\mathbf{K}}$) <cond($\mathbf{K}$).

(ii) A simple (and good) approximation for $\bar{\mathbf{K}}^{-1}$ will be $\mathbf{K}_1^{-1}$ with

$$\mathbf{K}_1 := \begin{pmatrix} \bar{\mathbf{K}}_1 & & & \\ & \bar{\mathbf{D}}_2 & 0 & \\ & 0 & \ddots & \\ & & & \bar{\mathbf{D}}_m \end{pmatrix} \tag{3.21}$$

and $\bar{\mathbf{K}}_1$ is the tangential stiffness matrix (Hessian) on the coarsest mesh, while the diagonal matrices $\bar{\mathbf{D}}_k$, $(k = 2, \ldots, m)$ are defined by:

$$\bar{\mathbf{D}}_k := \begin{pmatrix} \bar{K}_{n_{k-1},n_{k-1}} & & \\ 0 & \ddots & 0 \\ & & \bar{K}_{n_k,n_k} \end{pmatrix}. \tag{3.22}$$

(iii) $\mathbf{T}^T$ has a very simple structure which will be studied in more detail in chapter 3.3.

A proposition for the preconditioner would be

$$\tilde{\mathbf{K}}^{-1} := \mathbf{T}^T \mathbf{K}_1^{-1} \mathbf{T}, \tag{3.23}$$

which has many nice properties (Yserentant, 1983). Its only draw back is the complex datastructure.

3.1.4 Step size control

One can use all standard routines for minimizing a single variable function, as bisection, curve fitting, backtracking,..., which are explained in standard textbooks on optimization or numerical analysis, e.g. (Luenburger, 1984), (Schwetlick, 1978) (Denis,Schnabel, 1983).

3.2 Continuation Methods

Within nonlinear finite element formulation we start from the equilibrium equations, which will be denoted by

$$\mathbf{r}(\mathbf{v}, \Lambda) = 0. \tag{3.24}$$

$\mathbf{v}$ describes the deformation of the structure while Λ is the parameter by which the load can be changed. Eq. (3.24) represent N relations between $N+1$ degrees of freedom. In general, to specify a particular point on $\mathbf{r}$ we must introduce an extra equation to complement eq. (3.24), see e.g. Riks (1984), Wagner, Wriggers (1985), Kahn, Wagner (1986).

$$g(\mathbf{v}, \Lambda) - \gamma_i = 0. \tag{3.25}$$

From a geometrical point of view, this equation describe, in $\mathcal{R}_{N+1}$, intersections of the curve defined by eq. (3.24),see Fig.3.5.

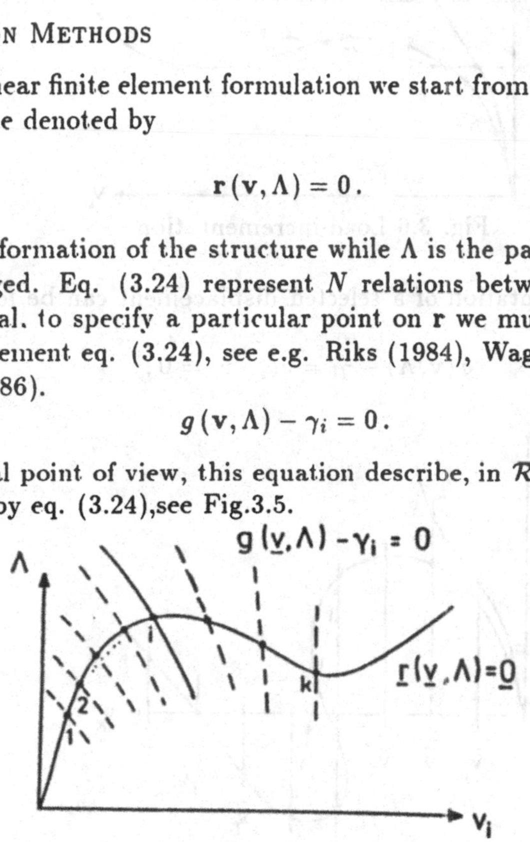

Fig. 3.5 Geometrical interpretation of continuation methods

The basic feature of a continuation method is the iterative computation of a set of points

$$\mathbf{y}(\gamma_i) = \begin{bmatrix} \mathbf{v}(\gamma_i) \\ \Lambda(\gamma_i) \end{bmatrix} \tag{3.26}$$

with

$$\gamma_i : \quad \gamma_0 < \gamma_1 < \gamma_2 < \ldots < \gamma_i < \gamma_k .$$

This leads to k solution points on the nonlinear equilibrium path, eq. (3.24) and may also called incremental-iterative solution strategy. Two classical approaches may be discussed. If we choose a load incrementation, the constraint condition, eq. (3.25), yields

$$g(\mathbf{v}, \Lambda) - \gamma_i = \Lambda - \bar{P}_i = 0 . \qquad (3.27)$$

where $\bar{P}_i$ is a prescribed value of the load at increment i.

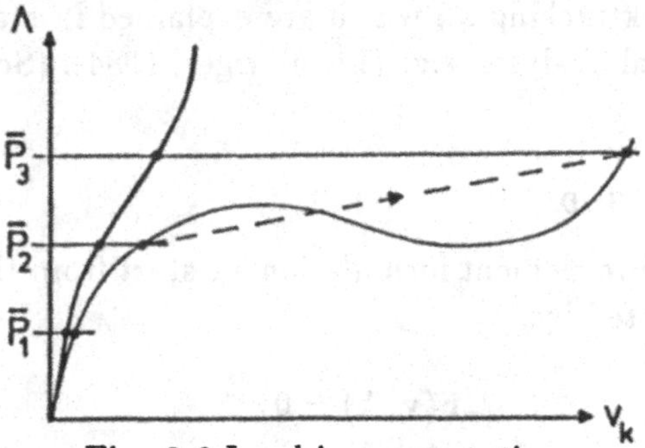

Fig. 3.6 Load incrementation

Similar an incrementation of a selected displacement can be formulated. Now eq. (3.25) leads to

$$g(\mathbf{v}, \Lambda) - \gamma_i = v_k - \bar{v}_i = 0 . \qquad (3.28)$$

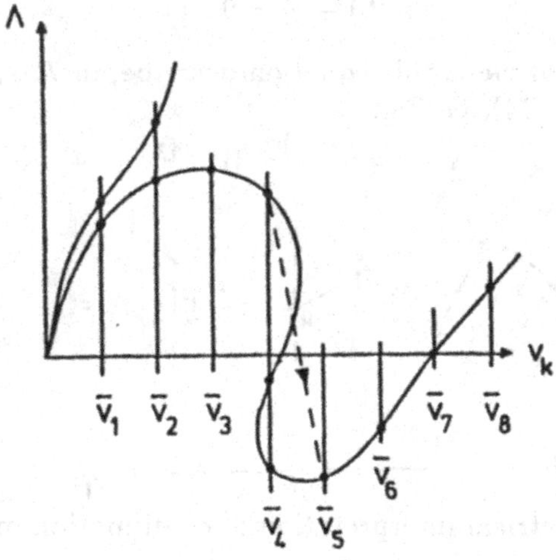

Fig. 3.7 Displacement incrementation

In Figure 3.6 and 3.7 the limitations of both approaches can be realized by the dash-dotted lines. Therefore we will treat the constraint condition in a more general way. If γ is specified as a path-parameter it is obvious to choose γ , thus $g(\mathbf{v},\Lambda)$, in such a way that the relationship between the natural measure of progress along the path s and γ is one-to-one. This condition corresponds to the requirement that $ds/d\gamma$ should keep its sign along the segment under consideration. Differentiation of the constraint condition with respect to γ along the path, see Figure 3.8, leads to

$$\frac{d}{d\gamma}[g(\mathbf{v},\Lambda)-\gamma=0] \quad\Longrightarrow\quad (\mathbf{n}^T\dot{\mathbf{y}})\frac{ds}{d\gamma}-1=0 \tag{3.29}$$

with

$$\mathbf{y}=\begin{pmatrix}\mathbf{v}\\ \Lambda\end{pmatrix}$$
$$\mathbf{n}=Grad\, g$$
$$\mathbf{t}=\frac{d}{ds}\mathbf{y}=\dot{\mathbf{y}}\,.$$

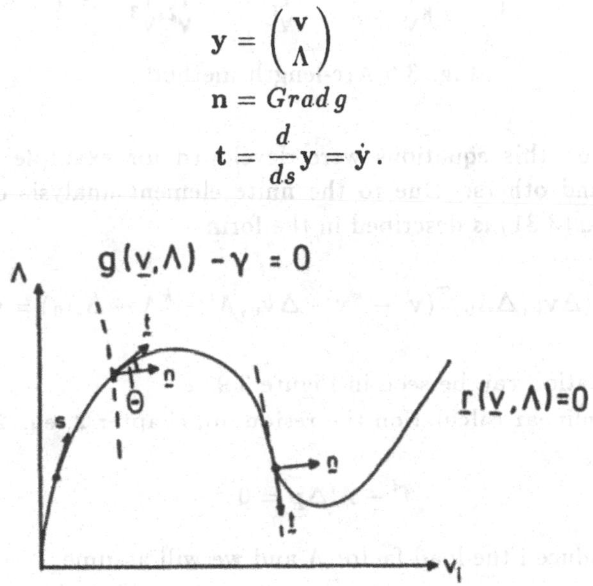

Fig. 3.8 Regular and singular intersections

We see that for $ds/d\gamma$ to keep its sign along the segment of interest, the angle between $\mathbf{n}$ and $\dot{\mathbf{y}}$ at the points of intersection should stay acute, and that the singular case occurs when

$$\mathbf{n}^T\dot{\mathbf{y}}=0\,;\quad \theta=\pi/2\,. \tag{3.30}$$

This point of the curve is a turning point, see Figure 3.8. Therefore the best possible choice in this sense would be $\theta=0$, which we achieve if we take $\mathbf{n}^T\dot{\mathbf{y}}=1$. This leads to $s=\gamma$. Although it is not possible to realize this idea in an exact manner there are several possibilities to do it approximately. Here we will restrict ourselves to one example for the choice of the constraint condition.

$$g=\mathbf{t}^T(\gamma_k)\,[\mathbf{y}(\gamma)-\mathbf{y}(\gamma_k)]-[\gamma-\gamma_k]=0\,. \tag{3.31}$$

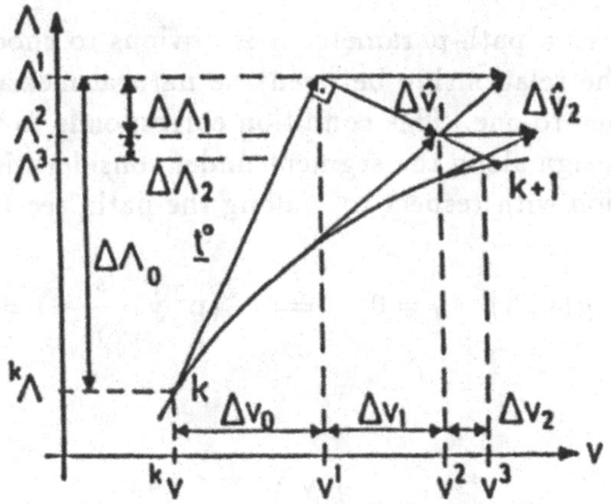

Fig. 3.9 Arc-length-method

Algorithms based on this equations were developed for example by Riks (1972), Wempner (1971) and others. Due to the finite element analysis of chapter 2 the constraint condition (3.31) is described in the form

$$g^i = (\Delta \mathbf{v}_0, \Delta\Lambda_0)^T(\mathbf{v}^i - {}^k\mathbf{v} - \Delta\mathbf{v}_0, \Lambda^i - {}^k\Lambda - \Delta\Lambda_0) = 0\,. \tag{3.32}$$

The appropiate notation can be seen in Figure 3.9.

Within the nonlinear calculation the residuum, chapter 2, eq. 2.11, is reformulated

$$\mathbf{f}^i - \Lambda^i \Delta\mathbf{p} = 0\,. \tag{3.33}$$

Here, we have introduced the load factor Λ and we will assume

$${}^k\mathbf{p} = {}^k\Delta\mathbf{p} \tag{3.34}$$

with an arbitrary load increment $\Delta\mathbf{p}$. The consistent linearization of the eq. (3.29) and (3.30) leads to an unsymmetric set of equations, see Wagner, Wriggers(1985).

$$\begin{bmatrix} \mathbf{K}_T & \Delta\mathbf{p} \\ \Delta\mathbf{v}_0^T & \Delta\Lambda_0 \end{bmatrix} \begin{Bmatrix} \Delta\mathbf{v}_i \\ \Delta\Lambda_i \end{Bmatrix} = - \begin{Bmatrix} \mathbf{f}^i - \Lambda^i\Delta\mathbf{p} \\ (\Delta\mathbf{v}_0, \Delta\Lambda_0)^T(\mathbf{v}^i - {}^k\mathbf{v} - \Delta\mathbf{v}_0, \Lambda^i - {}^k\Lambda - \Delta\Lambda_0) \end{Bmatrix} \tag{3.35}$$

In equation (3.35) $\mathbf{K}_T$ is the tangent stiffness matrix or Hessian. The numerical treatment of eq. (3.35) is based on a partitioning method. With the definition

$$\Delta\mathbf{v}_i = \Delta\Lambda_i \Delta\mathbf{v}_i^I + \Delta\mathbf{v}_i^{II} \tag{3.36}$$

the related algorithm, commonly called arc-length-method, can be derived. The main steps of this algorithm are stated below.

1. *Set initial values* $\mathbf{v}_0 = 0$

2. *For* $k = 1,\ldots,n$ *load – increments do* ${}^k\mathbf{p} = {}^{k-1}\mathbf{p} + \Delta\mathbf{p}_k$

3. *For* $i = 1,\ldots,m$ *until convergence do*

3.1) *Solve*
$$\mathbf{K}_T^i \Delta\mathbf{v}_i^I = \Delta\mathbf{p}$$
$$\mathbf{K}_T^i \Delta\mathbf{v}_i^{II} = -(\mathbf{f}^i - \Lambda^i \Delta\mathbf{p})$$

3.2) *Calculate the incremental load scaling factor*
$$\Delta\Lambda_i = -\frac{\Delta\mathbf{v}_0^T \Delta\mathbf{v}_i^{II}}{\Delta\mathbf{v}_0^T \Delta\mathbf{v}_i^I + \Delta\Lambda_o}$$

3.3) *Calculate the incremental displacements*
$$\Delta\mathbf{v}_i = \Delta\Lambda_i \Delta\mathbf{v}_i^I + \Delta\mathbf{v}_i^{II}$$

3.4) *Update displacements and incremental load scaling factor*
$$\mathbf{v}^{i+1} = \mathbf{v}^i + \Delta\mathbf{v}_i$$
$$\Lambda^{i+1} = \Lambda^i + \Delta\Lambda_i$$

3.5) *Check for convergence* $\qquad (3.37)$
$$\| \mathbf{r}^{i+1} \| \leq tol$$
Terminate goto 2
Else goto 3

The algorithm is of Newton type if $\mathbf{K}_T^i$ is the Hessian. With $\mathbf{K}_T^i = \mathbf{K}_T^0$ we have employed a modified Newton iteration. According to chapter 3.1.2 quasi- Newton-Methods can be introduced.

Examples in which continuation methods have to be used will be given in chapter 4.6.

3.3 Multi Grid Methods

3.3.1 Concept of Multi Grid Methods

Assume (at first) that two meshes are given, denoted by Ω_H (coarse mesh) and Ω_h (fine mesh). Let $\mathbf{v}_{h0}$ be the exact solution of an equation $\mathbf{f}_h(\mathbf{v}_h) = \mathbf{p}_h$ on the fine mesh and $\mathbf{v}_h$ an approximation of $\mathbf{v}_{h0}$.

IDEA:

(i) Smooth the errors $\mathbf{e}_h = \mathbf{v}_{h0} - \mathbf{v}_h$ on the fine mesh Ω_h.
(ii) Determine the smooth part of the errors on the coarse mesh Ω_H.

WANTED EFFECTS:

(i) An order of magnitude reduction in the error per calculation cycle.
(ii) The number of operations is $O(n)$ (n : number of variables) i.e. independent of h.

SIMPLEST VERSION OF A MG-ALGORITHM: TWO GRID CYCLE (Stüben, Trottenberg (1982)

STEP 1: "Smoothing of the Error" (Part I)

Determine $\bar{\mathbf{v}}_h$ from $\mathbf{v}_h$ by applying γ-times an iteration procedure with smoothing properties for the error:

$$\bar{\mathbf{v}}_h = (\mathbf{S}_h)^\gamma \mathbf{v}_h.$$

STEP 2: "Coarse grid correction"

a) determine the defect:

$$\mathbf{r}_h = -(\mathbf{f}_h(\mathbf{v}_{h0}) - \mathbf{f}_h(\bar{\mathbf{v}}_h))$$

b) transfer of the defect to the coarse grid:

$$\mathbf{r}_h = \mathbf{T}_h^H \mathbf{v}_h$$

c) transfer of the approximation $\bar{\mathbf{v}}_h$ to the coarse grid:

$$\bar{\mathbf{v}}_H = \hat{\mathbf{T}}_h^H \bar{\mathbf{v}}_h$$

d) solve on the coarse grid:

$$\mathbf{f}_H(\mathbf{w}_H) = \mathbf{f}_H(\bar{\mathbf{v}}_H) - \mathbf{r}_H$$

e) compute the correction on the coarse grid:

$$\mathbf{s}_H = \mathbf{w}_H - \bar{\mathbf{v}}_H$$

f) transfer of the correction to the fine grid:

$$\mathbf{s}_h = \mathbf{T}_H^h \mathbf{s}_H$$

g) correct $\bar{\mathbf{v}}_h$:

$$\tilde{\mathbf{v}}_h = \bar{\mathbf{v}}_h + \mathbf{s}_h$$

(for linear problems c), d), e) are equivalent to $\mathbf{f}_H \mathbf{s}_H = -\mathbf{r}_H$

STEP 3: "Smoothing" (Part II)

Determine a new approximate solution $\mathbf{v}_h$ from $\tilde{\mathbf{v}}_h$ by applying μ-times an iteration procedure with error smoothing properties:

$$\mathbf{v}_h = (\mathbf{S}_h)^\mu \tilde{\mathbf{v}}_h$$

REMARKS:

1) The operator for error smoothening $\mathbf{S}_h$ may be: SOR, Gauss-Seidel relaxation, conjugate gradient method,...

2) One has to establish:
   - discrete operator $\mathbf{f}_h$ on the coarse mesh
   - fine $\rightarrow$ coarse transfer operators $\mathbf{T}_h^H$, $\hat{\mathbf{T}}_h^H$
   - coarse $\rightarrow$ fine transfer operators $\mathbf{T}_H^h$.

3) Recursive application of the Two Grid Method leads to Multi Grid Methods (MGM).

3.3.2 Application of MGM to Finite Element discretizations

On the two meshes let the nodal points be denoted by $\{\mathbf{x}_i^H \mid i = 1,\ldots,n_H\}$ and $\{\mathbf{x}_j^h \mid j = 1,\ldots,n_h\}$. The basis for the corresponding shape functions are $\{N_h^j \mid j = 1,\ldots,m\}$. In general $\mathcal{V}_H$ is not a subset of $\mathcal{V}_h$ and then one has to approximate N_H^i by N_h^j:

$$N_H^i(\mathbf{x}) \sim \sum_j g_{ij} N_h^j(\mathbf{x}) \tag{3.38}$$

(with $g_{ij} := N_H^i(\mathbf{x}_j^h), i = 1,\ldots,M, j = 1,\ldots,m$ in the case of Lagrangian elements.)

One gets:

(i) Coarse - fine transfer for nodal displacements:
Since $\mathbf{v}_H(\mathbf{x}) = \sum_i \mathbf{v}_{Hi} N_H^i(\mathbf{x})$ is approximated by

$$\sum_j \left(\sum_i \mathbf{v}_{Hi} g_{ij}\right) N_h^j(\mathbf{x}) = \sum_j \mathbf{v}_{hj} N_h^j(\mathbf{x}) \tag{3.39}$$

$$\text{i. e.} \quad (\mathbf{v}_{hj}) = \mathbf{G}^T(\mathbf{v}_{Hi}) \qquad \text{with} \quad \mathbf{G} = (g_{ij})$$

it holds that

$$\mathbf{T}_H^h = \mathbf{G}^T \tag{3.40}$$

(ii) Fine - coarse transfer for nodal-forces:
$\mathbf{p}_{Hi} = \int_\Omega \mathbf{p}(\mathbf{x}) N_H^i(\mathbf{x})\, d\mathbf{x}$ is approximated by

$$\sum_j g_{ij} \int_\Omega \mathbf{p}(\mathbf{x}) N_h^j(\mathbf{x})\, d\mathbf{x} = \sum_j g_{ij} \mathbf{p}_{hj} \tag{3.41}$$

thus

$$\mathbf{T}_h^H = \mathbf{G} \tag{3.42}$$

(iii) Fine - coarse transfer for displacements:
With the presumption $\{\mathbf{x}_i^H \mid i = 1,\ldots,n_H\} \subseteq \{\mathbf{x}_j^h \mid j = 1,\ldots,n_h\}$ one possibility to define $\hat{\mathbf{T}}_h^H$ is

$$\hat{\mathbf{T}}_h^H = \mathbf{P} \qquad \text{where} \quad P_{ij} := \begin{cases} 1 & \text{if } \mathbf{x}_i^H = \mathbf{x}_j^h \\ 0 & \text{otherwise.} \end{cases} \tag{3.43}$$

This method is sometimes called "injection". Other possibilities are to define $\hat{\mathbf{T}}_h^H$ by an averaging process, that is to define $\mathbf{v}_{Hi}$ by

$$\mathbf{v}_{Hi} = \sum_{j \in I_i} \omega_{ij} \mathbf{v}_{hj} \qquad \text{where} \quad \sum_{j \in I_i} \omega_{ij} = 1 \tag{3.44}$$

and I_i contains the number of the variables in a neighbourhood of $\mathbf{x}_i^H$. This method seems to be advantageous for highly nonlinear problems (Brandt, 1982).

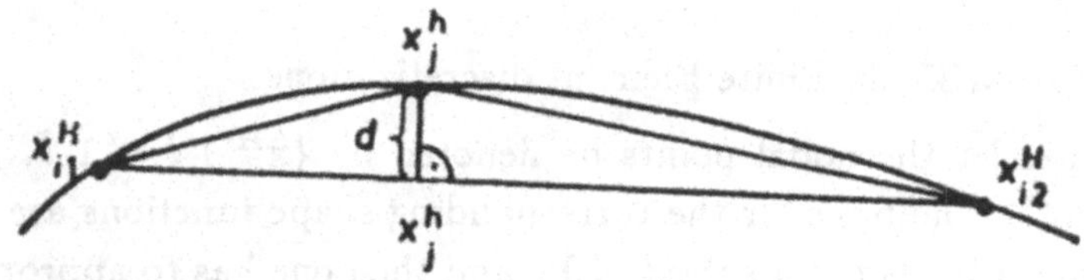

Fig. 3.10 Curved boundaries

REMARKS:

1) On curved boundaries (see Fig. 3.10) the shape functions on the coarse mesh are not neccessarily defined in nodal points $\mathbf{x}_j^h$ lying on the boundary of the fine mesh. If $d \ll H$ one possibility to circumvent this difficulty is to substitute $N_H^i(\mathbf{x}_j^h)$ by $N_H^i(\tilde{\mathbf{x}}_j^h)$, $(\tilde{\mathbf{x}}_j^h \in \partial\Omega_h)$.
2) The refinement can also be done in an adaptive way (Stein, Bischoff, Brand, Plank, 1985)

EXAMPLE: Transfer operator

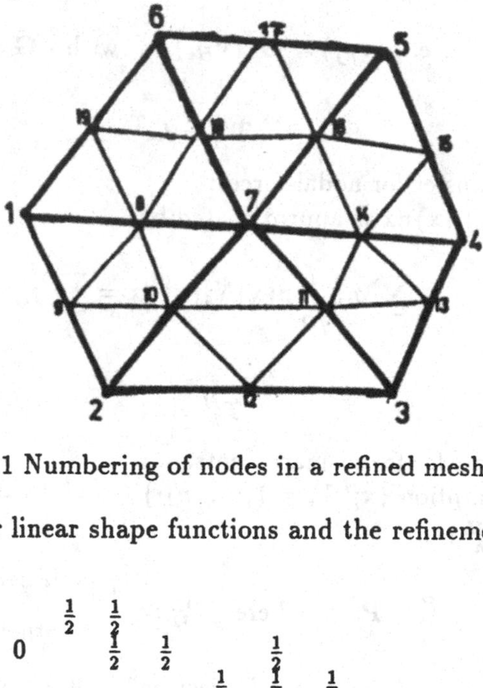

Fig. 3.11 Numbering of nodes in a refined mesh

Transfer operator for linear shape functions and the refinement in the example given above:

$$\mathbf{T}_H^h = \begin{pmatrix} 1 & & & & & & & \frac{1}{2} & \frac{1}{2} & & & & & & & & & & \frac{1}{2} \\ & & & & & & 0 & & \frac{1}{2} & \frac{1}{2} & & \frac{1}{2} & & & & & & & \\ & & \cdot & & & & & & & & \frac{1}{2} & \frac{1}{2} & \frac{1}{2} & & & & & & \\ & & & & & & & & & & & & \frac{1}{2} & \frac{1}{2} & \frac{1}{2} & & & & \\ & & & & \cdot & & & & & & & & & & \frac{1}{2} & \frac{1}{2} & \frac{1}{2} & & \\ & 0 & & & & & & & & & & & & & & & \frac{1}{2} & \frac{1}{2} & \frac{1}{2} \\ & & & & & & 1 & \frac{1}{2} & & \frac{1}{2} & \frac{1}{2} & & & \frac{1}{2} & & \frac{1}{2} & & \frac{1}{2} & \end{pmatrix}$$

Only the non-zero and non-unity elements of the transferoprator (and the stiffness matrix) need to be stored (together with their pointer fields). The data structure should be a dynamic one (Bank, 1983).

4 Nonlinear Shell Theory

4.1 Introduction

In this chapter we discuss some of the main aspects of a geometrically nonlinear shell theory and the numerical application. A characteristic of thin shell structures is the ability to sustain heavy loads given by their curvature. The curved middle surface produces a large variety of different response phenomena especially in the geometrically nonlinear range. For example we mention the sensivity against imperfections, buckling and postbuckling behaviour. Basically two concepts have been persued deriving thin shell finite elements. The classical concept starts with the derivation of a two dimensional nonlinear shell theory with all its problems to a consistent approximation of the shell equations. Based on this theory a lot of element formulations have been derived. For example we mention the curved shell elements *NACS*, Harte (1982), Harte,Eckstein (1986), mixed formulations, Harbord(1972), Noor et. al.(1984) and shallow shell elements, e.g. Wagner(1985).

On the other side exists the degeneration concept. Here we need no special shell theory. The formulation discretizes the 3D- field-equations directly applying simultaneously corresponding shell assumptions. This leads only to an approximate description of the shell geometry and the displacements. Other problems like "menbrane- and shear-locking" can be avoided by the choice of higher order elements. Therefore for accurate and good results we have to employ elements with up to 80 degress of freedom which leads to rather expensive calculations. Examples for this class of elements may be found in papers of Ramm (1976) or Huang, Hinton (1984). Figure 4.1 shows a classification of shell elements including element formulations for plane stress/strain and plates.

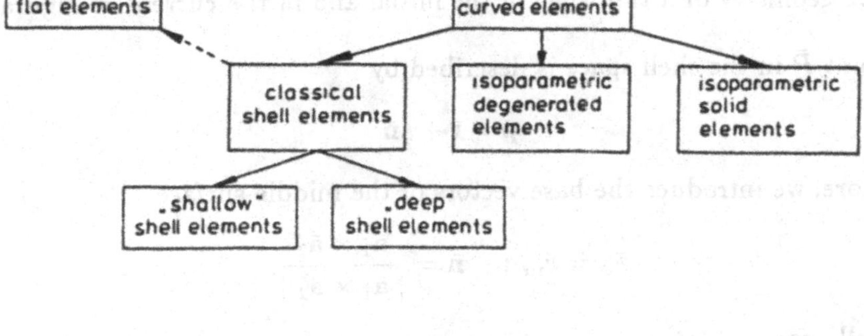

Fig. 4.1 Shell elements

Within the classical concept we can distinguish between a shell theory based on the well known Kirchhoff-Love-hypothesis and a shear elastic version of shell theory which then includes shear deformations. From a numerical point of view the latter one has advantages because the shape functions have to fulfil only C^0 - continuity. In this paper we discuss only some selected theoretical and numerical aspects out of the wide field of shell problems. Especially we will describe the main ideas of a shear elastic shell theory, finite element formulations for associated shallow shell and axisymmetric shell elements and some examples dealing with the postbuckling behaviour of shallow shells.

4.2 Derivation of nonlinear shell equations

Here we describe the main ideas and equations of a geometrically nonlinear shell theory including transverse shear. For a detailed description we refer to e.g. Wagner (1985) or to Başar, Krätzig (1985). We use a symbolic tensor notation because of its simplicity. In Figure 4.2 , the initial (i.c.) and the current configuration (c.c.) of the shell with the essential notations are shown. Quantities in the i.c. are marked by an upper bar

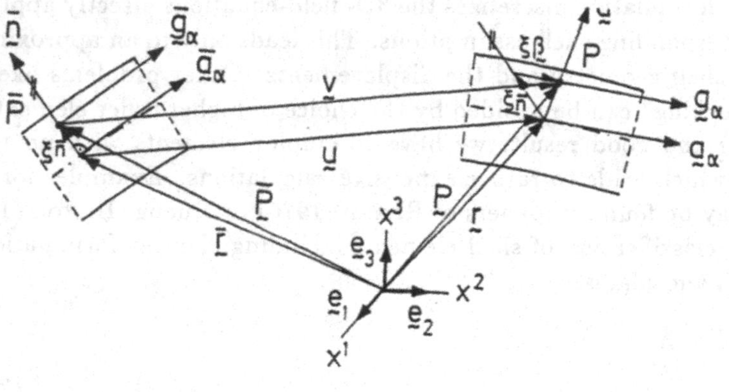

Fig. 4.2 geometry of a thin shell in the initial and in the current configuration

A point $\bar{P}$ in the shell space is described by

$$\bar{\mathbf{p}} = \bar{\mathbf{r}} + \xi \bar{\mathbf{n}} \tag{4.1}$$

Furthermore, we introduce the base vectors of the middle surface

$$\bar{\mathbf{a}}_\alpha = \bar{\mathbf{r}}_{,\alpha} \; ; \quad \bar{\mathbf{n}} = \frac{\bar{\mathbf{a}}_1 \times \bar{\mathbf{a}}_2}{|\, \bar{\mathbf{a}}_1 \times \bar{\mathbf{a}}_2 \,|} \tag{4.2}$$

and in shell space

$$\bar{\mathbf{g}}_\alpha = \bar{\mathbf{p}}_{,\alpha} + \xi \bar{\mathbf{n}}_{,\alpha} \; ; \quad \bar{\mathbf{g}}_3 = \bar{\mathbf{n}}. \tag{4.3}$$

The shifter tensor $\bar{\mathbf{Z}}$ combine the base vectors

$$\bar{\mathbf{g}}_i = \bar{\mathbf{Z}} \bar{\mathbf{a}}_i \tag{4.4}$$

with the definition of $\bar{\mathbf{Z}}$

$$\bar{\mathbf{Z}} = \bar{\mathbf{1}} - \xi \bar{\mathbf{B}} . \tag{4.5}$$

In the frequently used index notation eq.(4.4) and (4.5) leads to

$$\bar{\mathbf{g}}_\alpha = \bar{Z}_\alpha^\beta \bar{\mathbf{a}}_\beta ; \quad \bar{Z}_\alpha^\beta = (\delta_\alpha^\beta - \xi \bar{B}_\alpha^\beta).$$

In equation (4.5) we introduce the curvature tensor

$$\bar{\mathbf{B}} = -\bar{\mathbf{n}}_{,\alpha} \otimes \bar{\mathbf{a}}^\alpha \tag{4.6}$$

and the unit tensor in the initial configuration

$$\bar{\mathbf{1}} = \bar{\mathbf{a}}_i \otimes \bar{\mathbf{a}}^i = \bar{\mathbf{a}}_\alpha \otimes \bar{\mathbf{a}}^\alpha + \bar{\mathbf{n}} \otimes \bar{\mathbf{n}}. \tag{4.7}$$

The material deformation gradient $\mathbf{F}$ in shell space is derived for the introduced convected coordinates

$$\mathbf{F} = \mathbf{g}_i \otimes \bar{\mathbf{g}}^i . \tag{4.8}$$

According to Pietraszkiewicz (1977),(1979), the material deformation gradient of the middle surface is given by

$$\mathbf{G} = \mathbf{F}(\xi = 0) = \mathbf{a}_i \otimes \bar{\mathbf{a}}^i = \mathbf{a}_\alpha \otimes \bar{\mathbf{a}}^\alpha + \mathbf{d} \otimes \bar{\mathbf{n}}. \tag{4.9}$$

The relation between $\mathbf{F}$ and $\mathbf{G}$ can be introduced similar to eq.(4.4)

$$\mathbf{F} = \mathbf{Z}\mathbf{G}\bar{\mathbf{Z}}^{-1}. \tag{4.10}$$

With the above definitions one can describe the deformation gradient in a geometrical way as a composed mapping, see Figure 4.3.

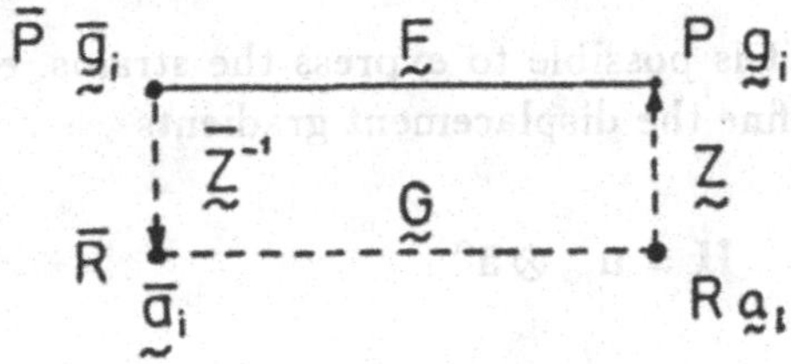

Fig. 4.3 Description of deformation gradient $\mathbf{F}$ as a composed mapping

With the material deformation gradient $\mathbf{F}$ the Green strain tensor $\mathbf{E}$ in shell space yields, see chapter 1, eq.(1.4)

$$\mathbf{E} = \frac{1}{2}(\mathbf{F}^T\mathbf{F} - \bar{\mathbf{1}}). \tag{4.11}$$

Introducing eq. (4.10), $\mathbf{E}$ can be represented in the form, see e.g. Pietraszkiewicz (1977)

$$\mathbf{E} = \bar{\mathbf{Z}}^{T-1}(\mathbf{E}^1 + \xi\mathbf{E}^2 + \xi^2\mathbf{E}^3)\bar{\mathbf{Z}}^{-1}. \tag{4.12}$$

$\mathbf{E}^1, \mathbf{E}^2$ and $\mathbf{E}^3$ are strain tensors defined on the middle surface of the shell

$$\begin{aligned} \mathbf{E}^1 &= \mathbf{E}^m + \mathbf{E}^{nl} + \mathbf{E}^s = \frac{1}{2}(\mathbf{G}^T\mathbf{G} - \bar{\mathbf{1}}), \\ \mathbf{E}^2 &= \mathbf{E}^b = -(\mathbf{G}^T\mathbf{D}\mathbf{G} - \bar{\mathbf{B}}), \\ \mathbf{E}^3 &= \frac{1}{2}(\mathbf{G}^T\mathbf{D}^T\mathbf{D}\mathbf{G} - \bar{\mathbf{B}}^T\bar{\mathbf{B}}). \end{aligned} \tag{4.13}$$

The curvature tensor $\mathbf{D}$ is defined in the current configuration similar to $\bar{\mathbf{B}}$, see eq. (4.6). The extraction of the shifter tensor in eq. (4.12) is advantageous within the formulation of the principle of virtual work on the middle surface of the shell.

To express the base vectors of the current configuration by vectors of the i.c. we introduce the displacement vetor

$$\mathbf{v} = \mathbf{u} + \xi\boldsymbol{\beta} \tag{4.14}$$

and the difference vector

$$\boldsymbol{\beta} = \mathbf{d} - \bar{\mathbf{n}} \tag{4.15}$$

see Figure 4.2. Similar to eq. (4.3) it holds

$$\mathbf{g}_\alpha = \mathbf{p}_{,\alpha} + \xi\mathbf{d}_{,\alpha}\ ; \quad \mathbf{g}_3 = \mathbf{d}. \tag{4.16}$$

With the eqs. (4.2) and (4.16) it is possible to express the strains, eqs. (4.13), by displacements. Previously we define the displacement gradients

$$\mathbf{H} = \mathbf{u}_{,\alpha} \otimes \bar{\mathbf{a}}^\alpha \tag{4.17}$$

and

$$\mathbf{V} = \boldsymbol{\beta}_{,\alpha} \otimes \bar{\mathbf{a}}^\alpha \tag{4.18}$$

At this stage of development it is necessary to introduce a classification of shell theories. Due to Koiter (1966), Simmonds and Danielson(1972), Reissner(1969), Pietraszkiewicz(1977), Başar and Krätzig(1977) and others the classification is performed with respect to the order of magnitude of rotations.

With the commonly introduced parameter

$$\vartheta = max\,(\frac{h}{L}, \frac{h}{d}, \sqrt{\frac{h}{R}}, \sqrt{\eta}) \tag{4.19}$$

(h : shell thickness, L : smallest wave length of deformation pattern, d : distance of considered point from lateral boundary, η : maximum eigenvalue of Green strain tensor , R : characteristic radius), a classification is shown in Figure 4.4.

$\omega = O(\vartheta^2)$	*small rotations*
$\omega = O(\vartheta)$	*moderate rotations*
$\omega = O(\sqrt{\vartheta})$	*large rotations*
$\omega = O(1)$	*finite rotations*

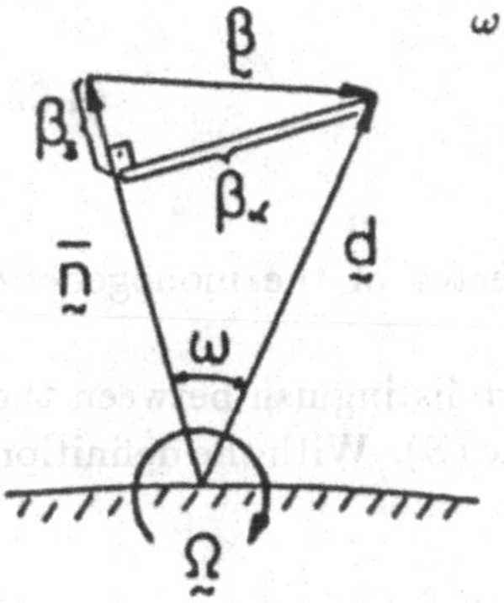

Fig. 4.4 Classification of magnitude of rotation angle ω in terms of ϑ

In the case of a "moderate rotation theory", it is possible to neglect the normal component of $\boldsymbol{\beta}$. A separation of the strains into linear membrane- (m), shear- (s), bending-(b) and nonlinear- (nl) terms leads to the following expressions

$$\begin{aligned}
\mathbf{E}^m &= \frac{1}{2}(\bar{\mathbf{1}}^T\mathbf{H} + \mathbf{H}^T\bar{\mathbf{1}}),\\
\mathbf{E}^{nl} &= \frac{1}{2}(\mathbf{H}^T\mathbf{P}^T\mathbf{P}\mathbf{H}),\\
\mathbf{E}^s &= (\boldsymbol{\beta}\otimes\bar{\mathbf{n}} + \mathbf{H}^T\mathbf{P}),\\
\mathbf{E}^b &= -\mathbf{H}^T\bar{\mathbf{B}} + \bar{\mathbf{1}}^T\mathbf{V})
\end{aligned} \tag{4.20}$$

For explanation of eq. (4.20) we express the nonlinear strain measures in index notation

$$\begin{aligned}
E^m_{\alpha\beta} &= \frac{1}{2}(u_{\alpha|\beta} + u_{\beta|\alpha} - 2\bar{B}_{\alpha\beta}u_3),\\
E^{nl}_{\alpha\beta} &= \frac{1}{2}(u_{3,\alpha} + u_{3,\beta} + u_{3,\alpha}\bar{B}^\lambda_\beta u_\lambda + u_{3,\beta}\bar{B}^\lambda_\alpha u_\lambda),\\
E^s_{\alpha 3} &= \frac{1}{2}(u_{3,\alpha} + \bar{B}^\lambda_\alpha u_\lambda + \beta_\alpha),\\
E^b_{\alpha\beta} &= \beta_{\alpha|\beta} - \bar{B}^\lambda_\beta u_{\lambda|\alpha} + \bar{B}^\lambda_\beta \bar{B}_{\lambda\alpha}u_3.
\end{aligned}$$

It can be seen that the strain tensor $\mathbf{E}^1$ remains nonlinear while the strain tensor $\mathbf{E}^2$ is now linear, see e.g. Pietraszkiewicz(1977). Within the shear elastic formulation only first derivatives due to the linear form of **H** and **V** exist. This is an advantage in the numerical treatment of shell problems. The orthogonal projection tensor **P** is introduced to consider only the normal components of the displacement gradient **H** within the nonlinear strain tensor **E** . **P** is defined as follows

$$\mathbf{P} = \bar{\mathbf{n}} \otimes \bar{\mathbf{n}} \tag{4.21}$$

The numerical treatment is based on the weak formulation in shell space. Here, a total Lagrangian formulation is adopted. It holds, see chapter 1, eq. (1.18)

$$g(\mathbf{v}, \boldsymbol{\eta}) = \int_{^\circ B} \mathbf{S} \cdot \delta \mathbf{E} \, dv - \int_{^\circ B_\sigma} \hat{\mathbf{t}} \cdot \boldsymbol{\eta} \, da \,. \tag{4.22}$$

S denotes the 2nd Piola-Kirchhoff-tensor and $\hat{\mathbf{t}}$ is the vector of the monogenetic surface loads.

In an analogous way to the strain tensor **E** we have to distinguish between the stress tensor in shell space (**S**) and on the shell middle surface ($\tilde{\mathbf{S}}$). With the definition of the shifter tensor Z defined in e.q. (4.5) we get

$$\mathbf{S} = \bar{\mathbf{Z}} \tilde{\mathbf{S}} \bar{\mathbf{Z}}^T . \tag{4.23}$$

The principle of virtual work in terms of the shell middle surface then yields

$$g(\mathbf{u}, \boldsymbol{\beta}, \delta\mathbf{u}, \delta\boldsymbol{\beta}) = \int_{^\circ M} [\mathbf{N} \cdot (\delta\mathbf{E}^m + \delta\mathbf{E}^{nl}) + \mathbf{M} \cdot \delta\mathbf{E}^b + \mathbf{Q} \cdot \delta\mathbf{E}^s]) \, da - \int_{^\circ M_\sigma} (\hat{\mathbf{f}} \cdot \delta\mathbf{u} + \hat{\mathbf{m}} \cdot \delta\boldsymbol{\beta}) \, da \,. \tag{4.24}$$

Assuming a homogeneous and isotropic elastic material, we introduce a linear hyperelastic material law which leads in the case of shear elastic shells to constitutive equations for the commonly used stress and couple resultant tensors.

$$\tilde{\mathbf{N}} = \frac{C}{1-\nu^2} \int_{-h/2}^{h/2} det \bar{\mathbf{Z}} \bar{\mathbf{Z}} \mathcal{H} [\mathbf{E}^m + \mathbf{E}^{nl} + \xi \mathbf{E}^b] \, d\xi \,,$$

$$\mathbf{M} = \frac{C}{1-\nu^2} \int_{-h/2}^{h/2} det \bar{\mathbf{Z}} \bar{\mathbf{Z}} \mathcal{H} \, \xi [\mathbf{E}^m + \mathbf{E}^{nl} + \xi \mathbf{E}^b] \, d\xi \,,$$

$$\mathbf{Q} = \chi G \int_{-h/2}^{h/2} det \bar{\mathbf{Z}} \bar{\mathbf{Z}} \mathcal{K} \, [\mathbf{E}^s] \, d\xi \,. \tag{4.25}$$

Analytical integration over the shell thickness and constant approximation of the determinant of the shifter tensor results in

$$\tilde{\mathbf{N}} = \frac{Ch}{1-\nu^2}\mathcal{H}\,[\mathbf{E}^m + \mathbf{E}^{nl}] - \frac{Ch^3}{12(1-\nu^2)}\bar{\mathbf{B}}\mathcal{H}\,[\mathbf{E}^b] = \mathbf{N} - \bar{\mathbf{B}}\mathbf{M}\,,$$
$$\mathbf{M} = \frac{Ch^3}{12(1-\nu^2)}\mathcal{H}\,[\mathbf{E}^b]\,,$$
$$\mathbf{Q} = \chi G\mathcal{K}\,[\mathbf{E}^s]\,. \tag{4.26}$$

with the constitutive tensors

$$\mathcal{H} = (1-\nu)\bar{\mathbf{a}}_\alpha \otimes \bar{\mathbf{a}}_\beta \otimes \bar{\mathbf{a}}^\alpha \otimes \bar{\mathbf{a}}^\beta + \nu\bar{\mathbf{a}}^\alpha \otimes \bar{\mathbf{a}}_\alpha \otimes \bar{\mathbf{a}}_\beta \otimes \bar{\mathbf{a}}^\beta\,,$$
$$\mathcal{K} = \bar{\mathbf{a}}_\alpha \otimes \bar{\mathbf{n}} \otimes \bar{\mathbf{a}}^\alpha \otimes \bar{\mathbf{n}}\,. \tag{4.27}$$

Within the finite element analysis, the solution of the nonlinear equation (4.24) is obtained by an iterative Newton scheme, see chapter 2. For this purpose, a sequence of linearized problems has to be solved which is stated below

$$\int_{{}^\circ M} \{\frac{Ch}{1-\nu^2}(\delta\mathbf{E}^m + \delta\mathbf{E}^{nl})\cdot\mathcal{H}\,[\Delta\mathbf{E}^m + \delta\mathbf{E}^{nl}]$$
$$+\frac{Ch^3}{12(1-\nu^2)}\delta\mathbf{E}^b\cdot\mathcal{H}\,[\Delta\mathbf{E}^b] + \chi Gh\delta\mathbf{E}^s\cdot\mathcal{K}\,[\Delta\mathbf{E}^s]\,\}\,da$$
$$+\int_{{}^\circ M}\mathbf{P}\delta\mathbf{H}\cdot\mathbf{N}\mathbf{P}\Delta\mathbf{H}\,da + g(\mathbf{u},\boldsymbol{\beta},\delta\mathbf{u},\delta\boldsymbol{\beta}) = 0. \tag{4.28}$$

For example, the variation $\delta\mathbf{E}^b$ and the linearization $\Delta\mathbf{E}^b$ of the bending strain tensor are defined as

$$\delta\mathbf{E}^b = -\delta\mathbf{H}^T\bar{\mathbf{B}} + \bar{\mathbf{1}}^T\delta\mathbf{V}\,;\quad \Delta\mathbf{E}^b = -\Delta\mathbf{H}^T\bar{\mathbf{B}} + \bar{\mathbf{1}}^T\Delta\mathbf{V}\,. \tag{4.29}$$

4.3 Special Theories

For plates and plane structures the curvature tensor $\bar{\mathbf{B}}$, eq. (4.6), vanishes. In the linear case the strain measures of eg. (4.20) reduces to

**plane stress*	$\mathbf{E}^m = \frac{1}{2}(\bar{\mathbf{1}}^T\mathbf{H} + \mathbf{H}^T\bar{\mathbf{1}})$
**plates*	$\mathbf{E}^b = -\mathbf{H}^T\bar{\mathbf{B}} + \bar{\mathbf{1}}^T\mathbf{V})$
	$\mathbf{E}^s = (\boldsymbol{\beta}\otimes\bar{\mathbf{n}} + \mathbf{H}^T\mathbf{P})$
**shallow shells*	$\mathbf{E}^m = \frac{1}{2}(\bar{\mathbf{1}}^T\mathbf{H} + \mathbf{H}^T\bar{\mathbf{1}})$
	$\mathbf{E}^b = -\mathbf{H}^T\bar{\mathbf{B}} + \bar{\mathbf{1}}^T\mathbf{V})$
	$\mathbf{E}^s = (\boldsymbol{\beta}\otimes\bar{\mathbf{n}} + \mathbf{H}^T\mathbf{P})$

(4.30)

4.4 Geometrically nonlinear shallow shell element

Starting from the strain tensors defined in eq. (4.20) and using the assumptions for shallow shells the physical strains may be expressed as follows

$$
\begin{aligned}
E_{xx} &= u_{,x} + z_{o,x}\, w_{,x} + \frac{1}{2} w_{,x}^{2} - z\beta_{x,x} \,, \\
E_{yy} &= v_{,y} + z_{o,y}\, w_{,y} + \frac{1}{2} w_{,y}^{2} - z\beta_{y,y} \,, \\
E_{xy} &= u_{,y} + v_{,x} + z_{o,x}\, w_{,y} + z_{o,y}\, w_{,x} + \frac{1}{2} w_{,x}\, w_{,y} - z(\beta_{x,y} + \beta_{y,x}) \,, \\
E_{x3} &= \frac{1}{2}(w_{,x} - \beta_{x}) \,, \\
E_{y3} &= \frac{1}{2}(w_{,y} - \beta_{y}) \,.
\end{aligned}
\tag{4.31}
$$

The eqs. (4.31) represent a shear-elastic version of the well known Donnell-Marguerre strain measures, Marguerre(1939). For the notation see Figure 4.5. The appropriate elasticity matrix **C** is defined in the usual way, see e.g. Hughes, Taylor, Kanoknukulchai (1977). The finite element formulation is based on an isoparametric displacement concept. Due to the shear elastic shell theory, only first derivatives exist in the strain measures.

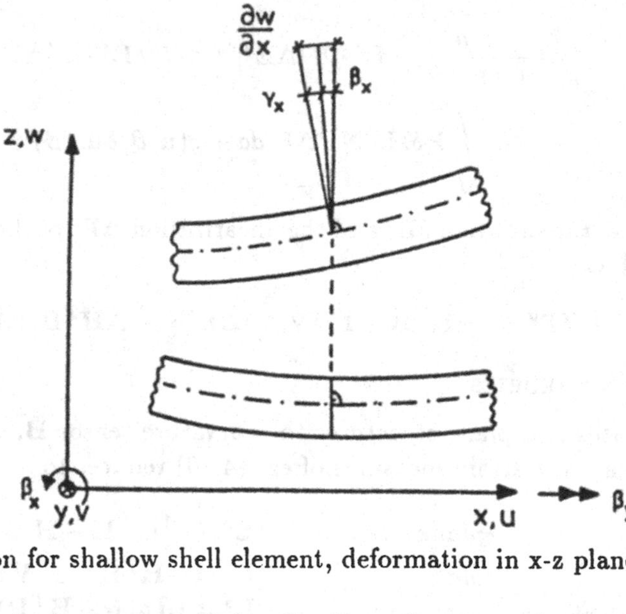

Fig. 4.5. Notation for shallow shell element, deformation in x-z plane

Therefore, C^0 - continuity is required for the shape functions. In the case of a straight element, bilinear shape functions are possible and the element formulation becomes extremely simple. Within the numerical integration of the element matrices we have to introduce reduced or selective integration schemes. Otherwise the well known effect of shear locking occurs, see e.g. Hughes, Taylor, Kanoknukulchai (1977).

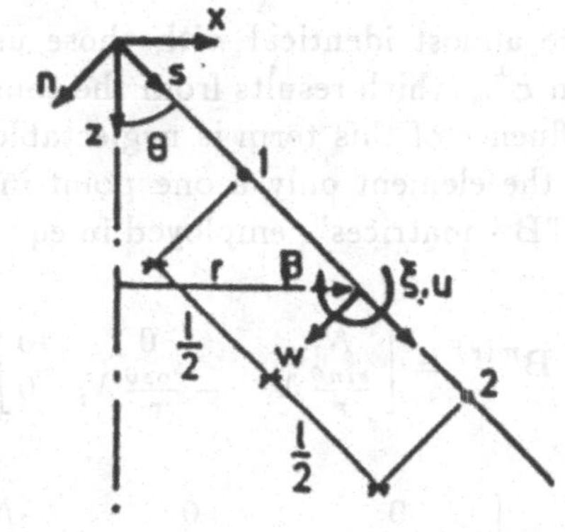

Fig. 4.6 Notation for axisymmetric shell element

Reduced or selective integrations and mixed methods are closely related, see Malkus, Hughes (1978). Another way to explain reduced integration schemes is the interpretation of the shear stiffness matrix as a penalty term for a Kirchhoff-type shell theory, Zienkiewicz (1977) . Within the reduced integration one has to observe possible zero energy modes. For a detailed description we refer to e.g. Hughes, Cohen, Haroun (1978).

The numerical finite element analysis is based on equation (4.28). For shear elastic shallow shells we reach at

$$\delta \mathbf{v}^T \{ \int_{^\circ M} [(\mathbf{B}^m + \mathbf{B}^{nl})^T \mathbf{C}^m (\mathbf{B}^m + \mathbf{B}^{nl}) + \mathbf{B}^{bT} \mathbf{C}^b \mathbf{B}^b + \mathbf{B}^s \mathbf{C}^s \mathbf{B}^s] \, da \, \Delta \mathbf{v}$$

$$+ \int_{^\circ M} \mathbf{G}^T \hat{\mathbf{N}} \mathbf{G} \, da \, \Delta \mathbf{v} + \mathbf{r} \} = \delta \mathbf{v}^T \{ \mathbf{K}_T \Delta \mathbf{v} + \mathbf{r} \} = 0 \, . \tag{4.32}$$

In eq. (4.32) the "B-matrices" are introduced which are described in chapter 2. They can be derived with respect to the above introduced strain measures, see eq. (4.31). A similar finite element formulation may be found in Pica, Wood (1980).

4.5 Geometrically nonlinear axisymmetric shell element

Starting from the strain tensor defined in equation (4.20) and using the conditions for axisymmetry

$$u_\varphi = 0 \, , \quad \beta_\varphi = 0 \, , \quad (\ldots)_{,\varphi} \tag{4.33}$$

the physical strains may be expressed as follows, notation see Figure 4.6

$$\begin{aligned} E^m_{ss} &= u_{,s} & E^m_{\varphi\varphi} &= \tfrac{1}{r}(u\sin\theta - w\cos\theta) \\ E^b_{ss} &= -\beta_{,s} & E^b_{\varphi\varphi} &= -\tfrac{1}{r}\beta\sin\theta - \tfrac{\cos\theta}{r^2}(u\sin\theta - w\cos\theta) \\ E^s_s &= \tfrac{1}{2}(w_{,s} - \beta) & E^{nl}_{ss} &= \tfrac{1}{2} w_{,s}^{\,2} \end{aligned} \tag{4.34}$$

The linear strain are almost identical with those used by Zienkiewicz (1977) except the second term in $E^b_{\varphi\varphi}$ which results from the consequently derived tensorial strain measures. The influence of this term is neglectable, Wagner (1985). For the successful application of the element only a one point integration is necessary, see Zienkiewicz (1977). The "**B** - matrices", employed in eq. (4.32), are stated below.

$$\mathbf{B}^{m(e)} = \begin{bmatrix} N_{i,s} & 0 & 0 \\ \frac{sin\theta}{r} N_i & -\frac{cos\theta}{r} N_i & 0 \end{bmatrix}$$

$$\mathbf{B}^{b(e)} = \begin{bmatrix} 0 & 0 & -N_{i,s} \\ \frac{sin\theta cos\theta}{r^2} N_i & -\frac{cos^2\theta}{r^2} N_i & -\frac{sin\theta}{r} N_i \end{bmatrix}$$

$$\mathbf{B}^{s(e)} = [0 \quad N_{i,s} \quad -N_i]$$

$$\mathbf{B}^{nl(e)} = \begin{bmatrix} 0 & w_{,s} N_{i,s} & 0 \\ 0 & 0 & 0 \end{bmatrix}$$

$$\mathbf{G}^e = \mathbf{B}^{nl(e)} \tag{4.35}$$

4.6 Examples

In the following three examples of postbuckling behaviour of thin shells, the above described shear elastic shell elements, chapter 4.4 and 4.5, and continuation (or path-following) algorithms, see chapter 3.2 are applied.

4.6.1 Example 1: Snap-through behaviour of a shallow cylindrical shell segment

The snap-through behaviour of a shallow cylindrical shell segment is considered as a first example. This problem is well known and was investigated by many authors. To calculate the nonlinear load displacement curve one one has to employ the arc-length-method because load and displacements can be reversible. Figure 4.7 shows the system and a reference solution, e.g. Ramm (1981). We choose as load increment $\Delta P = 0.25kN$ and as load factor $\Delta\Lambda = 1.0$. The results are obtained with 27 increments where only 3 or 4 iterations are necessary for convergence. Within the calculation we have tested two finite element meshes and elements with linear or quadratic shape functions. Figure 4.7 shows the results for different element/mesh combinations. It can be seen that even with the simple four node element and the coarse 2x2 mesh relatively good results can be obtained. Here, differences to the reference solution appear near the limit point and in the unstable branch of the load displacement curve. Rather good results are abtained by a mesh refinement (4x4 mesh) or with an quadratic element.

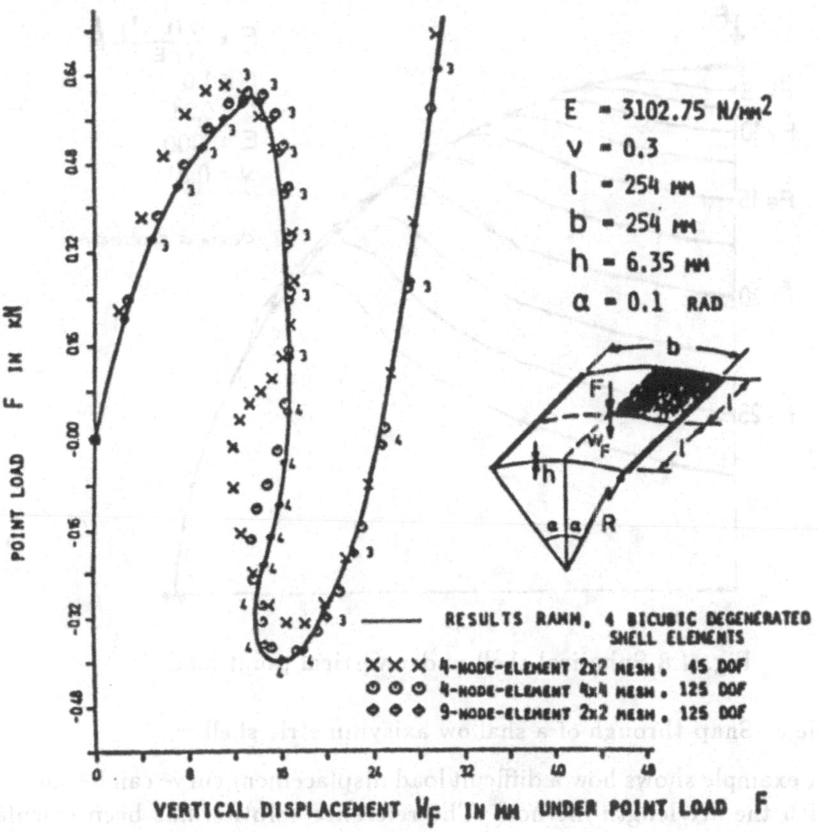

Fig. 4.7 Snap-through of a shallow cylindrical shell segment

4.6.2 Example 2: Bending of a spherical shell

The application of the element to an axisymmetric shell problem with finite deflections and rotations is considered as second example. For this purpose, we examine a spherical shell under point load at the apex. The system is shown in Figure 4.8 which also exhibits the deformed shapes for different load levels. The finite element mesh consist of 24 linear shell elements. This example was also treated by Fried (1985) who used the same number of elements but with cubic shape functions. Our solution is in very good agreement with that by Fried which furthermore shows the good performance of the simple and inexpensive element formulation given in section 4.5 Here, to include finite rotations and updated Lagrangian procedure has to be used.

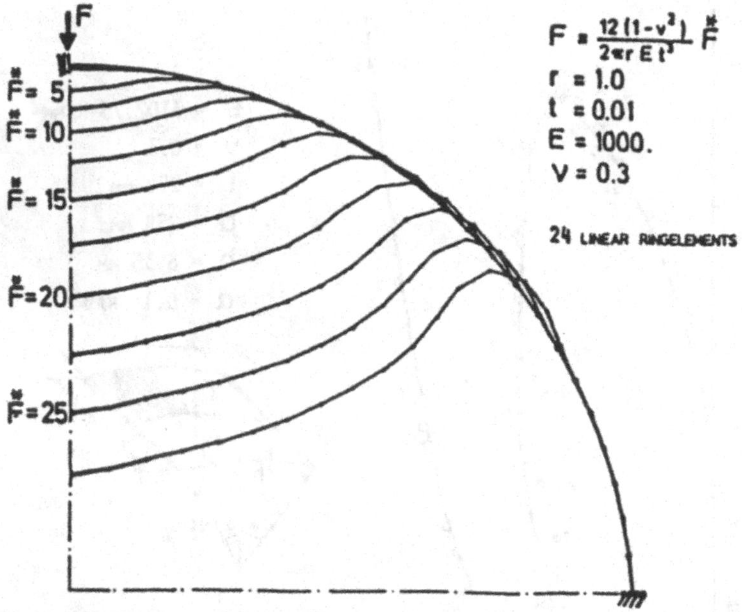

Fig. 4.8 Spherical shell under vertical point load

4.6.3 Example 3: Snap-through of a shallow axisymmetric shell

The next example shows how a difficult load displacement curve can be successful calculated with the arc-length-method. The reference solution has been calculated by Argyris et.al. (1980). This solution is restricted to that part of the load displacement curve which can be investigated without the arc-length-method. The total load displacement curve is calculated in 79 increments with an automatically reduction of the arc-length near limit points. Figure 4.9 shows system, load and system parameters. It should be mentioned that only axisymmetric buckling modes have been investigated. For the most part, the load displacement curve can only be obtained in a statically calculation because the state of equilibrium is unstable. The deformation pattern in several increments shows how the axisymmetric snapthrough of the shell takes place. The midpoint of the shell moves up and down, see Figure 4.9, while the wrinkling of the shell increases up to a turning point in which the deformed shell is nearly horizontal between the support points. The deformation of the shell is continued with a decreasing of the wrinkling. The snap-through behaviour is finished on the stable branch of the load displacement curve. For a detailed description we refer to Wagner (1985).

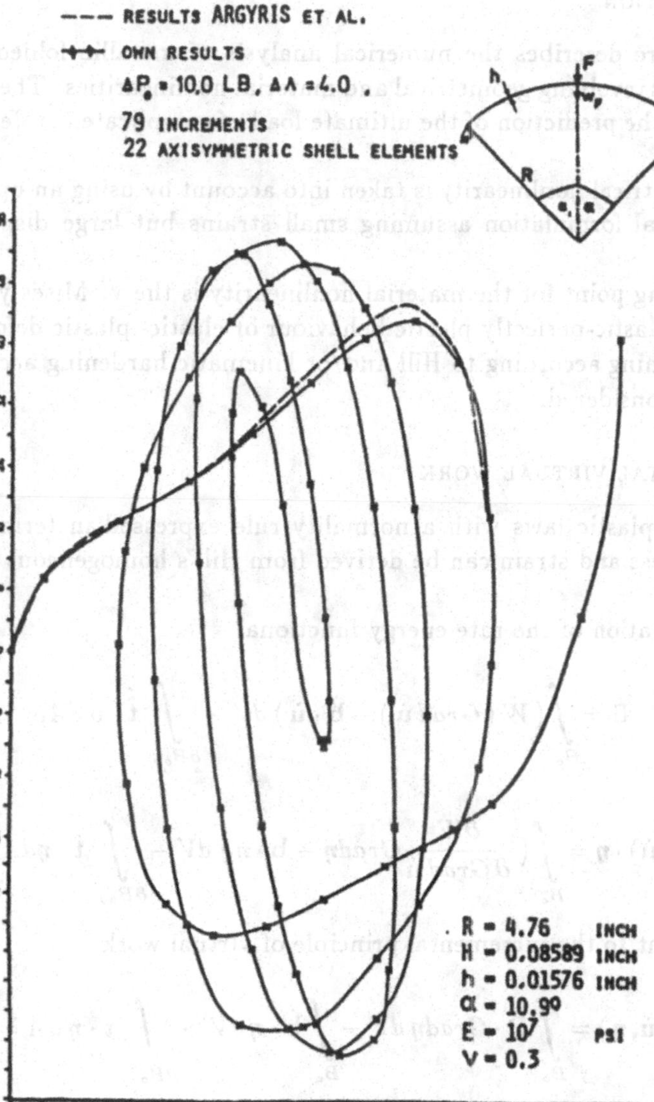

Fig. 4.9 Snap-through of a shallow axisymmetric shell

5 Elastic-Plastic Deformations of Plates and Shells

5.1 Introduction

This lecture describes the numerical analysis of metallic folded and stiffened plate structures involving geometrical and material nonlinearities. The purpose of the calculations is the prediction of the ultimate load of complicated stiffend engineering structures.

The geometrical nonlinearity is taken into account by using an updated Lagrangian incremental formulation assuming small strains but large displacements and rotations.

The starting point for the material nonlinearity is the v. Mises yield condition. Alternatively, elastic-perfectly plastic behaviour or elastic- plastic deformations with isotropic hardening according to Hill and/or kinematic hardening according to Prager/Ziegler is considered.

5.2 Incremental virtual work

All elastic-plastic laws with a normality rule expressed in terms of conjugate variables of stress and strain can be derived from Hill's homogeneous quadratic rate potential $\dot{W}$.

The 1. variation of the rate energy functional

$$\dot{\Pi} = \int_{B_o} (\dot{W}(Grad\,\dot{\mathbf{u}}) - \dot{\mathbf{b}} \cdot \dot{\mathbf{u}})\, dV - \int_{\partial B_{o\sigma}} \dot{\hat{\mathbf{t}}} \cdot \dot{\mathbf{u}}\, dA, \tag{5.1}$$

$$D\dot{\Pi}(\dot{\mathbf{u}}) \cdot \boldsymbol{\eta} = \int_{B_o} (\frac{\partial \dot{W}}{\partial\, Grad\,\dot{\mathbf{u}}} Grad\boldsymbol{\eta} - \dot{\mathbf{b}} \cdot \boldsymbol{\eta})\, dV - \int_{\partial B_{o\sigma}} \dot{\hat{\mathbf{t}}} \cdot \boldsymbol{\eta} dA = 0 \tag{5.2}$$

is equivalent to the incremental principle of virtual work

$$\dot{g}(\dot{\mathbf{u}}, \boldsymbol{\eta}) = \int_{B_o} \dot{\mathbf{P}} \cdot Grad\boldsymbol{\eta}\, dV - \int_{B_o} \dot{\mathbf{b}} \cdot \boldsymbol{\eta}\, dV - \int_{\partial B_{o\sigma}} \dot{\hat{\mathbf{t}}} \cdot \boldsymbol{\eta}\, dA = 0 \tag{5.3}$$

with

$$\dot{\mathbf{P}} = \frac{\partial \dot{W}}{\partial\, Grad\,\dot{\mathbf{u}}} \tag{5.4}$$

and the test function

$$\boldsymbol{\eta} = \delta \dot{\mathbf{u}}. \tag{5.5}$$

The incremental principle of virtual work leads to a symmetric stiffness matrix and is equivalent to the time derivative of the linearized principle of virtual work with respect to a space and time fixed configuration, denoted with "-". In material description, eq. (1.20) yields

$$\frac{d}{dt}\left(Dg(\bar{\mathbf{u}},\boldsymbol{\eta})\cdot\Delta\mathbf{u}\right) = \int_{B_o} tr\left(Grad\,\dot{\mathbf{u}}\,\bar{\mathbf{S}}\,Grad^T\boldsymbol{\eta}\right)dV + \int_{B_o} tr\left(\bar{\mathbf{F}}\bar{\mathcal{A}}\left(Grad^T\dot{\mathbf{u}}\bar{F} + \bar{F}^T Grad\,\dot{\mathbf{u}}\right)Grad^T\boldsymbol{\eta}\right)dV = -\frac{d}{dt}g(\bar{\mathbf{u}},\boldsymbol{\eta}), \tag{5.6}$$

with $\mathcal{A}$ being a material property tensor in material description, which will be specified later.

In spatial description, eq. (1.28) leads to

$$\frac{d}{dt}\left(Dg(\bar{\mathbf{u}},\boldsymbol{\eta})\cdot\Delta\mathbf{u}\right) = \int_{B} tr\left(grad\dot{\mathbf{u}}\,\bar{\sigma}\,grad^T\boldsymbol{\eta}\right)dv + \int_{B} tr\left(\bar{a}\frac{1}{2}\left(grad^T\dot{\mathbf{u}} + grad\dot{\mathbf{u}}\right)grad^T\boldsymbol{\eta}\right)dv = -\frac{d}{dt}g(\bar{\mathbf{u}},\boldsymbol{\eta}), \tag{5.7}$$

with the material property tensor a in spatial description, the rate of deformation tensor

$$\mathbf{d} = \frac{1}{2}\left(grad^T\dot{\mathbf{u}} + grad\dot{\mathbf{u}}\right) \tag{5.8}$$

and the Lie derivative of Cauchy stresses

$$\check{\sigma} = a\,\mathbf{d}. \tag{5.9}$$

For small strains and moderate rotations per increment the spatial description leads to the following kind of updated Lagrangian formulation (Bathe, Ramm, Wilson 1975), which is used for the computation of the examples. Therefore the fixed configuration "-" is identified with the converged configuration "1" of the last increment.

The assumption of small strains yields for the Lie derivative of Cauchy stresses

$$\check{\sigma} = \mathbf{R}\dot{\mathbf{S}}\mathbf{R}^T. \tag{5.10}$$

The rotation tensor can be decomposed as

$$\mathbf{R} = {}^2_1\mathbf{R}\,{}^1_0\mathbf{R} \tag{5.11}$$

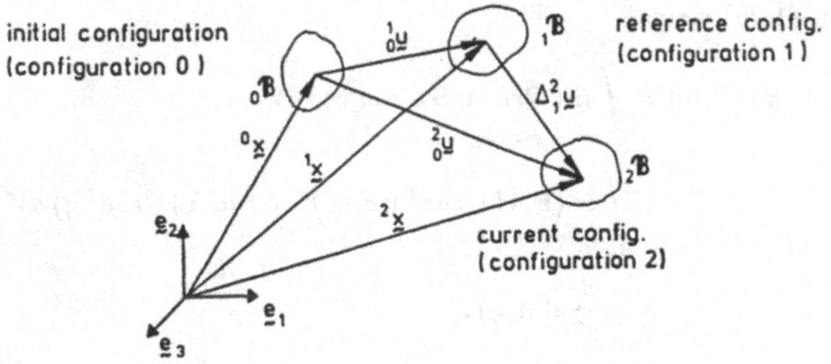

Fig. 5.1 Three configurations of a solid body

so that

$$\breve{\boldsymbol{\sigma}} = {}^2_1\mathbf{R}^1_0\mathbf{R}^2_0\dot{\mathbf{S}}^1_0\mathbf{R}^{T2}_{\;\;1}\mathbf{R}^T = {}^2_1\mathbf{R}^2_1\dot{\mathbf{S}}^2_1\mathbf{R}^T. \tag{5.12}$$

The condition of moderate rotations per increment leads to ${}^2_1\mathbf{R} \simeq \mathbf{1}$, so that

$$\breve{\boldsymbol{\sigma}} \simeq {}^2_1\dot{\mathbf{S}}. \tag{5.13}$$

Neglecting body forces and replacing time derivatives by finite increments eq. (5.7) leads to

$$\begin{aligned} \Delta g(\,\Delta\mathbf{u}, \delta\Delta\mathbf{u}\,) = & \int_{B_1} tr\left({}_1Grad\Delta^2_1\mathbf{u}\,{}^1_1\sigma\,(\,{}_1Grad\delta\Delta^2_1\mathbf{u}\,)^T\right) dV \\ & + \int_{B_1} tr\,(\,\Delta^2_1\mathbf{S}\,{}_1Grad\delta\Delta^2_1\mathbf{u}\,)\,dV \\ & - \int_{\partial B_{1\sigma}} \Delta^2_1\hat{\mathbf{t}}\,\delta\Delta^2_1\mathbf{u}\,dA. \end{aligned} \tag{5.14}$$

Now problem (5.14) is solved until the residual vanishes within a prescribed tolerance.

5.3 Material laws

The well known rate constitutive equation for linear hypoelastic behaviour given by Hooke's law is

$$\dot{\mathbf{S}} = \mathcal{C}\,\dot{\mathbf{E}} \tag{5.15}$$

with $\mathcal{C}$ being the material property tensor.

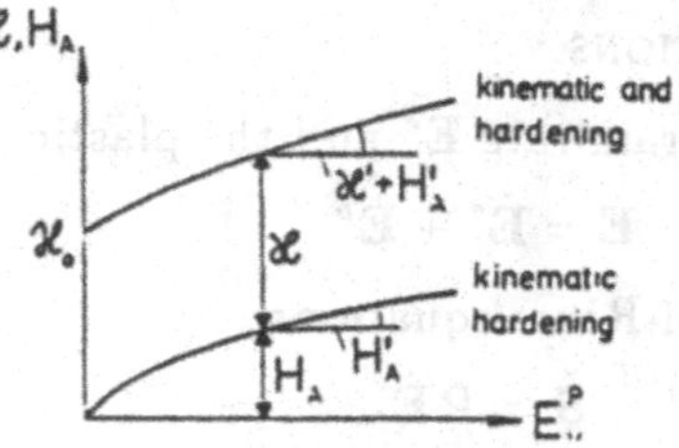

Fig. 5.2 Nonlinear kinematic and isotropic hardening function

For the calculation of the inelastic deformations especially for metals, the v. Mises yield condition

$$f = (\mathbf{S} - \mathbf{A}) \cdot \mathcal{D} (\mathbf{S} - \mathbf{A}) - \frac{2}{3}\kappa^2 = 0 \tag{5.16}$$

$$\mathcal{D} := \mathcal{I} - \frac{1}{3}\mathbf{1} \otimes \mathbf{1}$$

with the fourth order unit tensor $\mathcal{I}$, the second order unit tensor $\mathbf{1}$ and the associated flow rule

$$\dot{\mathbf{E}}^p = \frac{3}{2}\dot{\lambda}\mathcal{D}(\mathbf{S} - \mathbf{A}) \tag{5.17}$$

holds. $\dot{\lambda}$ is the yield parameter and κ is the yield stress, which in the case of isotropic hardening according to Hill is a function of the comparative plastic strain E_v^p

$$\kappa = \kappa(E_v^p). \tag{5.18}$$

$\mathbf{A}$ is the back stress tensor, which in the case of kinematic hardening according to Prager/Ziegler is also a function of comparative plastic strain E_v^p ,

$$\mathbf{A} = \mathbf{A}(E_v^p). \tag{5.19}$$

For the back stress tensor $\mathbf{A}$ the evolution equation

$$\dot{\mathbf{A}}^D = \frac{2}{3}H'_A\dot{\mathbf{E}}^p \tag{5.20}$$

holds. The index D signifies the deviator of a tensor. The connexion between the plastic stain rate tensor $\dot{\mathbf{E}}^p$ and the equivalent plastic strain rate $\dot{E}_v^p$ is given by

$$\dot{E}_v^p = \sqrt{\frac{2}{3}\dot{\mathbf{E}}^p \cdot \dot{\mathbf{E}}^p}. \tag{5.21}$$

A nonlinear hardening function for stable material is assumed as follows

$$\kappa' = \frac{d\kappa}{dE_v^p}, \; H'_A = \frac{dH_A}{dE_v^p}.$$

5.4 Elastic-plastic deformations

The addition of the elastic strain rate $\dot{\mathbf{E}}^e$ and the plastic strain rate $\dot{\mathbf{E}}^p$

$$\dot{\mathbf{E}} = \dot{\mathbf{E}}^e + \dot{\mathbf{E}}^p \tag{5.22}$$

yields the well known Prandtl-Reuss equations

$$\dot{\mathbf{S}} = \mathcal{P}\,\dot{\mathbf{E}} \tag{5.23}$$

with the elastic-plastic material property tensor

$$\mathcal{P} = \mathcal{C} - \frac{\mathcal{C}\,(\mathbf{S}^D - \mathbf{A}^D) \otimes \mathcal{C}\,(\mathbf{S}^D - \mathbf{A}^D)}{\frac{4}{9}\kappa^2(\kappa' + H'_A) + (\mathbf{S}^D - \mathbf{A}^D) \cdot \mathcal{C}\,(\mathbf{S}^D - \mathbf{A}^D)}. \tag{5.24}$$

An equivalent formulation is given with the variational inequality

$$\dot{\mathbf{E}}^p \cdot (\tilde{\mathbf{S}} - \mathbf{S}) - \kappa' \dot{E}^p_v(\tilde{E}^p_v - E^p_v) - \dot{\mathbf{E}}^p \cdot (\tilde{\mathbf{A}} - \mathbf{A}) \leq 0 \tag{5.25}$$

$\forall(\tilde{\mathbf{S}}, \tilde{\mathbf{A}}, \tilde{E}^p_v)$ with $f(\tilde{\mathbf{S}}, \tilde{\mathbf{A}}, \tilde{E}^p_v) \leq 0$ with the constraint $f(\mathbf{S}, \mathbf{A}, E^p_v) \leq 0$.

5.5 Elastic-plastic algorithms

For numerical treatment of eqs. (5.23) and (5.25) the rates are replaced by finite increments. Recall that small strains and moderate rotations are assumed, so $\mathbf{E}$ is replaced by $\mathbf{E}^{lin}$ in the sequel.

5.5.1 Explicit integration

For an explicit integration method, version (5.23) of the Prandtl-Reuss equations is used

$$^2_1\mathbf{S} = {}^1_1\boldsymbol{\sigma} + \Delta^2_1\mathbf{S} \tag{5.26}$$

with

$$\Delta^2_1\mathbf{S} = \mathcal{P}\,({}^1_1\boldsymbol{\sigma}, {}^1_1\mathbf{A}, {}^1E^p_v)\;\Delta^2_1\mathbf{E}, \tag{5.27}$$

$$\Delta^2_1\mathbf{E} = \Delta^2_1\mathbf{E}^{lin} = \frac{1}{2}\,(\,{}_1Grad\Delta^2_1\mathbf{u} + {}_1Grad^T\,\Delta^2_1\mathbf{u}\,). \tag{5.28}$$

This integration method describes an orthogonal projection of inadmissible stresses onto a hyperplane tangential to the yield surface.

Shortcomings of this method:

- the yield condition usually is not fulfilled

$$f\,({}^1_1\boldsymbol{\sigma}^D + \Delta^2_1\mathbf{S}\,) \neq 0$$

and has to be corrected by iteration

$$f\,({}^1_1\boldsymbol{\sigma}^D + \Delta^2_1\mathbf{S}^D_c\,) = 0,$$

- a modification is required for

$$f\,({}^1_1\boldsymbol{\sigma}^D\,) < 0,$$

- this method is not stable for large increments.

Remark: When using an explicit integration method for the constitutive equations the classical Prandtl-Reuss tensor is the consistent tangent tensor of the problem.

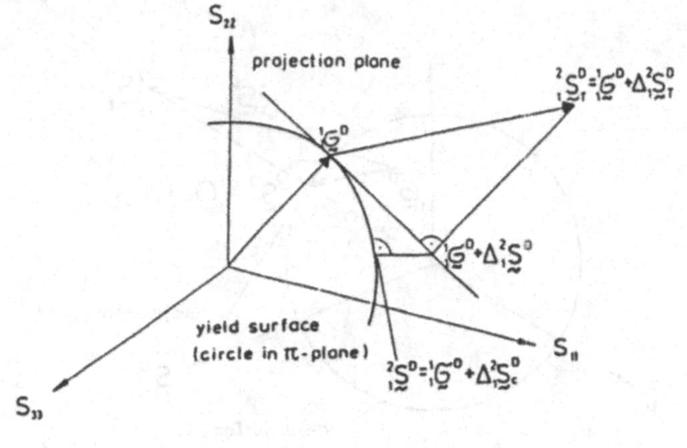

Fig. 5.3 Orthogonal projection of inadmissible stresses onto a tangential hyperplane of the yield surface

5.5.2 Implicit integration

To derive the implicit integration method, version (5.25) of the Prandtl- Reuss equations is used, for simplicity $\kappa' = 0$ and $H'_A = 0$ is assumed (Stein, Lambertz and Plank 1985). Therefore eq. (5.25) reduces to

$$\Delta_1^2\mathbf{E}^p \cdot ({}_1^2\tilde{\mathbf{S}} - {}_1^2\mathbf{S}) \leq 0 \tag{5.29}$$

$\forall {}_1^2\tilde{\mathbf{S}}$ with $f({}_1^2\tilde{\mathbf{S}}) \leq 0$ with the constraint $f({}_1^2\mathbf{S})$.

From eq. (5.22)

$$\Delta_1^2\mathbf{E}^p = \Delta_1^2\mathbf{E} - \Delta_1^2\mathbf{E}^e \tag{5.30}$$

with the elastic stresses

$$\Delta_1^2\mathbf{S} = \mathcal{C}\,\Delta_1^2\mathbf{E}^e \tag{5.31}$$

and the fictitious (trial) stresses

$$\Delta_1^2\mathbf{S}_T = \mathcal{C}\,\Delta_1^2\mathbf{E}, \tag{5.32}$$

follows equivalently to eq. (5.29)

$$(\Delta_1^2\mathbf{S}_T - \Delta_1^2\mathbf{S}) \cdot \mathcal{C}^{-1}({}_1^2\tilde{\mathbf{S}} - {}_1^2\mathbf{S}) \leq 0 \tag{5.33}$$

$\forall {}_1^2\tilde{\mathbf{S}}$ with $f({}_1^2\tilde{\mathbf{S}}) \leq 0$ with the constraint $f({}_1^2\mathbf{S}) \leq 0$.

The associated minimal problem is

$$J({}_1^2\mathbf{S}) := ({}_1^2\mathbf{S}_T - {}_1^2\mathbf{S}) \cdot \mathcal{C}^{-1}({}_1^2\mathbf{S}_T - {}_1^2\mathbf{S}) \rightarrow min \tag{5.34}$$

with the constraint $f({}_1^2\mathbf{S}) = 0$.

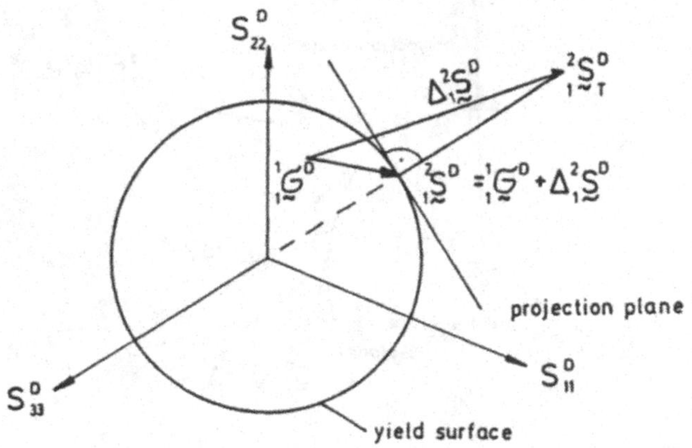

Fig. 5.4 Radial return of inadmissible stresses onto the yield surface

This integration method describes an orthogonal projection of the inadmissible stresses

$$ {}^2_1\mathbf{S}_T = {}^1_1\sigma + \Delta^2_1\mathbf{S}_T $$

onto the yield surface.

Advantages of this method:

- the yield condition is fulfilled without additional iterations,
- no modifications are necessary for

$$ f\,(\,{}^2_1\mathbf{S}\,) \leq 0, $$

- this method is unconditionally stable
- and applicable for yield conditions with edges.

Instead of the minimal problem (5.34), the saddle point problem

$$ L(\,{}^2_1\mathbf{S},\lambda\,) := J\,(\,{}^2_1\mathbf{S}\,) + \lambda f\,(\,{}^2_1\mathbf{S}\,) \underset{{}^2_1\mathbf{S},\,\lambda}{\rightarrow stat} \tag{5.35} $$

with the Lagrangian multiplier λ is solved.

The directional derivative of eq. (5.36) leads to the algebraic equations

$$ \frac{\partial L(\,{}^2_1\mathbf{S},\lambda\,)}{\partial^2_1\mathbf{S}} = -2\mathcal{C}^{-1}(\,{}^2_1\mathbf{S}_T - {}^2_1\mathbf{S}\,) + 3\lambda\mathcal{D}\,{}^2_1\mathbf{S} = 0 \tag{5.36a} $$

$$ \frac{\partial L(\,{}^2_1\mathbf{S},\lambda\,)}{\partial\lambda} = (\,{}^2_1\mathbf{S}\,)\cdot\mathcal{D}(\,{}^2_1\mathbf{S}\,) - \frac{2}{3}\kappa_o^2 = 0. \tag{5.36b} $$

Eq. (5.36a) is the flow rule $\Delta^2_1\mathbf{E}^p = \frac{2}{3}\lambda\mathcal{D}\,{}^2_1\mathbf{S}$ and eq. (5.36b) is the yield condition.

Remark: When using an implicit integration method for the constitutive equations the classical Prandtl-Reuss tensor is not the tangent tensor of the problem. The tangent tensor consistent with the implicit return mapping algorithm will be given in the sequel.

5.6 CONSISTENT TANGENT TENSOR

The consistent tangent tensor for an implicit integration method is achieved by differentation of

$$\frac{\partial_1^2\mathbf{S}(\,_1^2\mathbf{E},\lambda)}{\partial_1^2\mathbf{E}} = \frac{\partial_1^2\mathbf{S}(\,_1^2\mathbf{E},\lambda)}{\partial\Delta_1^2\mathbf{E}}, \tag{5.37}$$

$$_1^2\mathbf{S}(\,_1^2\mathbf{E},\lambda) = \bar{\mathcal{C}}(\lambda)\,_1^2\mathbf{E} \tag{5.38}$$

with the definitions

$$_1^2\mathbf{E} := \mathcal{C}^{-1}\,_1^1\sigma + \Delta_1^2\mathbf{E}, \tag{5.39}$$

$$\bar{\mathcal{C}}(\lambda) := (\mathcal{C}^{-1} + \frac{3}{2}\lambda\mathcal{D})^{-1}. \tag{5.40}$$

λ is a function of $_1^2\mathbf{E}$ but cannot be described explicitly. Therefore, $\frac{\partial\lambda}{\partial_1^2\mathbf{E}}$ is calculated by implicit differentation of the yield condition eq. (5.36b), expressed in terms of $_1^2\mathbf{E}$

$$f(\,_1^2\mathbf{E},\lambda) = \frac{3}{2}(\bar{\mathcal{C}}\,_1^2\mathbf{E})\cdot\mathcal{D}(\bar{\mathcal{C}}\,_1^2\mathbf{E}) - \kappa_o^2 = 0. \tag{5.41}$$

Then eq. (5.38) yields

$$\frac{\partial_1^2\mathbf{S}}{\partial_1^2\mathbf{E}} = \bar{\mathcal{C}} - \frac{\frac{\partial_1^2\mathbf{S}}{\partial\lambda}\otimes\frac{\partial f}{\partial_1^2\mathbf{E}}}{\frac{\partial f}{\partial\lambda}} \tag{5.42}$$

or after differentation of all terms the consistent tangent tensor without hardening effects

$$\bar{\mathcal{P}} = \bar{\mathcal{C}} - \frac{\bar{\mathcal{C}}\,_1^2\mathbf{S}^D \otimes \bar{\mathcal{C}}\,_1^2\mathbf{S}^D}{_1^2\mathbf{S}^D\cdot\bar{\mathcal{C}}\,_1^2\mathbf{S}^D}. \tag{5.43}$$

5.6.1 Consistent tangent tensor for isotropic and kinematic hardening

If isotropic and/or kinematic hardening is assumed the eqs.(5.17), (5.20) and (5.19), in increments

$$\Delta_1^2\mathbf{E}^p = \frac{3}{2}\lambda\mathcal{D}(\,_1^2\mathbf{S} - \,_1^2\mathbf{A}), \tag{5.44}$$

$$\Delta_1^2\mathbf{A} = \frac{3}{2}H'_A\Delta_1^2\mathbf{E}^p, \tag{5.45}$$

$$_1^2\mathbf{A} = \,_1^1\mathbf{A} + \Delta_1^2\mathbf{A}, \tag{5.46}$$

have to be solved for the stresses.
The result is

$$_1^2\mathbf{S}(\,_1^2\mathbf{E},\lambda) = \,_1^1\mathbf{A} + \bar{\mathcal{C}}(\lambda)\,_1^2\mathbf{E}, \tag{5.47}$$

$$_1^2\mathbf{A}(\,_1^2\mathbf{E},\lambda) = \,_1^1\mathbf{A} + \frac{\lambda H'_A}{1+\lambda H'_A}\bar{\mathcal{C}}(\lambda)\,_1^2\mathbf{E} \tag{5.48}$$

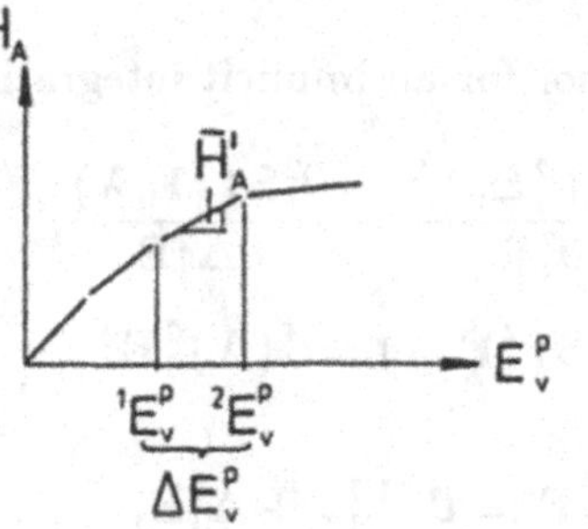

Fig. 5.5 Approximation of the kinematic hardening function by piecewise linear polynomials

with the definitions

$$ {}^2_1\mathbf{E} := \mathcal{C}^{-1}(\,{}^1_1\sigma - {}^1_1\mathbf{A}\,) + \Delta{}^2_1\mathbf{E}, \tag{5.49}$$

$$\bar{\mathcal{C}}(\,\lambda\,) := (\,\mathcal{C}^{-1} + \frac{3\lambda}{2(\,1+\lambda H'_A\,)}\mathcal{D}\,)^{-1}. \tag{5.50}$$

The yield condition (5.41) including hardening effects expressed in terms of ${}^2_1\mathbf{E}$ is

$$f(\,{}^2_1\mathbf{E},\lambda\,) = \frac{3}{2}(\,\bar{\mathcal{C}}\,{}^2_1\mathbf{E}\,)\cdot\mathcal{D}(\,\bar{\mathcal{C}}\,{}^2_1\mathbf{E}\,) - \kappa^2(\,{}^2E^p_v\,) = 0. \tag{5.51}$$

The connexion between ${}^2E^p_v$ and ${}^2_1\mathbf{E}$ is given by the equivalent plastic strain eq. (5.21), formulated in increments with the help of eqs. (5.44), (5.47) and (5.48)

$$\Delta{}^2E^p_v = \frac{1}{1+\lambda H'_A}\sqrt{\frac{3}{2}(\,\bar{\mathcal{C}}\,{}^2_1\mathbf{E}\,)\cdot\mathcal{D}(\,\bar{\mathcal{C}}\,{}^2_1\mathbf{E}\,)} = \lambda\kappa. \tag{5.52}$$

Differentation according to eq. (5.37) or eq. (5.42) leads to the consistent tangent tensor

$$\bar{\mathcal{P}} = \bar{\mathcal{C}} - \frac{\bar{\mathcal{C}}(\,{}^2_1\mathbf{S}^D - {}^2_1\mathbf{A}^D\,)\otimes\bar{\mathcal{C}}(\,{}^2_1\mathbf{S}^D - {}^2_1\mathbf{A}^D\,)}{\frac{4}{9}\kappa^2\frac{(\,H'_A+\kappa'+\lambda H''_A\kappa\,)(\,1+\lambda H'_A\,)}{(\,1-\lambda\kappa'-\lambda^2 H''_A\kappa\,)} + (\,{}^2_1\mathbf{S}^D - {}^2_1\mathbf{A}^D\,)\cdot\bar{\mathcal{C}}(\,{}^2_1\mathbf{S}^D - {}^2_1\mathbf{A}^D\,)} \tag{5.53}$$

For $\lambda \rightarrow d\lambda$ or $\dot{\lambda}$ (infinitesimal load increments) the consistent tangent tensor becomes the classical Prandtl-Reuss tensor (5.24). When the load increments increase, the differences between the two tensors grow.

5.6.2 Three dimensional stress case

A constant slope of the kinematic hardening function $H'_A(\,E^p_v\,)$ in the increment is assumed

$$\bar{H}'_A = \frac{{}^2H_A - {}^1H_A}{{}^2E^p_v - {}^1E^p_v} = \frac{\Delta H'_A}{\Delta E^p_v}. \tag{5.54}$$

Therefore $\bar{H}''_A = 0$ holds.
With the definitions

$$K = \frac{2G(1+\nu)}{3(1-2\nu)}, \tag{5.55}$$

G: shear modulus, ν: Poisson ratio,

$$\beta = \frac{1+\lambda \bar{H}'_A}{1+\lambda \bar{H}'_A + 3\lambda G}, \tag{5.56}$$

$$\gamma = \frac{1}{1+\frac{\kappa'+\bar{H}'_A}{3G}} + \beta - 1, \tag{5.57}$$

$$\mathbf{n} := \frac{{}^2_1\mathbf{S}^D - {}^2_1\mathbf{A}^D}{\sqrt{({}^2_1\mathbf{S}^D - {}^2_1\mathbf{A}^D)\cdot({}^2_1\mathbf{S}^D - {}^2_1\mathbf{A}^D)}}, \tag{5.58}$$

the following result for the three dimensional stress case (Simo and Taylor 1985) is obtained:

$$\bar{\mathcal{P}} = K\mathbf{1}\otimes\mathbf{1} + 2G\beta(\mathcal{I} - \frac{1}{3}\mathbf{1}\otimes\mathbf{1}) - 2G\gamma\mathbf{n}\otimes\mathbf{n}. \tag{5.59}$$

5.6.3 Two-dimensional stress case

For simplicity a matrix notation is used. The consistent tangent matrix (Gruttmann 1985) and (Simo and Taylor 1986) for the two-dimensional stress state has the same form as the tensor (5.53)

$$\bar{\mathcal{P}} = \bar{\mathcal{C}} - \frac{\bar{\mathcal{C}}({}^2_1\mathbf{S}^D - {}^2_1\mathbf{A}^D)({}^2_1\mathbf{S}^D - {}^2_1\mathbf{A}^D)^T\bar{\mathcal{C}}^T}{\frac{4}{9}\kappa^2\frac{(H'_A+\kappa'+\lambda H''_A\kappa)(1+\lambda H'_A)}{(1-\lambda\kappa'-\lambda^2 H''_A\kappa)} + ({}^2_1\mathbf{S}^D - {}^2_1\mathbf{A}^D)^T\bar{\mathcal{C}}({}^2_1\mathbf{S}^D - {}^2_1\mathbf{A}^D)} \tag{5.60}$$

but $\bar{\mathcal{C}}$ contains the following material property tensor for the plane stress case

$$\bar{\mathcal{C}}^{-1} = \breve{\mathcal{C}}^{-1} + \frac{3\lambda}{2(1+\lambda H'_A)}\mathcal{D} \tag{5.61}$$

with

$$\breve{\mathcal{C}}^{-1} = \frac{1}{E}\begin{pmatrix} 1 & -\nu & 0 \\ -\nu & 1 & 0 \\ 0 & 0 & 2(1+\nu) \end{pmatrix},$$

$$\mathcal{D} = \frac{1}{3}\begin{pmatrix} 2 & -1 & 0 \\ -1 & 2 & 0 \\ 0 & 0 & 6 \end{pmatrix}.$$

5.7 FE-DISCRETIZATION

5.7.1 Isoparametric 4-node two-dimensional element

For the computations of the examples in 5.8.1 an isoparametric 4-node two dimensional element was used with 2 DOF's at each corner. For a detailed description of this element see chapter 1. The integration was carried out in 2 Gaussian points in each direction.

5.7.2 The extended DKT-element

For the computations of the examples in 5.8.2 and 5.8.3 a plane triangular element with the 6 primar DOF's at each corner was applied. The element stiffness matrix is given as a linear superposition of the following independent parts:
- membrane stiffness matrix $\mathbf{K}_M$
- bending stiffness matrix $\mathbf{K}_B$
- formal rotational matrix $\mathbf{K}_{\theta z}$

for rotations with respect to the normal direction of the element.

$\mathbf{K}_B$ is the stiffness matrix of the DKT-element (DKT means discrete Kirchhoff theory), investigated by Bathe and Ho (1981) and Batoz, Bathe and Ho (1980). $\mathbf{K}_M$ is the constant strain plane stiffness matrix of a three node element.

$$\mathbf{V} = \left\{ u_1\, u_2\, u_3\, v_1\, v_2\, v_3\, w_1\, \vartheta_{x1}\, \vartheta_{y1}\, w_2\, \vartheta_{x2}\, \vartheta_{y2} \atop w_3\, \vartheta_{x3}\, \vartheta_{y3}\, \vartheta_{z1}\, \vartheta_{z2}\, \vartheta_{z3} \right\}$$

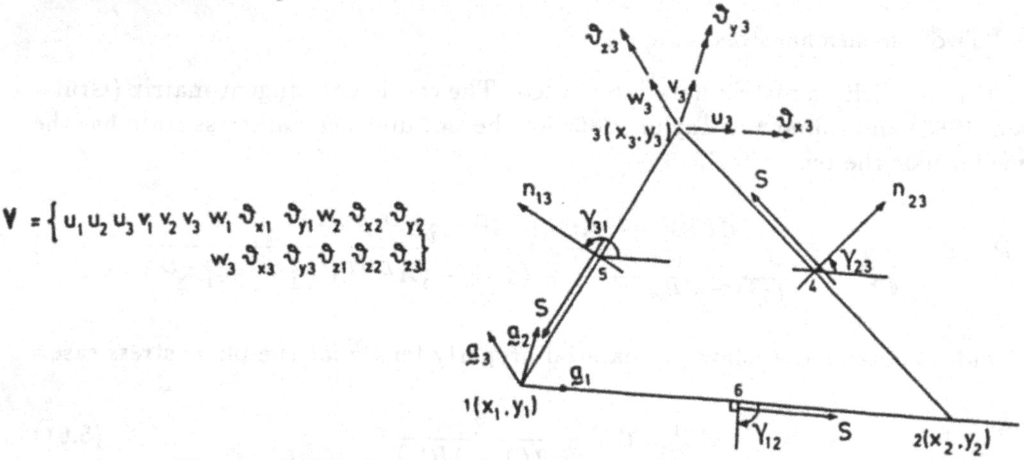

Fig. 5.6 DKT-element and element displacement vector

5.7.3 Layer models

For progressing plastic zones over the thickness two models are used, which are based on the introduction of so called layer points distributed over the element thickness (Stein and Paulun 1981).

In model 1, the same distance is chosen between each pair of adjacent layer points, introduced in each Gaussian point of the element. The strain increment is calculated assuming the Kirchhoff-Love hypothesis and the yield condition is fullfilled

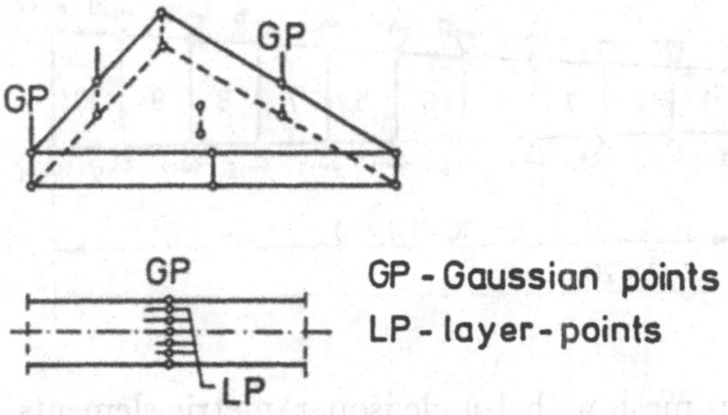

Fig. 5.7 Gaussian points and layer points

in each layer point by using the projection method described in chapter 5.2. The integration of the calculated stresses is done with Newton-Cotes formulas, e. g. Simpson's rule. The disadvantage of this model is a low accuracy if a nonlinear stress distribution is integrated with only few layer points, e. g. 5 layer points. In this case, discontinious changes in the stiffness occur.

This disadvantage may be treated either by using a larger number of layer points (which requires more CP-time) or by application of the layer model 2. This model admits the variable position of the layer points corresponding to the expansion of the plastic zones.

The positions of the variable layer points denote the boundaries of the elastic region. Assuming a piecewise linear distribution, the stresses are integrated exactly.

5.8 Examples

5.8.1 Plane strain problems

The progressing plastifying of two examples is investigated using 4-node isoparametric elements in a 2 dimensional displacement field.

Example 1.

Thick-walled cylinder under internal pressure, a plane strain, geometrically linear problem. The rate of convergence is given for the Prandtl-Reuss tensor and the consistent tangent tensor.

System and material data:

$p_i = 22.918\, kN/cm^2$ in 5 load increments

$E = 21000\, kN/cm^2$

$\nu = 0.3$

$R_a = 20\, cm$

$R_i = 10\, cm$

$\kappa_o = 24\, kN/cm^2$

$\kappa' = 200\, kN/cm^2$ linear isotropic hardening

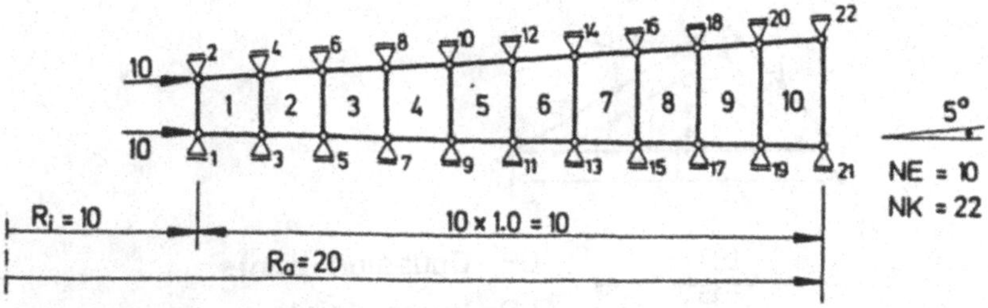

Fig. 5.8 Finite element mesh with 4-node isoparametric elements, sector of degrees, 10 elements, 22 nodes; units in (cm) and (kN).

Load increment		1	2	3	4	5
Number of plastified elements, beginning with the inner edge		0	0	2	7	10
Number of iterations	a) Prandtl-Reuss	2	2	6	8	10
	b) konsistent tangential-tensor	2	2	5	6	5

Tab. 5.1 Comparison of the number of iterations for a) Prandtl-Reuss law b) consistent tangential tensor

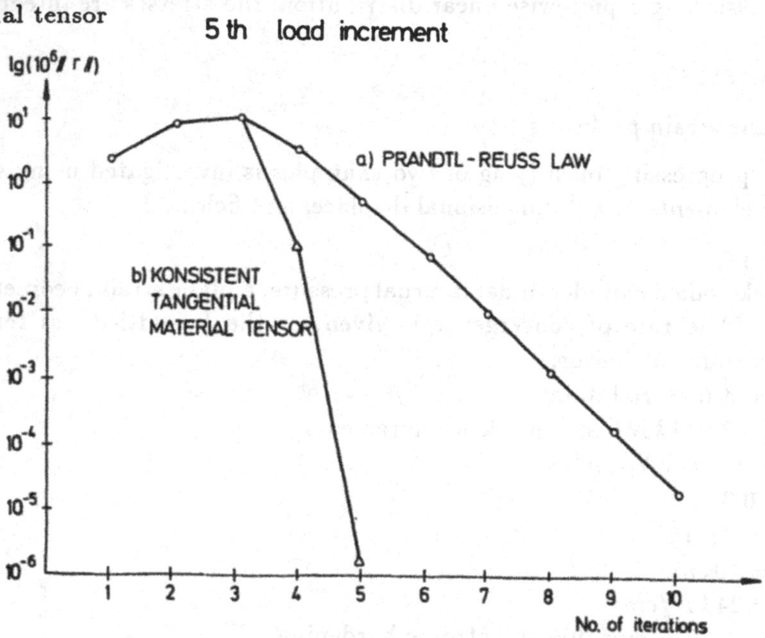

Fig. 5.9 Euclidian norm of the residual force vector r (logarithmic) for each iteration step

It's evident that the use of consistent elasto-plastic tangents within the implicit integration of the flow rule leads to about the half of the number of iterations in comparison with the use of the classical Prandtl-Reuss tensor.

Example 2.

Perforated strip under uniaxial tension, a plane strain, geometrically linear problem. The rate of convergence is given, when using the Prandtl- Reuss tensor and the consistent tangent tensor.
System and material data:

$p = 17.6\,kN/cm^2$ in 8 load increments
$E = 21000\,kN/cm^2$
$\nu = 0.3$
$\kappa_o = 24\,kN/cm^2$
$\kappa' = 200\,kN/cm^2$ linear isotropic hardening

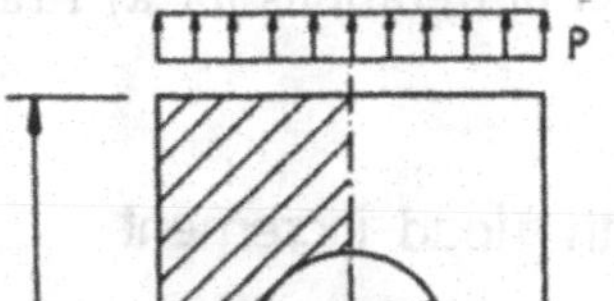

Fig. 5.10 System and loading; units in (cm) and (kN).

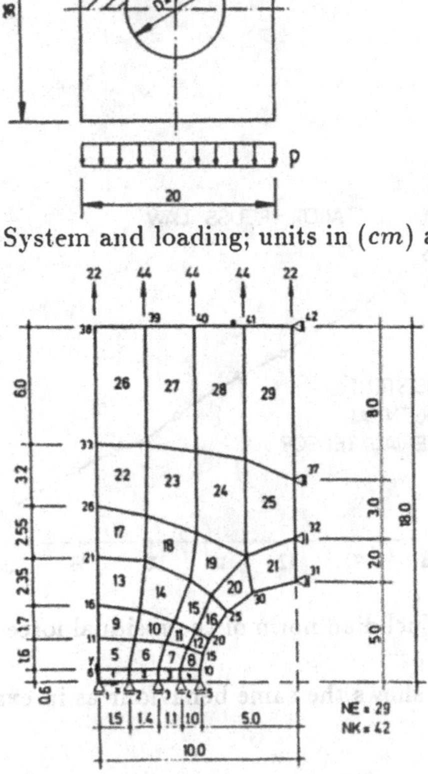

Fig. 5.11 Finite element mesh with 4-node isoparametric elements

Load increment		1	2	3	4	5	6	7	8
Number of plastified elements		0	0	0	0	2	4	6	13
Number of iterations	a) Prandtl-Reuss	2	2	2	2	10	13	11	20
	b) konsistent tangential-tensor	2	2	2	2	5	5	5	6

Tab. 5.2 Comparison of the number of iterations for a) Prandtl-Reuss law b) consistent tangent tensor

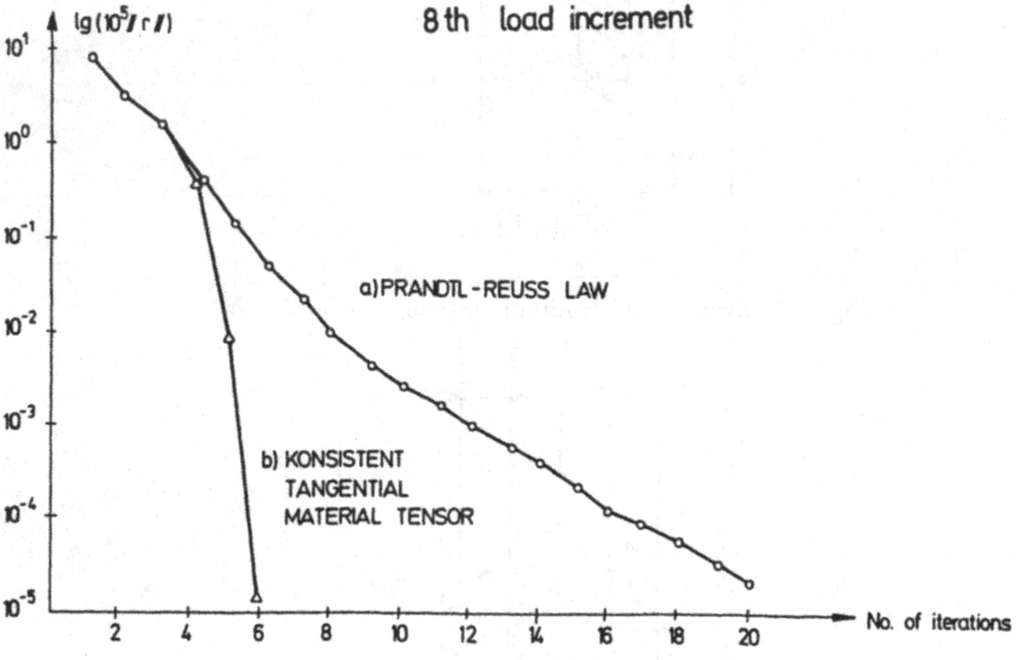

Fig. 5.12: Euclidian norm of the residual force vector r

The iteration process shows the same behaviour as in example 1.

5.8.2 Plane stress problems

To calculate the plastifying of the following two examples, the DKT-element described in chapter 5.7.2 was used.

Example 3.

Perforated strip under uniaxial tension, a plane stress geometrically linear problem. The rate of convergence is shown for the Prandtl-Reuss matrix and the consistent tangent matrix.

System and material data:

$p = 12.88\, kN/cm^2$
$E = 7000\, kN/cm^2$
$\nu = 0.2$
$\kappa_o = 24.34\, kN/cm^2$
$\kappa' = 200.5\, kN/cm^2$ linear isotropic hardening
$t = 1mm$
units in (mm)

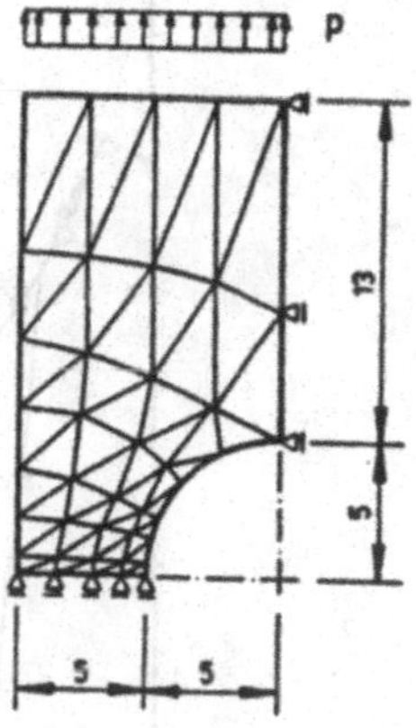

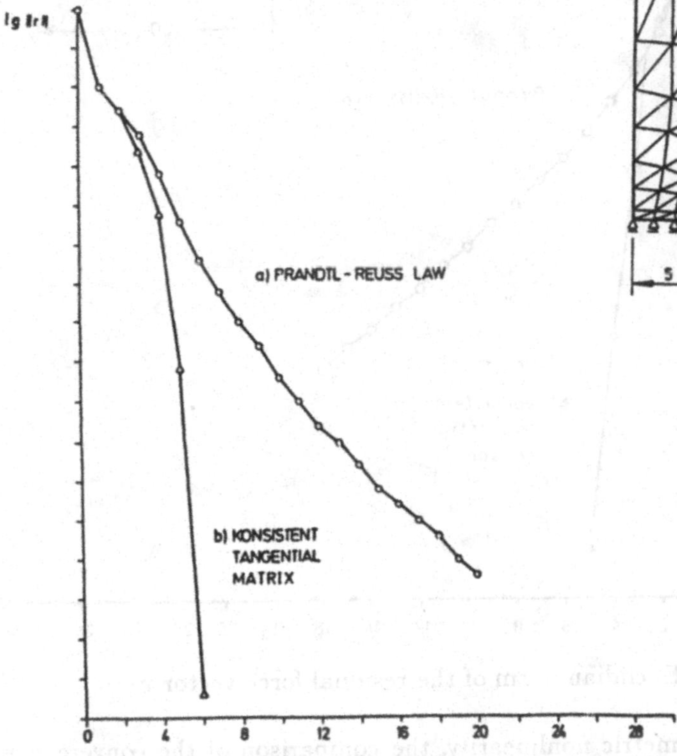

Fig. 5.13 Euclidian norm of the residual force vector **r**

Also in the plane stress case the use of the consistent elasto-plastic tangent shows the same behaviour as in the examples before.

Example 4.

Rectangular plate with hinged boundaries, a plane stress geometrically nonlinear problem under the assumption of small strains. The rate of convergence is given for the Prandtl-Reuss matrix and the consistent tangent matrix.

System and material data:

$p = 0.025\, kN/cm^2$
$E = 19600\, kN/cm^2$
$\nu = 0.33$
$\kappa_o = 19.6\, kN/cm^2$
$t = 5mm$
$l = 400mm$

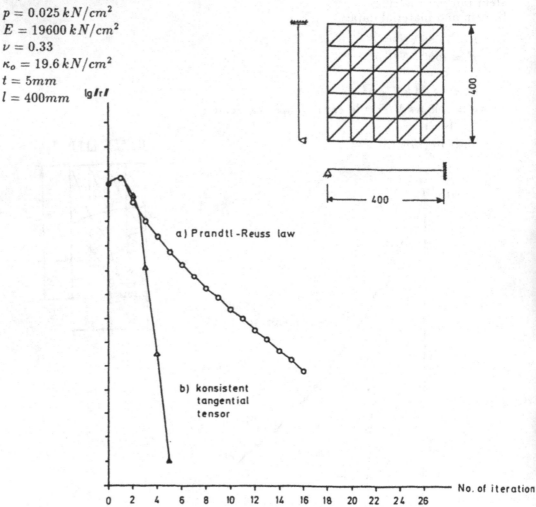

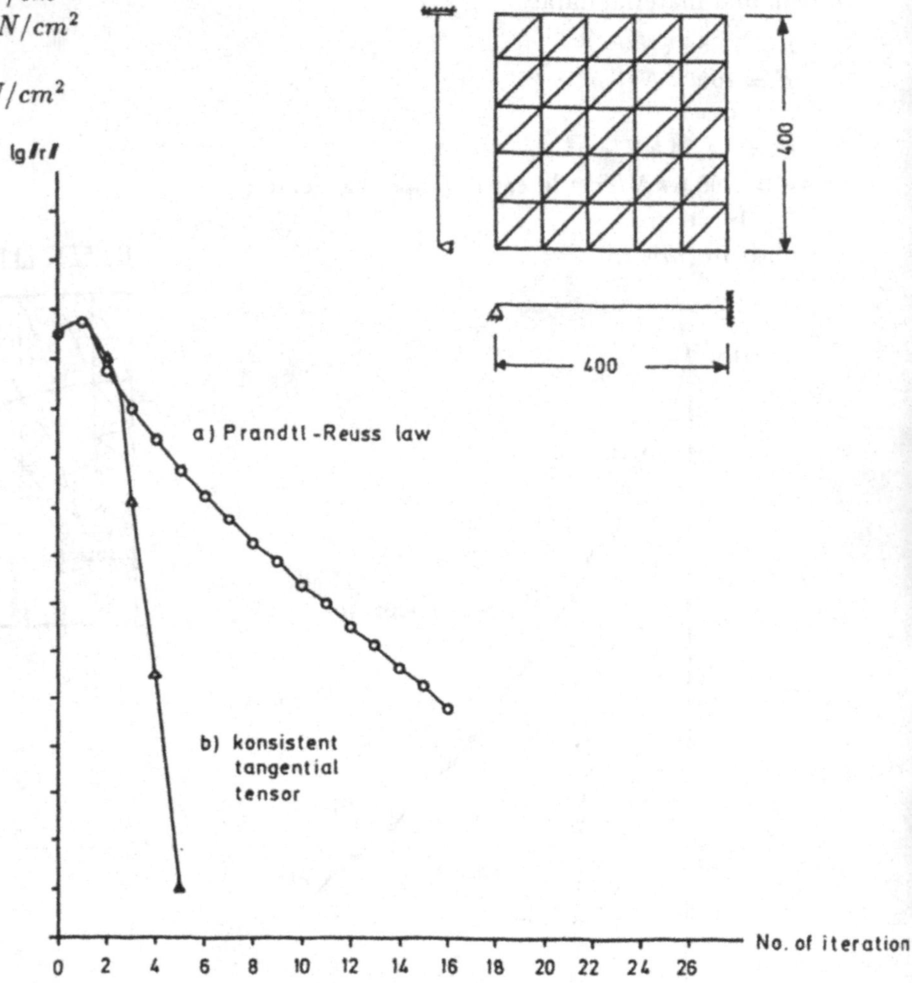

Fig. 5.14 Euclidian norm of the residual force vector **r**

In spite of the geometric nonlinearity, the comparison of the convergence rate shows the same behaviour as before, when using the consistent elasto- plastic tangents and the Prandtl-Reuss law.

5.8.3 Ultimate load of a steel girder with thin web

The failure of the girder was computed using the DKT-element described in chapter 5.7.2.

Example 5.

Steel girder with extremly thin web, where plastifying and buckling interacts until crippling of the web happens and the girder fails. This problem is nonlinear in its geometric and material behaviour. To compute the material nonlinearity the projection method described in chapter 5.5.2 was used. Within the calculation of the geometric nonlinearity the Prandtl-Reuss law was applied. A comparison with an experiment is given.
System and material data:

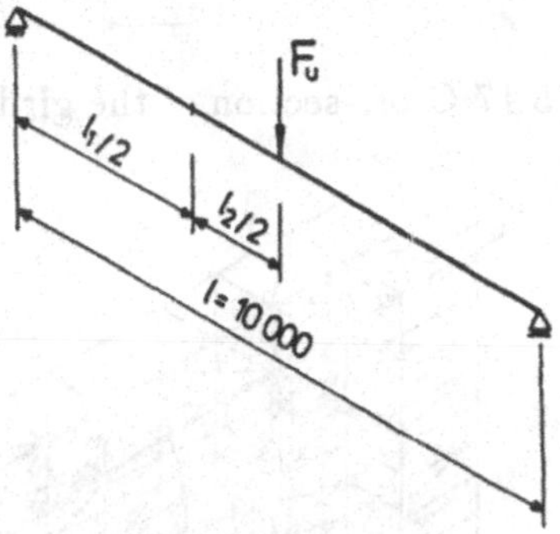

Fig. 5.15 System and loading (unit mm)

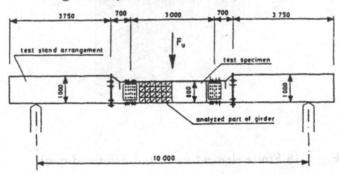

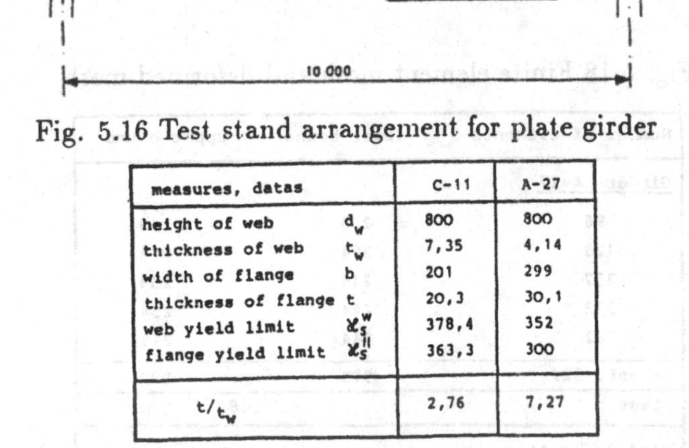

Fig. 5.16 Test stand arrangement for plate girder

measures, datas		C-11	A-27
height of web	d_w	800	800
thickness of web	t_w	7,35	4,14
width of flange	b	201	299
thickness of flange	t	20,3	30,1
web yield limit	β_S^w	378,4	352
flange yield limit	β_S^{II}	363,3	300
t/t_w		2,76	7,27

Tab. 5.3 Material datas (units in N and mm)

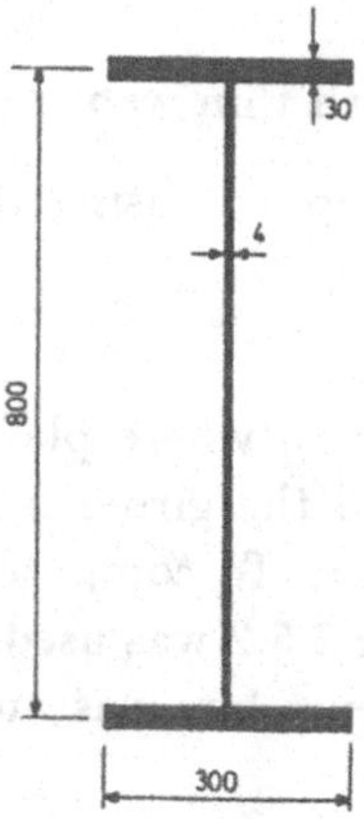

Fig. 5.17 Cross-section of the girder (unit mm)

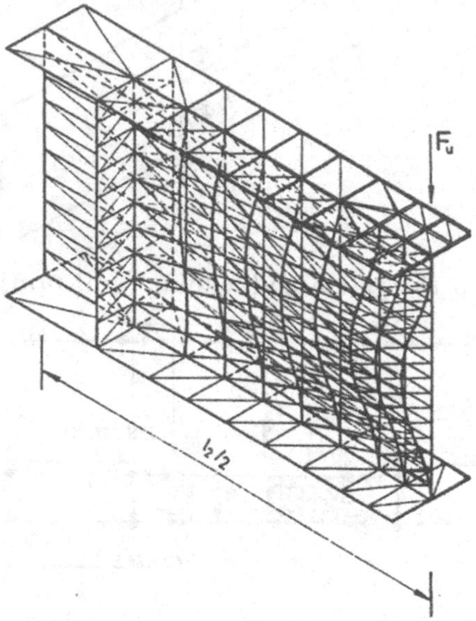

Fig. 5.18 Finite element mesh and deformed mesh

Number of elements	lower bound	upper bound
Girder A-27		
96	220	244
128	284	294
257	244	254
322	244	254
362	244	249
elast. 322	970	980
Test	248	
Girder C-11		
96	492	520
Test	535	

Tab. 5.4 Calculated ultimate loads $\mathbf{F}_u$ (unit kN)

Failure mode girder A-27: web $t_w = 4mm$
Elastic buckling
$\mathbf{F} = 24.75\,kN$
$w = 1.7mm$
Web-crippling
Yielding at $\mathbf{F} = 120\,kN$
$w = 6.8mm$
Plastic buckling
Large yield zone and failure at $\mathbf{F}_u = 249\,kN$
$w = 18.1mm$

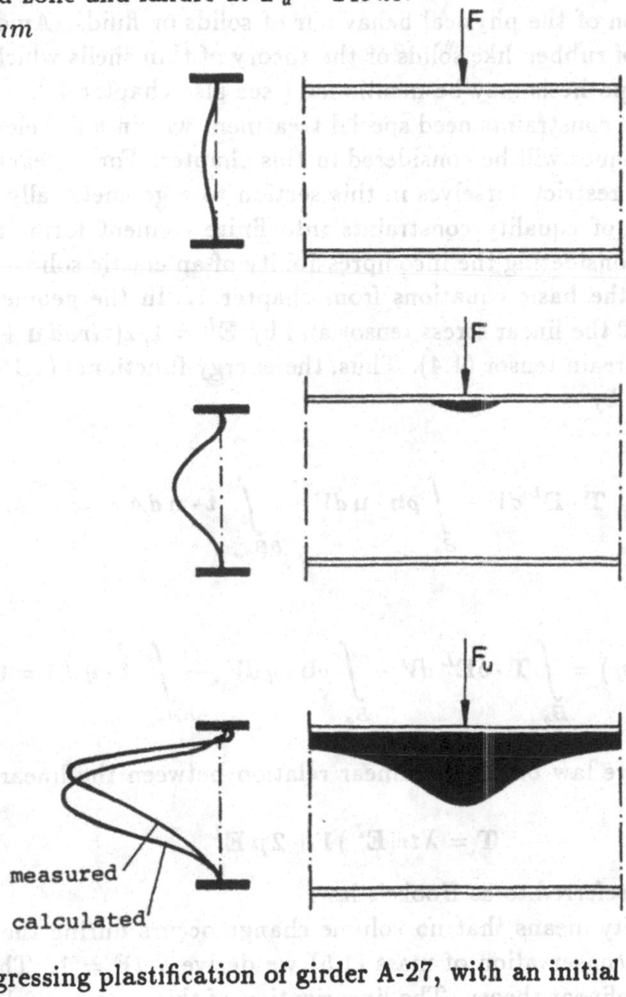

Fig. 5.19 Progressing plastification of girder A-27, with an initial web deflection of max $w_v = 5mm$

The computation of the girder shows a good corresponding to the experiment, which was carried out in Braunschweig, FRG, 1983 (Scheer 1983).

6 FE–Formulation for problems with constraints

Within a wide range of applications constraint conditions have to be incorporated into the formulation of the continuum mechanics of solids. These constraints can occur in equality or inequality form and will be considered within the next two sections.

6.1 Equality constraints

There are many examples for the occurance of constraints within the mathematical formulation of the physical behaviour of solids or fluids. Among others the incompressibility of rubber–like solids or the theory of thin shells which bases on the Kirchhoff–Love hypothesis may be mentioned (see also chapter 4).

Problems with constraints need special treatment within finite element formulations. Some techniques will be considered in this chapter. For a clear description of the basic ideas we restrict ourselves in this section to a geometrically linear theory. The incorporation of equality constraints into finite element formulations will be demonstrated by considering the incompressibility of an elastic solid.

Let us recall the basic equations from chapter 1. In the geometrically linear case we denot by $\mathbf{T}$ the linear stress tensor and by $\mathbf{E}^L = 1/2(Grad\,\mathbf{u} + Grad^T\mathbf{u})$ the linearized Green streain tensor (1.4). Thus, the energy functional (1.15) and its first variation are given by

$$\Pi^L(\mathbf{u}) = \frac{1}{2}\int_{B_o} \mathbf{T}\cdot\mathbf{E}^L\,dV - \int_{B_o} \rho\mathbf{b}\cdot\mathbf{u}\,dV - \int_{\partial B_{o_\sigma}} \hat{\mathbf{t}}\cdot\mathbf{u}\,dA \quad\Rightarrow\quad Min. \tag{6.1}$$

and

$$g^L(\mathbf{u},\boldsymbol{\eta}) = \int_{B_o} \mathbf{T}\cdot\delta\mathbf{E}^L\,dV - \int_{B_o} \rho\mathbf{b}\cdot\boldsymbol{\eta}\,dV - \int_{\partial B_{\sigma_o}} \hat{\mathbf{t}}\cdot\boldsymbol{\eta}\,dA = 0 \tag{6.2}$$

The constitutive law becomes a linear relation between the linear stresses and strains

$$\mathbf{T} = \lambda\,tr(\mathbf{E}^L)\mathbf{1} + 2\,\mu\,\mathbf{E}^L \tag{6.3}$$

which is often referred to as Hooke's law.

Incompressibility means that no volume change occurs during the deformation process. From the conservation of mass (1.5) we derive $det\mathbf{F} = 1$. This relation is valid within the nonlinear theory. The linearization of this eqation yields

$$Div\,\mathbf{u} = tr\,\mathbf{E}^L = 0. \tag{6.4}$$

For the incorporation of constraint equation (6.4) into a finite element method several formulations are possible. Two of the commonly used methods are the

— Penalty formulation and
— Lagrangian multiplier method.

The derivation of these methods is based on a split of strains and stresses into bulk and deviatoric parts. This is shown in detail in e.g. Wriggers (1985). We obtain

$$\mathbf{E}^L = \frac{1}{3} Div\, \mathbf{u}\, \mathbf{1} + \mathbf{E}^D$$

$$\mathbf{T} = p\, \mathbf{1} + \mathbf{T}^D \tag{6.5}$$

where $(\ldots)^D$ denotes the deviatoric part. With (6.5) Hooke's law yields

$$\mathbf{T} = (\,\lambda + \frac{2}{3}\mu\,)\, Div\, \mathbf{u}\, \mathbf{1} + 2\,\mu\, \mathbf{E}^D \tag{6.6}$$

Finally, equation (6.6) can be split with $(6.5)_2$ and $K = \lambda + \frac{2}{3}\mu$ into

$$p = K\, Div\, \mathbf{u},$$

$$\mathbf{T}^D = 2\,\mu\, \mathbf{E}^D. \tag{6.7}$$

With equation (6.7) the first term in the weak form (6.2) yields

$$\int_{B_o} \mathbf{T} \cdot \delta \mathbf{E}^L\, dV = \int_{B_o} (\, K\, Div\, \mathbf{u}\, Div\, \boldsymbol{\eta} + 2\,\mu\, \mathbf{E}^D \cdot \delta \mathbf{E}^D\,)\, dV. \tag{6.8}$$

Note, that the constitutive law (6.7) for the pressure term p cannot be used for incompressibility since $Div\, \mathbf{u} \to 0$ leads to $K \to \infty$. In this case, the Lagrangian multiplier method may be used to determine the unknown pressure field p.

On the other hand, the weak form (6.8) may be used to approximate condition (6.4). By choosing a large number for K nearly incompressible behaviour is obtained. This approach is equivalent to a penalty method. The penalty term then is the first term on the left side of (6.8). A formulation from which both methods, described above, may be deduced is given by a mixed model for the pressure.

$$\Pi^L(\mathbf{u}, p) = \frac{1}{2} \int_{B_o} \mathbf{T}^D \cdot \mathbf{E}^D\, dV + \int_{B_o} p\,(\, Div\, \mathbf{u} - \frac{1}{2K}\, p\,)\, dV$$

$$- \int_{B_o} \rho \mathbf{b} \cdot \mathbf{u}\, dV - \int_{\partial B_{o\sigma}} \hat{\mathbf{t}} \cdot \mathbf{u}\, dA \;\Rightarrow\; Stat. \tag{6.9}$$

This formulation is often referred to as *perturbed* Lagrangian formulation. The first variation of (6.9) leads to

$$g^L(\mathbf{u}, p, \boldsymbol{\eta}) = \int_{B_o} 2\,\mu\, \mathbf{E}^D \cdot \delta \mathbf{E}^D \, dV - \int_{B_o} p\, Div\, \boldsymbol{\eta}\, dV$$

$$- \int_{B_o} \rho \mathbf{b} \cdot \boldsymbol{\eta}\, dV - \int_{\partial B_{\sigma o}} \hat{\mathbf{t}} \cdot \boldsymbol{\eta}\, dA = 0$$

$$g^L(\mathbf{u}, p, \delta p) = \int_{B_o} \delta p (\, Div\, \mathbf{u} - \frac{1}{K} p\,)\, dV = 0. \tag{6.10}$$

Solving equation $(6.10)_2$ for p we recover the constitutive relation $(6.7)_1$. Inserting this into $(6.10)_1$ leads to a pure displacement (or penalty) formulation (6.8). Furthermore for $K \rightarrow \infty$ the classical Lagrangian multiplier formulation is obtained.

The finite element discretization leads for a two–dimensional plane strain problem to

$$Div\, \mathbf{u}_e^h = \sum_{i=1}^{n} \mathbf{B}_{Vi} \mathbf{v}_i = \sum_{i=1}^{n} [\, N_{i,1} \;\; N_{i,2}\,]\, \mathbf{v}_i \tag{6.11}$$

and

$$\mathbf{E}_e^{D\,h} = \sum_{i=1}^{n} \mathbf{B}_{Di}\, \mathbf{v}_i \tag{6.12}$$

with

$$\mathbf{B}_{Di} = \frac{1}{3} \begin{bmatrix} 2\,N_{i,1} & -N_{i,2} \\ -N_{i,1} & 2\,N_{i,2} \\ -N_{i,1} & -N_{i,2} \\ 3\,N_{i,2} & 3\,N_{i,1} \end{bmatrix}$$

where the notation of chapter 2 and equation (6.5) have been used. The introduction of (6.11) and (6.12) into the linearization of (6.8) leads to the element stiffness matrix for the penalty approach

$$\mathbf{K}^P = \sum_{i=1}^{n} \sum_{k=1}^{n} \int_{\Omega_e^h} [\, \mathbf{B}_{Vi}^T\, K\, \mathbf{B}_{Vk} + \mathbf{B}_{Di}^T\, 2\,\mu \mathbf{I}\, \mathbf{B}_{Dk}\,]\, d\Omega. \tag{6.13}$$

An integration of **K** with 2 quadrature points in all directions shows the well known locking behaviour of the mesh, see e.g. Hughes et. al. (1976). This phenomenon can be avoided by a selective–reduced integration, see e.g. Malkus, Hughes (1977). Selective–reduced integration means a reduced order of integration, $(1\, x\, 1)$, for the pressure term in (6.13). This procedure yields good results for plane strain problems but not for axisymmetric applications. For those cases the mixed model provides a reasonable approach. Here we have to approximate **u** and p. The LBB–condition (Ladyshenkaya–Babuska–Brezzi—condition) for mixed approximations tells us to use within the shape functions for p one order of degree of polynominal less than for **u**. In the case where a linear polynominal is used for **u** we approximate p by a constant value p_o.

With these shape functions the pressure terms in (6.10) are approximated with (6.11) by

$$\int_{B_o} p \, Div \, \boldsymbol{\eta} \, dV = \sum_{i=1}^{n} \boldsymbol{\eta}_i^T \int_{\Omega_e^h} \mathbf{B}_{Vi}^T \, d\Omega \; p_o,$$

$$\int_{B_o} \delta p \, (Div \, \mathbf{u} - \frac{1}{K} p) \, dV = \sum_{i=1}^{n} \int_{\Omega_e^h} (\mathbf{B}_{Vi} \, \mathbf{v}_i - \frac{1}{K} p_o) \, d\Omega \, \delta p_o . \tag{6.14}$$

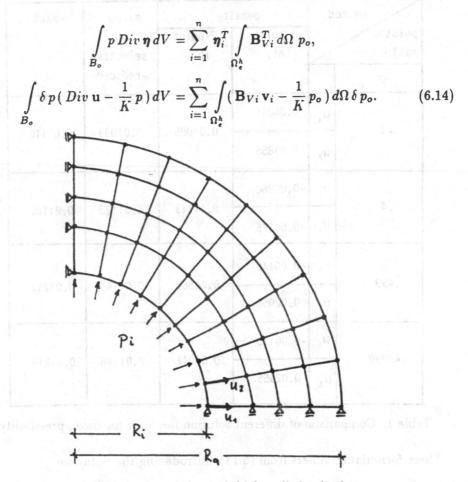

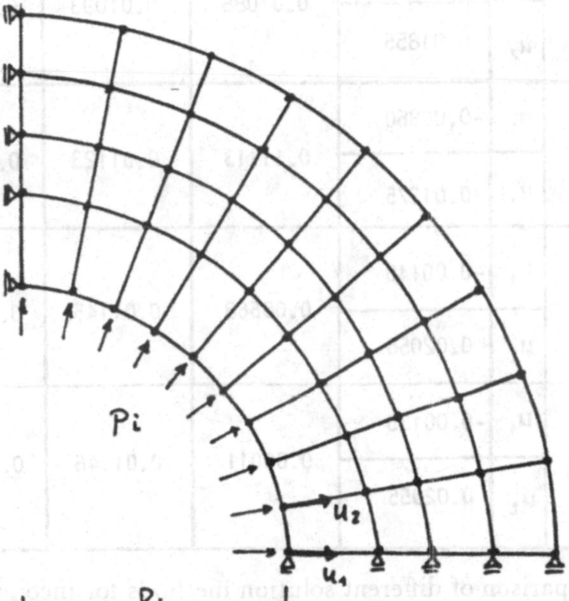

Fig. 6.1 FE–mesh and data of thick walled cylinder

From a computational standpoint it is more convenient to work with displacements as only unknowns. Therefore equation $(6.14)_2$ may be solved for p_o

$$p_o = \frac{K}{\Omega_e^h} \int_{\Omega_e^h} \mathbf{B}_{Vi} \, d\Omega \, \mathbf{v}_i \tag{6.15}$$

and inserted into $(6.14)_1$ which yields a pure displacement formulation

$$\mathbf{K}^M = \sum_{i=1}^{n} \sum_{k=1}^{n} \frac{K}{\Omega_e^h} \int\limits_{\Omega_e^h} \mathbf{B}_{Vi}^T \, d\Omega \int\limits_{\Omega_e^h} \mathbf{B}_{Vk} \, d\Omega + \int\limits_{\Omega_e^h} \mathbf{B}_{Di}^T \, 2\,\mu \mathbf{I} \, \mathbf{B}_{Dk} \, d\Omega. \tag{6.16}$$

poisson ratio ν \ method		penalty 1-point int.	penalty 2-point int.	mixed or selective -reduced	exact
.3	u_1	-0.00815	0.01088	0.01093	0.01110
	u_2	0.01855			
.4	u_1	-0.00860	0.11113	0.01123	0.01163
	u_2	0.01275			
.499	u_1	-0.00146	0.00568	0.01145	0.01213
	u_2	0.02056			
.49999	u_1	-0.00135	0.00011	0.01146	0.01219
	u_2	0.02055			

Table 1: Comparison of different solution methods for incompressibility

These formulation differs from (6.13). Introducing the notation

$$\mathbf{G}_i = \int\limits_{\Omega_e^h} \mathbf{B}_{Vi}^T \, d\Omega \tag{6.17}$$

and

$$\mathbf{K}^D = \sum_{i=1}^{n} \sum_{k=1}^{n} \int\limits_{\Omega_e^h} \mathbf{B}_{Di}^T \, 2\,\mu \mathbf{I} \, \mathbf{B}_{Dk} \, d\Omega. \tag{6.18}$$

we obtain

$$\mathbf{K}^M = \mathbf{K}^D + \sum_{i=1}^{n}\sum_{k=1}^{n} \frac{K}{\Omega_e^h}\, \mathbf{G}_i\, \mathbf{G}_k^T \tag{6.19}$$

which denotes a *rank-one-update* of $\mathbf{K}^D$.

It can be shown, that for plane problems the matrices in (6.13) and (6.16) are equivalent when selective-reduced integration is used for $\mathbf{K}^P$, see e.g. Malkus, Hughes (1977). However for axisymmetric situations this does not hold.

The different formulations stated above are compared by means of an plane strain examples. We consider a thick walled cylinder shown in Fig. 6.1. Table 1 contains the displacements u_1 and u_2 obtained for different values of the Poisson ratio ν. The information contained in table 1 may be interpreted as follows. The locking phenomenon for two-point quadrature can be observed for increasing ν. A one-point quadrature for all terms does not help. Locking vanishes but now a mesh instability is introduced by the underintegration of $\mathbf{K}^D$. This is due to rank deficiency which is imposed on $\mathbf{K}^D$ by one-point integration.

The mixed-or selective-reduced integration yield the same result which is in good agreement with the exact solution.

6.2 Inequality constraints

6.2.1 Continuous contact problem

In the case of contact between deformable bodies we have to augment the principle of minimal potential energy by inequality constraints for the displacements (socalled unilateral constraints). These are given by the directional distance $g : \Gamma_C \rightarrow \mathcal{R}$ which measures the distance of points on the contact surface to the obstacle and is positive if penetration occures. Thus we have to add the constraints

$$g(\mathbf{X} + \mathbf{u}(\mathbf{X})) \le 0, \qquad \text{for all } \mathbf{X} \in \Gamma_C \tag{6.20}$$

to the energy principle (1.15).

Necessary for equilibrium states is, that the virtual work of all nonreaction forces $D\Pi(\mathbf{u})$ are greater or equal than zero for all kinematically admissible virtual displacements, i. e.

$$\langle D\Pi(\mathbf{u}), \eta\rangle \ge 0 \qquad \text{for all } \eta \text{ with } \quad g(\mathbf{X} + \mathbf{u}(\mathbf{X}) + \eta(\mathbf{X})) \le 0. \tag{6.21}$$

For plastic contact problems the same ideas can be applied to the rate potential (Bischoff, 1987). The calculation then must be an incremental one (see chapter 5).

6.2.2 Discretization

6.2.2.1 General considerations

Additionally to problems without constraints one has to discretize the distance-function

$$d : \begin{cases} \Gamma_C \to \mathcal{R}_+ \\ \mathbf{X} \mapsto \min_{\mathbf{X}_C \in \Gamma_{C\mathbf{X}}} |\mathbf{X} + \mathbf{u}(\mathbf{X}) - (\mathbf{X}_C + \mathbf{u}(\mathbf{X}_C))| \end{cases} \tag{6.22}$$

by means of which we can compute the constraints given by g ($\Gamma_{C\mathbf{X}}$ is the obstacle relevant for the point $\mathbf{X}$).

If the obstacle for a contact surface is given by an analytical function, the distance and its first and second derivative can be computed in an explicit form in general.

Here we will be concerned with the case that both, the contacting body and the obstacle are discretized by means of finite elements (this is unavoidingly the case for multi-body contact or self-contact). Thus, the distancefunction d has to be changed to

$$\bar{d}_h : \begin{cases} \Gamma_{hC} \to \mathcal{R}_+ \\ \mathbf{X} \mapsto \min_{\mathbf{X}_C \in \Gamma_{hC\mathbf{X}}} |\mathbf{X} + \mathbf{u}_h(\mathbf{X}) - (\mathbf{X}_C + \mathbf{u}_h(\mathbf{X}_C))| \end{cases} . \tag{6.23}$$

For practical consideration this expression is computed only in nodal points and thus reduces to

$$d_h : \begin{cases} \Gamma_{hC} \to \mathcal{R}_+ \\ \mathbf{X}_h \mapsto \min_{\mathbf{X}_C \in \Gamma_{hC\mathbf{X}_h}} |\mathbf{X}_h + \mathbf{u}_h(\mathbf{X}_h) - (\mathbf{X}_C + \mathbf{u}_h(\mathbf{X}_C))| \end{cases} . \tag{6.24}$$

which says, that we will compute only the distance of the nodal points to the discretized obstacle surface

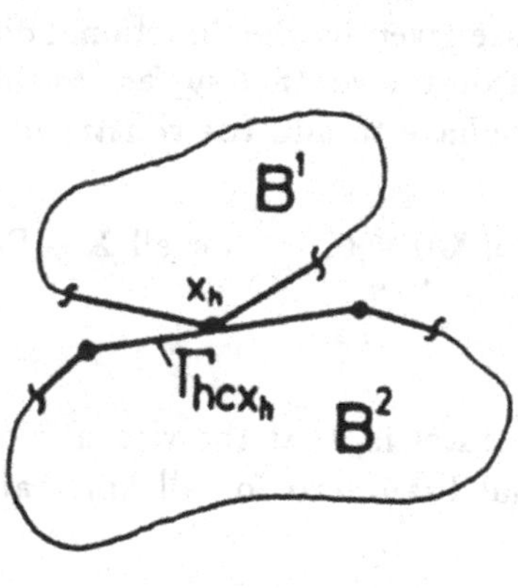

Fig. 6.2 Node-surface contact

As we will see, the quantities neeeded in contact algorithms are the directed distance and its first and second derivatives.

6.2.2.2 Determination of the directed distance

The directed distance g at a point $\mathbf{x}_h = \mathbf{X}_h + \mathbf{u}_h(\mathbf{X}_h)$ of the contact surface from the discretized obstacle will be determined in two steps. Since the obstacle is defined piecewise by finite elements one first has to select the right portion of the obstacle, belonging to the considered contact node and then to compute the distance.

The first step is performed by search over all obstacle nodes. In order to accelerate this procedure the search could be done in a hierarchical way, using the multi-grid concept e.g. or only local in the neighbourhood of the last node with minimal distance. In the last case one has to add global search steps after convergence in order to avoid convergence to local minima.

In the second step the distance on the surface patch belonging to the node with minimal distance will be computed and compared. How this can be done (for example) is shown in Fig. 6.3, (see Wriggers, Simo, 1985, Hallquist, 1979), where the node $\mathbf{x}_s$ is contacting a surface given by nodes $\mathbf{x}_1$ and $\mathbf{x}_2$.

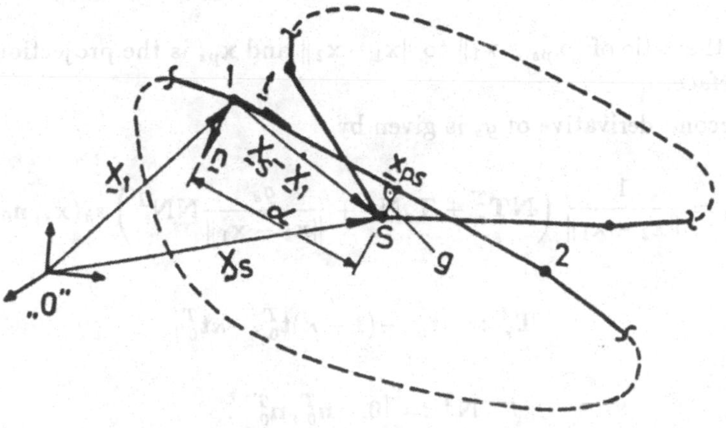

Fig. 6.3 Contact formulation

Let

$$g_s(\mathbf{x}_s, \mathbf{x}_1, \mathbf{x}_2) := \mathbf{n}_0^T(\mathbf{x}_s - \mathbf{x}_1)s_h(\mathbf{x}_s, \mathbf{n}_0) \tag{6.25}$$

with

$$\mathbf{n}_0 := \begin{pmatrix} 0 & 1 \\ -1 & 0 \end{pmatrix} \mathbf{t}_0, \tag{6.26}$$

$$\mathbf{t}_0 := \frac{\mathbf{x}_2 - \mathbf{x}_1}{\|\mathbf{x}_2 - \mathbf{x}_1\|} = \frac{\mathbf{t}}{\|\mathbf{t}\|}, \tag{6.27}$$

$$s_h(\mathbf{x}_s, \mathbf{n}_0) := \begin{cases} -1 & \mathbf{x}_s \text{ is a feasible point and } \mathbf{n}_0 \text{ is the outer normal on the} \\ & \text{obstacle or } \mathbf{x}_s \text{ is an unfeasible point and } \mathbf{n}_0 \text{ is the} \\ & \text{inner normal on the obstacle} \\ 1 & \text{otherwise,} \end{cases} \tag{6.28}$$

be the directed distance of node $\mathbf{x}_s$ from the obstacle. Penetration or gap of the node $\mathbf{x}_s$ is detected then by the sign of g_s:

$$g_s > 0 \longrightarrow \text{penetration}$$
$$g_s = 0 \longrightarrow \text{contact}$$
$$g_s < 0 \longrightarrow \text{gap}$$

and the constraints are:

$$\mathbf{g}(\mathbf{x}_h) \leq 0. \tag{6.29}$$

6.2.2.3 Determination of the derivatives

For the 2d-problem the first variation of the function g_s given above is given by

$$\nabla g_s = \mathbf{N}_s^T = \left[\mathbf{n}_0^T, -(1-\kappa)\mathbf{n}_0^T, -\kappa\mathbf{n}_0^T\right] s_h(\mathbf{x}_s, \mathbf{n}_0) \tag{6.30}$$

where κ is the ratio of $\|\mathbf{x}_{ps} - \mathbf{x}_1\|$ to $\|\mathbf{x}_1 - \mathbf{x}_2\|$ and $\mathbf{x}_{ps}$ is the projection of $\mathbf{x}_s$ on the contact surface.

The second derivative of g_s is given by

$$\nabla^2 g_s = \frac{1}{\|\mathbf{x}_2 - \mathbf{x}_1\|}\left(\mathbf{N}\mathbf{T}_s^T + \mathbf{T}_s\mathbf{N}^T + \frac{g_s}{\|\mathbf{x}_2 - \mathbf{x}_1\|}\mathbf{N}\mathbf{N}^T\right) s_h(\mathbf{x}_s, \mathbf{n}_0) \tag{6.31}$$

with

$$\mathbf{T}_s^T := \left[\mathbf{t}_0^T, -(1-\kappa)\mathbf{t}_0^T, -\kappa\mathbf{t}_0^T\right] \tag{6.32}$$

and

$$\mathbf{N}^T := \left[0, -\mathbf{n}_0^T, \mathbf{n}_0^T\right]. \tag{6.33}$$

This derivation is due to Wriggers, Simo (1985).

6.2.3 Algorithms

Algorithms for contact problems which are on the safe side usually rest on descent methods for constrained problems. The great variety of possible algorithms is restricted by the large dimension of the problems. There are essentially four types of algorithms which are suited to fit into a finite element code:

- Active Set Strategie
- Dual Methods
- Penalty Methods
- Newton-Methods (SQP)

which will be characterized shortly in the following chapters. A detailed discussion of these algorithms can be found in (Bischoff, 1985), (Bischoff, Mahnken, 1984), (Wriggers, Nour-Omid, 1984), (Bischoff, 1986). An extension to nonlinear stability problems is given in (Wriggers, Wagner, Stein, 1987).

6.2.3.1 Active Set Strategies

The idea underlying active set methods is to partition inequality constraints into two groups: those that are to be treated as active and those that are to be treated as inactive. At each phase of the algorithm a working set is chosen to be a subset of the nodes that are actually in contact for the current displacement. The algorithm then proceeds to move on with prescribed displacements for the nodes defined by the working set to an improved (with respect to the energy) displacement. At this new point the working set may be changed.

Convergence can be shown if the changes in the working set are made in the following way: One starts with a given working set and begins minimizing the energy with corresponding fixed working nodes. If a new node touches the obstacle it may be added to the working set, but no constraints are dropped from the working set. Finally, a point is obtained that minimizes with respect to the current working set. The corresponding reaction forces (i.e. the Lagrange multipliers) are determined and if they are all nonnegative, the solution is optimal. Otherwise, one or more constraints with negative reaction forces are dropped from the working set. The procedure is reinitiated with this new working set, and will strictly decrease on the next step.

6.2.3.2 Dual methods

Dual methods do not attack the original constrained problem directly but instead attack an alternate problem, the dual problem, whose unknowns are the Lagrange multipliers (here the reaction-forces) of the primal problem.

Such alternate problems can be constructed by means of a saddle function $L : \mathcal{R}^n \times \mathcal{R}^{n_C} \rightarrow \mathcal{R}$ (Ekeland, Temam, 1976) which characterises the optimal displacements $\mathbf{v}_0$ and the corresponding reaction-forces λ_0, i.e.:

$$L(\mathbf{v}_0, \lambda) \leq L(\mathbf{v}_0, \lambda_0) \leq L(\mathbf{v}, \lambda_0) \qquad \text{for all } \mathbf{v} \in \mathcal{R}^n, \lambda \in \mathcal{R}_+^{n_C}, \tag{6.34}$$

where n_C is the number of the constraints.

The dual function which has to be maximized then will be given by

$$\Phi(\lambda) := \min_{\mathbf{v} \in \mathcal{R}^n} L(\mathbf{v}, \lambda). \tag{6.35}$$

Possible choices for L are the Lagrangian

$$L(\mathbf{v}, \lambda) := \Pi(\mathbf{v}) + \sum_{i=1}^{n_C} \lambda_i g_i(\mathbf{v}) \tag{6.36}$$

or an augmented Lagrangian:

$$L(\mathbf{v}, \lambda) := \Pi(\mathbf{v}) + \sum_{i=1}^{n_C} \lambda_i \max\left\{g_i(\mathbf{v}), -\frac{\lambda_i}{\epsilon}\right\} + \frac{1}{2}\epsilon \sum_{i=1}^{n_C} \left(\max\left\{g_i(\mathbf{v}), -\frac{\lambda_i}{\epsilon}\right\}\right)^2. \tag{6.37}$$

In these cases the gradient of Φ is given by $\mathbf{g}(\mathbf{v}_0(\lambda))$, where $\mathbf{v}_0(\lambda)$ solves the minimization problem

$$L(\mathbf{v}_0(\lambda),\lambda) = \min_{\mathbf{v}\in\mathcal{R}^n} L(\mathbf{v},\lambda). \tag{6.38}$$

A widely used ascent method to solve the dual problem

$$\Phi(\lambda_0) = \max_{\lambda\in\mathcal{R}_+^{nC}} \Phi(\lambda) \tag{6.39}$$

is the UZAWA - Algorithm (Arrow, Hurwitz, Uzawa 1958), which in a first step of the i-th iteration cycle solves the minimization problem

$$L(\mathbf{v}^k,\lambda^{k-1}) = \min_{\mathbf{v}\in\mathcal{R}^n} L(\mathbf{v},\lambda^{k-1}), \tag{6.40}$$

then determines a preliminary new value for the Lagrangian multipliers:

$$\tilde{\lambda}^k = \lambda^{k-1} + \alpha_k g(\mathbf{v}^k) \tag{6.41}$$

with an appropriate chosen α_k and finally projects $\tilde{\lambda}^k$ on the feasible set:

$$\lambda^k = \max\{\tilde{\lambda}^k, 0\}. \tag{6.42}$$

Usually it is a problem to choose α_k in a right way. In the case, that the dual function is constructed by means of the augmented Lagrangian, one can show that an optimal choice for α_k will be ϵ (Großmann, Kaplan, 1979).

6.2.3.3 Penalty Methods

Penalty methods are procedures for approximating constrained optimization problems by unconstrained ones. The approximation is accomplished by adding to the energy a term that penalizes the violation of the obstacle; e.g. by

$$\beta : \begin{cases} \mathcal{R}^n \to \mathcal{R}_+ \\ \mathbf{v} \mapsto \frac{1}{2}\epsilon \sum_{i=1}^{nC} \left(\max\{g_i(\mathbf{v}),0\}\right)^2 . \end{cases} \tag{6.43}$$

This term can be interpreted as a spring acting only in the case of penetration. The spring coefficient is given by ϵ and by letting ϵ tend to infinity one can simulate a spring with arbitrary stiffness.

The approximate problem is then given by

$$\Pi_\epsilon(\mathbf{v}_\epsilon) = \min_{\mathbf{v}\in\mathcal{R}^n} \Pi_\epsilon(\mathbf{v}) \tag{6.44}$$

with

$$\Pi_\epsilon(\mathbf{v}) = \Pi(\mathbf{v}) + \frac{1}{2}\epsilon\beta(\mathbf{v}), \tag{6.45}$$

and one gets for the solution v_ϵ:

$$\lim_{\epsilon\to\infty} \mathbf{v}_\epsilon = \mathbf{v}_0. \tag{6.46}$$

But numerically stability problems arise for $\epsilon \to \infty$.

6.2.3.4 Newton methods (SQP)

In analogy to Newton's method for problems without constraints a correction $(\Delta \mathbf{v}^k, \Delta\lambda^k)$ of a current iteration point $(\mathbf{v}^k, \lambda^k)$ is calculated by solving the linearized system of Kuhn-Tucker-inequalities in the point $(\mathbf{v}^k, \lambda^k)$:

$$\begin{aligned} \nabla L(\mathbf{v}^k,\lambda^k)^T + \nabla^2 L(\mathbf{v}^k,\lambda^k)\Delta\mathbf{v}^k + \nabla \mathbf{g}(\mathbf{v}^k)^T\Delta\lambda^k &= 0 \\ \nabla\mathbf{g}(\mathbf{v}^k)\Delta\mathbf{v}^k + \mathbf{g}(\mathbf{v}^k) &\leq 0 \\ \lambda^k + \Delta\lambda^k &\geq 0 \\ \left(\nabla\mathbf{g}(\mathbf{v}^k)\Delta\mathbf{v}^k + \mathbf{g}(\mathbf{v}^k)\right)^T(\lambda^k + \Delta\lambda^k) &= 0. \end{aligned} \tag{6.47}$$

This can be done by solving the quadratic optimization problem

$$Q(\Delta\mathbf{v}^k) = \min\left\{Q(\Delta\mathbf{v}) \mid \nabla\mathbf{g}(\mathbf{v}^k)\Delta\mathbf{v} + \mathbf{g}(\mathbf{v}^k) \leq 0, \Delta\mathbf{v} \in \mathcal{R}^n\right\} \tag{6.48}$$

with

$$Q(\Delta\mathbf{v}) := \frac{1}{2}\Delta\mathbf{v}^T\nabla^2 L(\mathbf{v}^k,\lambda^k)\Delta\mathbf{v} + \nabla L(\mathbf{v}^k,\lambda^k)\Delta\mathbf{v}. \tag{6.49}$$

The new iteration point $(\mathbf{v}^{k+1}, \lambda^{k+1})$ is given then by

$$\begin{aligned} \mathbf{v}^{k+1} &= \mathbf{v}^k + \alpha_k\Delta\mathbf{v}^k \\ \lambda^{k+1} &= \lambda^k + \alpha_k\Delta\lambda^k. \end{aligned} \tag{6.50}$$

To determine α_k a line search with a merit function must be used, since no objective function is available. There are several proposals for a merit function (Han, 1977), (Luenburger, 1984); the one which seem to perform best, both in theoretical and numerical point of view is the augmented Lagrangian given by (Schittkowski, 1981).

The efficiency of the algorithm highly depends on the choice of the quadratic subproblem solver. The use of standard QP-solvers is restricted again by the large dimension. We have tried SOR, Penalty, Lagrange and a special kind of active set strategy basing on ideas of Bertsekas (Bertsekas, 1982). In combination with the last, SQP-methods showed an excellent convergence behaviour which differed only with an factor of about 1.5 from problems without constraints (Bischoff, 1986).

6.2.4 Example: Contact of a plastifying bar

For numerical tests we considered a parabolically symmetric bar with width 1 cm (see Fig. 6.4), whose material properties are given by the Young modulus E = 21000 kN/cm^2, the Poisson ratio $\nu = 0.3$ and the yield limit $k_0 = 40$ kN/cm^2 The bar was discretized by 20 4-node, 20 8-node and 80 4-node plane elements, respectively, while the load of 50 kN was applied in 1, 2 and 4 increments

The results of the calculation with the SQP-method (for the discretization with 20 8-node elements and 4 load increments) are shown in Fig. 6.5 and a comparison

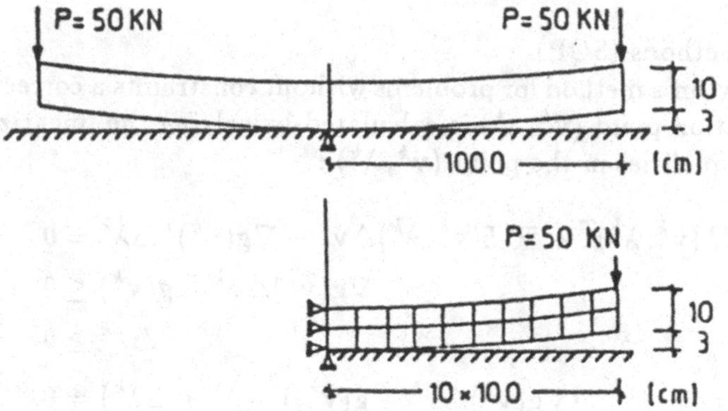

Fig. 6.4 Parabolically curved bar on a rigid plane with two point loads at the end

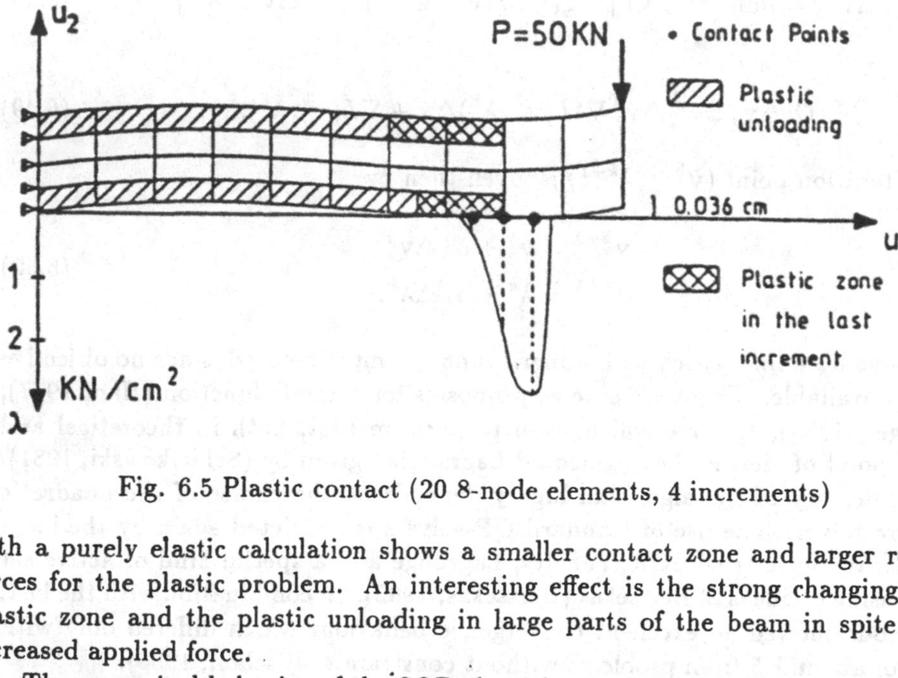

Fig. 6.5 Plastic contact (20 8-node elements, 4 increments)

with a purely elastic calculation shows a smaller contact zone and larger reaction forces for the plastic problem. An interesting effect is the strong changing of the plastic zone and the plastic unloading in large parts of the beam in spite of the increased applied force.

The numerical behavior of the SQP algorithm is given in table 6.1 which shows the number of Newton steps and in every Newton-step the number of active set steps times the number of Armijo-line search steps. As seen in the table, this line search is called only very rarely and thus doesn't consume almost any time - although it is essential for the convergence. The determination of the exact contact zone is indicated by a dashed line. One can see that this zone is always reached within the first three iteration steps, i.e. before the plastic iteration is converged, which shows the excellent behavior of this algorithm. We have compared the 4-node and the 8-node element for the bar discretized with 20 elements and the load applied within 1 and 4 increments. (The results for other incrementations or discretizations show no

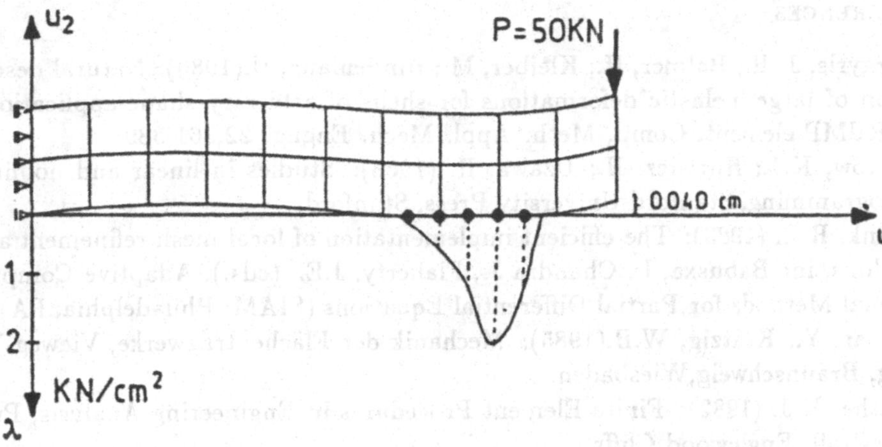

Fig. 6.6 Elastic contact (20 8-node elements)

significant differences.)

Table 6.2: Numerical behavior of the algorithm

Newton steps	1 increment	4 increments				1 incr.	4 increments			
		I	*II*	*III*	*IV*		*I*	*II*	*III*	*IV*
1	7x1	1x1	2x1	2x1	2x1	8x1	7x1	7x1	4x1	4x1
2	1x1	2x1	3x1	1x1	1x1	4x1	1x5+ 1x2	6x1	2x1	3x1
3	1x1	1x1	1x1	1x1	1x1	2x1	1x1	1x1	1x1	1x1
4	1x1	1x1	1x1	1x1	1x1	1x1	1x1	1x1	1x1	1x1
5	1x1	1x1	1x1	1x1	1x1	1x1	1x1	1x1	1x1	1x1
6			1x1	1x1		1x1	1x1	1x1		
7							1x1			

A comparison of the calculation with a prescribed contact zone shows that the algorithm needs exactly the same number of Newton-steps as he does with a variable zone. The CPU time for the contact problem increases about the factor 1.5 due to the condensation and the optimization algorithm. This is just the amount which was recognized too for the purely elastic problem.

7 REFERENCES

1. Argyris, J. H.; Balmer, H.; Kleiber, M.; Hindenlang, U.(1980): Natural description of large inelastic deformations for shells of arbitrary shape-application of TRUMP element, Comp. Meth. Appl. Mech. Engng. 22,361-389.
2. Arrow, K.J.; Hurwicz, H.; Uzawa, H. (1958): Studies in linear and nonlinear programming, Stanford University Press, Stanford.
3. Bank, R.E. (1983): The efficient implementation of local mesh refinement algorithms, in: Babuska, I., Chandra J., Flaherty, J.E., (eds.), Adaptive Computational Methods for Partial Differential Equations (SIAM, Philadelphia, PA).
4. Basar, Y.; Krätzig, W.B.(1985): Mechanik der Flächentragwerke, Vieweg-Verlag, Braunschweig,Wiesbaden.
5. Bathe, K.J. (1982): Finite Element Procedures in Engineering Analysis, Prentice-Hall, Englewood Cliffs.
6. Bathe, K.J.; Ramm, E.; Wilson, E.L. (1975): Finite element formulation for large deformation analysis, Int. J. Num. Meth. Eng. 9, 353-386.
7. Bathe, K.J.; Ho, L.W. (1981): A simple and effective element for analysis of general shell structures, Comp. & Struct. 13, 673-681.
8. Batoz, J.L.; Bathe, K.J.; Ho, L.W. (1980): A study of three-node triangular plate bending elements, Int. J. Num. Meth. Eng. 15, 1771-1812.
9. Bertsekas, D.P. (1982): Constrained optimization and Lagrange multiplier methods, Academic Press, New York.
10. Bischoff, D. (1985): Indirect optimization algorithms for nonlinear contact problems, in: Whiteman, J.R. (ed.): The mathematics of finite elements and applications V (MAFELAP 1984), Acad. Pr., London, 533-545.
11. Bischoff, D. (1986): Generalization of Newton-type methods to inelastic contact problems, in: el Piero, G.: Unilateral problems in structural analysis, to appear.
12. Bischoff, D.; Mahnken, R. (1984): Zur Konvergenz von Kontaktalgorithmen die auf Active Set Strategien beruhen, in: Bufler, H., Gwinner, J. (Eds.): Unilaterale Probleme (GAMM-Seminar, Stuttgart), Bi 1-16.
13. Brandt, A. (1982): Guide to multi-grid development, in: Hackbusch, W., Trottenberg, U., eds., Multigrid Methods, Springer, Berlin, 220-312.
14. Ciarlet, P.G. (1983): Lectures on three-dimensional elasticity, Springer, Berlin.
15. Dennis, J.E.; Schnabel, R. (1982): Numerical methods for unconstrained optimization and nonlinear equations, Pr. Hall, Englewood Cliffs, N.J.
16. Ekeland, I.; Temam, R. (1976): Convex analysis and variational problems, North Holland, Amsterdam (1976).
17. Fried, I.(1985): Nonlinear Finite Element Analysis of the Thin Elastic Shell of Revolution. Comp. Meth. Appl. Mech. 48, 283-299.
18. Grossmann, C.; Kaplan, A.A. (1979): Strafmethoden und modifizierte Lagrangefunktionen in der nichtlinearen Optimierung, B.G. Teubner, Leipzig.
19. Gruttmann, F. (1985): Konsistente Steifigkeitsmatrizen in der Elasto- Plastizitätstheorie, Workshop: Diskretisierungen in der Kontinuumsmechanik - Finite

Elemente und Randelemente, Bad Honnef, Gr1-Gr14.
20. Hallquist, J.Q. (1979): NIKE 2D: an implicit finite-element code for analysing the static and dynamic response of two-dimensional solids, Rept. UCRL-52678, Univ. of Calif.
21. Han, S.-P. (1977): A globally convergent method for nonlinear programming, J. Optim. Theory Appl., 22, 297-309.
22. Harbord, R.(1972): Berechnungen von Schalen mit endlichen Verschiebungen-gemischte finite Elemente- Bericht Nr. 72-7 Inst. f. Statik, TU Braunschweig.
23. Harte, R.(1982): Doppeltgekrümmte finite Dreieckelemente für die lineare und geometrisch nichtlineare Berechnung allgemeiner Flächentragwerke, Inst.KIB, Techn. Bericht Nr. 82-10, Ruhr-Universität Bochum.
24. Harte, R.; Eckstein, U.(1986): Derivation of geometrically nonlinear finite shell elements via tensor notation, Int. J. Num. Meth. Engng. 23,p.367-384.
25. Huang, H.C.; Hinton E.(1984): A nine node degenerated shell element with enhanced shear interpolation, Research Report, Dept. of Civil Engng.,Swansea, U.K.
26. Hughes, T.J.R.; Taylor, R.L.; Kanoknukulchai, W.(1977): A simple and efficient finite element for plate bending. Int. J. Num. Meth. Engng. 11, 1529- 1543.
27. Hughes, T.J.R.; Cohen, M.; Haroun, M.(1978): Reduced and selective integration techniques in the finite element analysis of plates, Nucl. Engng. Des. 46, 203-222.
28. Kahn, R.; Wagner, W. (1986): Überkritische Berechnung ebener Stabtragwerke unter Berücksichtigung einer vollständig geometrisch nichtlinearen Theorie, ZAMM, 67, T197-199.
29. Koiter, W.T.(1966): On the Nonlinear Theory of thin elastic Shells. Proc. Koninl. Ned. Akad. Wetenshap Serie B69 1-54.
30. Luenburger, D.G. (1984): Introduction to linear and nonlinear programming, Addison - Wesley, Reading.
31. Malkus, D.S.; Hughes, T.R.J.(1978): Mixed finite element methods - reduced and selective integration techniques: a unification of concepts. Comp. Meth. Appl. Mech. Engng. 15, 63-81.
32. Malvern, L.E. (1969): Introduction to the Mechanics of a Continuous Medium, Prentice-Hall, Englewood Cliffs.
33. Marguerre, K. (1939): Jahrbuch der deutschen Luftfahrtforschung 1939,413.
34. Marsden, J.E.; Hughes, T.R.J. (1983): Mathematical Foundations of Elasticity, Prentice Hall, Englewood Cliffs.
35. Matthies, H.; Strang, G. (1979): The solution of nonlinear finite element equations, Int. J. for Num. Meths. in Eng. 14, 1613-1626.
36. Noor, A. K.; Peters, J. M.; Andersen, C., M.(1984): Mixed Models and Reduction Techniques for Large-Rotation Nonlinear Problems. Comp. Meth. Appl. Mech. Eng., 44, 67-89.
37. Pica, A.; Wood, R. D.(1980): Postbuckling behaviour of Plates and Shells using a Mindlin Shallow Shell Formulation, Comp. & Struct. 12, 759-768.

38. Pietraszkiewicz, W.(1977): Introduction to the Nonlinear Theory of Shells. Mitt. aus dem Inst. f. Mechanik, Nr. 10, Ruhr-Universität Bochum.
39. Pietraszkiewicz, W.(1979): Finite Rotations and Lagrangean Description in the Non-linear Theory of Shells. Polish Academy of Sciences, Institute of fluid-flow machinery, Polish Scientific Publishers, Warszawa-Poznan.
40. Ramm, E.(1976): A plate/shell element for large deflections and rotations, in: US-Germany Symp. on " Formulations and Computational Algorithms in Finite Element Analysis", MIT, MIT-Press.
41. Ramm, E.(1981): Strategies for Tracing the Nonlinear Response Near Limit Points. in: Nonlinear Finite Element Analysis in Structural Mechanics, ed. Wunderlich, Stein, Bathe, Springer Verlag Berlin-Heidelberg-New-York.
42. Riks, E. (1972): The application of Newtons Method to the problem of Elastic stability, J. Appl. Mech., 39, p. 1060-1066.
43. Riks, E., (1984): Some computational aspects of the stability analysis of nonlinear structures, Comp. Meth. Appl. Mech. Engng. 47, p. 219-259.
44. Schittkowski, K. (1981): The nonlinear programming method of Wilson, Han and Powell with an augmented Lagrangian type line search function, part I, II, Numerische Mathematik, 39, 83-127.
45. Schwetlick, H. (1979): Numerische Lösung nichtlinearer Gleichungen, R. Oldenbourg, München.
46. Simmonds, J.G.; Danielson, D.A.(1972): Nonlinear Shell Theory with Finite Rotations and Stress-Function Vectors. J. Appl. Mech. Trans. ASME Serie E 39 1085-1090.
47. Simo, J.C; Taylor, R.L. (1985): Consistent tangent operators for rate- independent elasto-plasticity, Comp. Meth. Appl. Mech. Eng. 48, 101-118.
48. Stein, E.; Bischoff, D.; Brand, G.; Plank, L. (1985): Adaptive multi-grid methods for finite element systems with bi- and unilateral constraints, Comp. Meths. in Appl. Mech. Eng. 52, 873-884.
49. Stein, E.; Lambertz, K.H.; Plank, L. (1985): Ultimate load analysis of folded plate structures with large elastoplastic deformations - theoretical and practical comparisons of different FE-algorithms, Numeta 85, Swansea, U.K.
50. Stein, E.; Paulun, J. (1981): Incremental elastic-plastic deformations of stiffened plates in compression and bending, in: Nonlinear finite element analysis in Structural mechanics, Springer.
51. Stüben, K.; Trottenberg, U. (1982): Multigrid methods: Fundamental algorithms, model problem analysis and applications, in: Hackbusch, W., Trottenberg, U., eds., Multigrid Methods, Springer, Berlin, 1-176.
52. Truesdell, C.; Noll, W. (1965): The Non-Linear Field Theories of Mechanics, in Handbuch der Physik, Band III/3, ed. S. Flügge, Springer-Verlag, Berlin.
53. Wagner, W.; Wriggers, P. (1985): Kurvenverfolgungsalgorithmen in der Strukturmechanik, in: GAMM-Seminar: Geometrisch/physikalisch nichtlineare Probleme, Forschungs- und Seminarberichte aus dem Bereich der Mechanik der Universität Hannover, S 85/3.

54. Wagner, W.; Wriggers, P. : A Simple Method for the Calculation of Postcritical Branches, to appear in Engineering Computations.
55. Wagner, W. (1985): Eine geometrisch nichtlineare Theorie schubelastischer Schalen mit Anwendung auf Finite-Element- Berechnungen von Durchschlag- und Kontaktproblemen. Forschungs-und Seminarberichte aus dem Bereich der Mechanik der Universität Hannover, Nr. F 85/2.
56. Wempner, G.A.(1971): Discrete approximations related to nonlinear theories of solids, Int. J. Solids Structures 7, 1581-1599.
57. Wriggers, P.; Nour-Omid, B.(1984): Solution Methods for Contact Pronlems, Rept. UCB/SESM 84/09, Dept. of Civil Engng. UC Berkeley.
58. Wriggers, P.; Simo, J.C. (1985): A note on tangent stiffnesses for fully nonlinear contact problems, Comm. Appl. Num. Meths., 1, 199-203.
59. Wriggers, P. (1985): Die Behandlung der Inkompressibilität mit gemischten finiten Elementen oder der selektiv-reduzierten Integration, in Seminarbericht: Gemischte finite Elemente, Herausgeber: W. Wunderlich, Bochum.
60. Wriggers, P.; Wagner, W.; Stein, E. (1987): Algorithms for nonlinear Contact Constraints with Application to Stability Problems of Rods and Shells, Computational Mechanics, **2** (1987),p. 215–230.
60. Yserentant, H. (1983): On the multi-level splitting of FE-spaces, Bericht Nr. 21, RWTH Aachen, Inst. f. Geometrie u. Prakt. Mathematik.
61. Zienkiewicz, O.C.; Bauer, J.; Morgan, K.(1977): A Simple and Efficient Element for Axisymmetric Shells. Int. J. Num. Meth. Engng.11, 1545-1558.

BOUNDARY ELEMENT TECHNIQUE IN ELASTOSTATICS AND LINEAR FRACTURE MECHANICS

Theory and Engineering Applications

G. Kuhn
Universität Erlangen, Nuremberg, F.R.G.

Abstract

The theoretical fondations and some engineering applications of the direct Boundary Element Method (BEM) are presented for the case of linear elastostatics including thermal loading and commonly encountered body forces like gravity and centrifugal forces. In plane case problems the classical formulation contains two eigenvalues, hence the solution is non-unique. Two different solution techniques are proposed to avoid this difficulty.
The computation of stress intensity factors (SIF) represents one of the most attractive application areas for BEM. A variety of interesting techniques have been developed in order to derive accurate results. After a brief introduction to the basic SIF-concept of linear Fracture Mechanics some of these techniques are discussed.

1. INTRODUCTION

Due to their complexity, practical engineering problems can be solved often only in an approximate manner. Among numerical procedures applied in continuum mechanics we can distinguish three basic methods. These are the well-known

i) **Finite Difference Method (FDM)**
ii) **Finite Element Method (FEM)**
iii) **Boundary Element Method (BEM)**

Historically, the first widely known procedure was the FDM. Within this method, the differential equation describing a physical problem is reduced to a discrete difference version, by defining a series of nodes and by using a local expansion for the variables. Thus, the infinite number of degrees of freedom of the continuous system are reduced to a finite set. By doing this the problem can be solved numerically. The finite difference solution satisfies the discrete version of the differential equation at nodal points.

The FEM is the best known and indeed most commonly used procedure. The method is based on variational principles or, sometimes more generally, on weighted residual expressions. The differential equation or its weak formulation is satisfied in an average sense over a region or element. The method is well established. A great number of FEM-program systems are available on the software market specialized in structural analysis, fluid flow problems and other problems from various fields of applied physics. In continuum mechanics the application ranges from the analysis of simple linear statics up to complex non-linear dynamic problems. With continuous progress in computer technique, more and more complex structures can be analysed to a sufficiently exact degree. However, for complex problems the necessary number of degrees of freedom may become immense. The immediate consequence of this would be increased expenditures and costs of data pre- and postprocessing.

In this field one finds the essential advantage of the BEM. The basic idea is to transform the differential description, usually given in terms of partial differential equations, into a corresponding integral description of the boundary effects, in the form of boundary integral equations. This integral description leads to a formulation of the problem on a lower dimensional level. Therefore only the boundary needs to be discretized. A direct consequence of this is a substantially reduced set of equations, containing fewer degrees of freedom. Immediately connected with this is the simplification of the whole data processing. However, a shortened CPU time cannot be automatically deduced from this. The reason for this lies in the final matrix of the set of algebraic equations, which is fully occupied, non-symmetrical and not positively definite.

Speaking of the BEM, we have to distinguish the so-called "direct" and "indirect" methods. In the case of the "direct" method the corresponding physical variables are directly coupled with each other. In the "indirect" method we have to calculate an auxillary singularity distribution along the boundary at first, to derive then, in a more indirect way, the physical unknowns sought for.

In this part of the course we will chiefly deal with the direct BEM. Within the past few years this formulation has clearly prevailed over the indirect formulation, especially in the field of elastostatics, where the direct collocation method is less sensitive to geometrical singularities, when complex boundaries with corners and edges are considered.

1.1 SOME HISTORICAL REMARKS

More than 150 years ago the English mathematician **G. GREEN** showed how to pass from a differential domain formulation to a boundary integral description. His idea was introduced into the mechanics of continuum for the first time by **FREDHOLM** [1] at the turn of the century.

It is due to the Soviet mathematicians led by **MUSKHELISHVILI** [2], **MIKHLIN** [3], **SMIRNOV** [4] and **KUPRADZE** [5], that the theory of singular integral equations and the development of integral equation methods made a decisive step forward. These methods, at first hardly recognized by engineers, were more and more introduced into applied physics. Under the names like "source method" or "indirect method" these techniques were applied in fluid mechanics and in potential theory. Very extensive investigations of the "indirect methods" were performed by **KELLOG** [6], **JASWON** [7], **SYMM** [8] and **MASSONET** [9], not to mention many others.

It is not easy to say, who first formulated the "direct" BEM, although in another representation one can find it already in **KUPRADZE**'s book [5]. But from the engineering point of view, the beginnings of the direct BEM applications in elastostatics have to be dated back to the papers of **RIZZO** [10] and **CRUSE/ RIZZO** [11]. Also, the further development of the method throughout the past two decades is closely associated with the names of **RIZZO** and **CRUSE** as well as **SHIPPY, LACHAT, BREBBIA, BANERJEE, MUKHERJEE, WROBEL, TELLES, TANAKA, WENDLAND** and others, to list only some of them.

In the seventies this method, which at first sight seemed to be promising only for linear problems, was also successfully applied for non-linear and time depending problems. Various, very interesting solution techniques have consequently developed. For some specific applications, like problems characterized by infinite elastic regions, the BEM is undoubtedly superior to other numerical techniques. In spite of the remarkable success of the BEM in nonlinear mechanics of solids, one should, however, not forget that the main advantages of the BEM lie in the linear field of application. The treatment of nonlinearities as well as that of inhomogeneties and anisotropy remains one of the domains of the FEM.

An interesting alternative, which attracts more and more the interest of scientists, is to combine the two methods, in order to exploit their advantages, while evading their disadvantages. Apart from obtaining the description of the problem by one lower dimension and the merits connected with this, the advantages of the BEM are certainly in the simple modelling of infinite elastic regions and in the excellent approximation of stress concentration problems, whereas the FEM promises advantages, when dealing with nonlinearities or else time depending problems.

1.2 A SHORT REMARK TO THE "INDIRECT" BEM

Without going into detail, the basic procedure of the "indirect" method shall be shown briefly for the linear plane problem.

We consider the linear-elastic domain Ω^+ with the boundary Γ^+ shown left on Fig. 1a). On the boundary Γ^+ a generally mixed boundary value problem is given and we are interested in the complete state of stresses and displacements in the domain Ω^+ and on Γ^+. Within the indirect method the actual interior domain Ω^+, which is under consideration, is completed by the exterior domain Ω^- to the infinite basic domain Ω^0. The boundary of the outer domain Ω^- is denoted Γ^-. The boundary coordinates s^+ or s^- as well as the according unit outward normal vectors n^+ or n^- are defined in Fig. 1. For the infinite plane (basic domain Ω^0) a number of fundamental solutions are known analytically. In the indirect method the

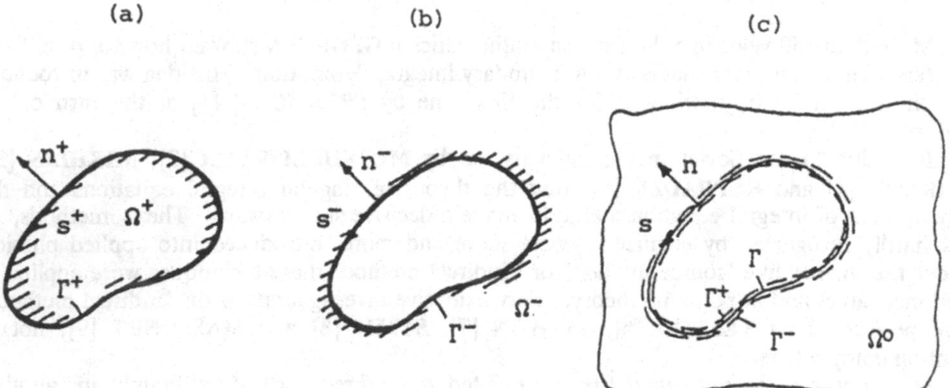

Fig. 1 (a) actual domain Ω^+, (b) complementary domain Ω^-, (c) infinite basic domain Ω^0

fundamental solution of a unit load F (or a unit displacement b) applied to an arbitrary point P in the unbounded plane is considered.

In order to find the solution of the boundary value problem one has to calculate first an auxiliary distribution of fictitious forces or displacements along the contour Γ lying between Γ^+ and Γ^-. This auxiliary singularity distribution has to be calculated in such a way, that in the domain Ω^+ and on Γ^+ the boundary value problem under consideration is satisfied. When the necessary singularity distribution is determined the stresses and displacements in the domain Ω^+ as well as on the boundary Γ^+ can be calculated in a second step. The separation of the boundary into Γ^+ and Γ^- is necessary, because stresses and displacements show a discontinuity as a result of the singularity distribution along Γ.

It is generally known that the indirect method provides bad results when boundaries with corners are considered. A simple physical consideration should illustrate the reason for this. It is well-known that in the simple rectangle plane under tension, shown in Fig. 2, there is a constant elongation in length and according to the POISSON ratio a constant contraction in width.

In the indirect method this very simple state of deformation must appear within the considerably stiffer infinite basic domain Ω^0 as a result of the singularity distribution along Γ. Considering the stiffness of the exterior domain Ω^- it is easy to understand that especially in regions with corners a very high intensity of the singularity distribution may be necessary which then leads to the well-known difficulties during the numerical procedure. Practically the indirect method leads to acceptably exact results only for regions with a smooth boundary.

2. ELASTOSTATICS WITH BODY FORCES AND THERMAL LOADING

In this chapter the theoretical foundations of the direct BEM are presented for the case of linear elastostatics, including thermal loading and commonly encountered body forces like gravity and centrifugal forces. Throughout this chapter the Cartesian tensor notation is used

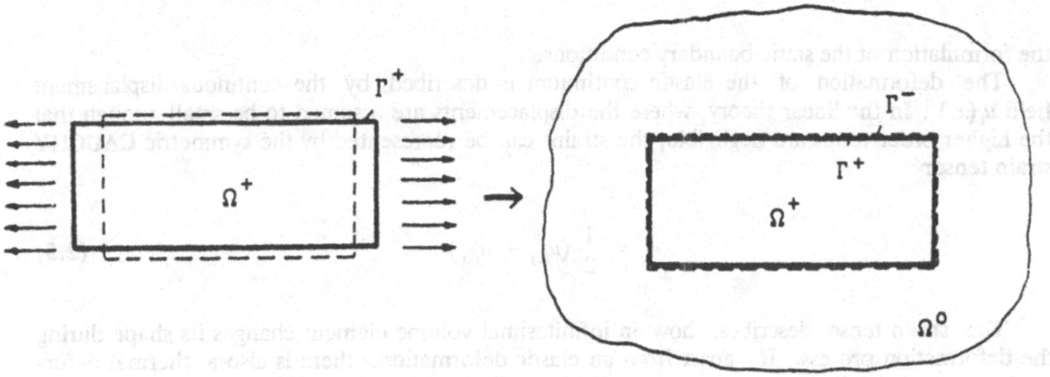

Fig. 2

and EINSTEIN's summation convention is valid. The indices i,j,k=1,2,3 represent the Cartesian coordinates x,y,z (or generally x_i).

2.1 BASIC DIFFERENTIAL EQUATIONS

For an infinitesimal volume element dV=dxdydz within an elastic continuum, the equilibrium conditions of forces yield

$$\sigma_{ij,i} + b_j = 0 , \tag{2.1}$$

where the components of the stress tensor are represented by σ_{ij} and the components of the body force vector by b_j . The partial derivative with regard to the Cartesian coordinate x_i is indicated by the abbreviation

$$\frac{\partial (...)}{\partial x_i} = (...),i , \tag{2.2}$$

which is usual in tensor notation. For the same infinitesimal volume element the equilibrium conditions of moments provide

$$\sigma_{ij} = \sigma_{ji} \tag{2.3}$$

and causes the components of the stress tensor to be symmetric. The components of the stress tensor σ_{ij} at a certain boundary point are related to the traction components t_i, applied to this point by the relation

$$t_i = \sigma_{ij} \, n_j , \tag{2.4}$$

where n_j represents the unit vector to the boundary in this point. Relation (2.4) is needed for the formulation of the static boundary conditions.

The deformation of the elastic continuum is described by the continous displacement field $u_i(x_i)$. In the linear theory, where the displacements are assumed to be small enough that the higher order terms are negligible, the strains can be represented by the symmetric **CAUCHY** strain tensor

$$\epsilon_{ij} = \frac{1}{2}(u_{i,j} + u_{j,i}) \,. \tag{2.5}$$

The strain tensor describes, how an infinitesimal volume element changes its shape during the deformation process. If, apart from an elastic deformation, there is also a thermal deformation, corresponding to a temperature field loading, the entire strain tensor consists of the two parts

$$\epsilon_{ij} = \epsilon_{ij}^{el} + \epsilon_{ij}^{th} \,. \tag{2.6}$$

For linear isotropic thermal expansion one obtains the thermal strains in the form

$$\epsilon_{ij}^{th} = \alpha\, \Theta\, \delta_{ij} \,, \tag{2.7}$$

with the linear thermal expansion coefficient α , the difference of temperature Θ related to a reference temperature and the **KRONECKER** symbol δ_{ij} , according to

$$\delta_{ij} = \begin{cases} 1 & \text{for } i=j \,, \\ 0 & \text{for } i \neq j \,. \end{cases} \tag{2.8}$$

If the displacement field is known, the strain tensor can be determined, according to Eq. (2.5). In the inverse problem, the strain field must satisfy the following compatibility conditions

$$\epsilon_{ij,kl} + \epsilon_{kl,ij} - \epsilon_{ik,jl} - \epsilon_{jl,ik} = 0 \,, \tag{2.9}$$

in order to derive a compatible displacement field, satisfying Eq. (2.5). For an isotropic, linear-elastic material the generalized **HOOKE**'s law

$$\epsilon_{ij}^{el} = \frac{1}{2G}\left[\sigma_{ij} - \frac{\nu}{1+\nu}\delta_{ij}\sigma_{kk}\right] , \tag{2.10}$$

with the shear modul G and the **POISSON**'s ratio ν provides a relation between stresses and strains. Substituting Eqs. (2.7) and (2.10) into (2.6) one obtains the generalized thermoelastic **HOOKE**'s law

$$\epsilon_{ij} = \frac{1}{2G}\left[\sigma_{ij} - \delta_{ij}\left(\frac{\nu}{1+\nu}\sigma_{kk} - 2G\,\alpha\,\Theta\right)\right] , \tag{2.11a}$$

which can also be expressed in the form

$$\sigma_{ij} = 2G \{\epsilon_{ij} + \frac{1}{1-2\nu} \delta_{ij} [\nu\epsilon_{kk} - (1+\nu)\, \alpha\, \Theta]\} . \tag{2.11b}$$

Equilibrium conditions (2.1), strain-displacement relations (2.5) and constitutive equations (2.11) present a set of 15 scalar partial differential equations for the determination of the 6 components of the symmetric stress tensor, the 6 components of the symmetric strain tensor and the three components of the displacement vector.

Eliminating the stress and strain components, one obtains the well-known **NAVIER's** equation in the form

$$u_{j,ii} + \frac{1}{1-2\nu} u_{k,kj} + \frac{1}{G} b_j^* = 0 , \tag{2.12}$$

with the generalized body force vector

$$b_j^* = b_j - 2G \frac{1+\nu}{1-2\nu} \alpha\, \Theta_{,j} = b_j - \gamma\, \Theta_{,j} . \tag{2.13}$$

It is interesting that for this process of elimination the compatibility conditions (2.9) are not needed. **NAVIER's** equation (2.12) presents the equilibrium conditions, expressed in the displacement components. This form of presentation has its advantages when the displacements are prescribed over the whole boundary (**DIRICHLET** boundary conditions). If the boundary tractions are prescribed (**NEUMANN** boundary conditions) or if a mixed boundary value problem is considered, the traction components can also be represented by the displacements. Substituting Eqs. (2.5) and (2.11b) into Eq. (2.4), one obtains

$$t_i = \{2G[\frac{1}{2}(u_{i,j} + u_{j,i}) + \frac{\nu}{1-2\nu} \delta_{ij}\, u_{k,k}] - \gamma\, \delta_{ij}\, \Theta\}\, n_j . \tag{2.14}$$

The displacement field $u_i(x_i)$ is solved from the **NAVIER** equation to satisfy the boundary conditions. After the displacements have been found, the strains can be obtained by Eq. (2.5) and the stresses can be determined by **HOOKE's** law (2.11b).

2.2 BOUNDARY INTEGRAL FORMULATION

As we have already mentioned, the BEM is not based on a differential problem description (eg. **NAVIER's** equation) but rather on an integral problem formulation transformed to the boundary. This boundary integral formulation can be deduced in different ways. Within linear elastistatics the derivation can be performed by ...

o ... **GREEN**'s third identity
o ... **BETTI**'s reciprocal work theorem
o ... **SOMIGLIANA**'s identity
o ... weighted residual method.

First, the "classical" way of derivation, based on the **BETTI**'s theorem shall be presented. Afterwards, following **BREBBIA** et al. [12] a more general alternative derivation by means of weighted residual statements shall be considered. This way of derivation gives a better insight into the approximative character of the BEM and permits a straightforward extension to more complex differential equations.

Although thermoelasticity has been considered so far, thermal loading will not be included in the following sections for reasons of simplicity. Thermoelastic problems and their detailed treatment will be presented in chapter 2.4.

2.2.1 BETTI'S THEOREM AND SOMIGLIANA'S IDENTITY

We consider the elastic domain Ω with the boundary Γ, shown on the left in Fig. 3. On the boundary Γ, a general mixed boundary value problem is given, with tractions t_j prescribed on the boundary part Γ_σ and with displacements u_j known on the part Γ_u.

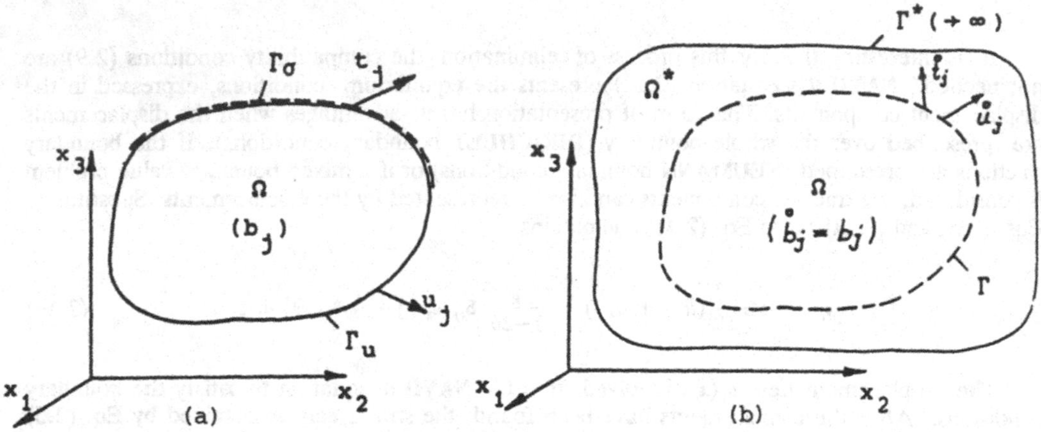

Fig. 3 (a) Elastic body with volume Ω and boundary Γ
(b) General region Ω^* with Γ^* containing the actual body

Besides the actual problem we consider a reference problem for the general region Ω^* and its boundary Γ^*, containing the body Ω and having the same elastic properties.

BETTI's theorem, which can be based on the principle of virtual work or which can be derived by applying **GREEN**'s third idendity to the **NAVIER**'s equation, couples the actual problem with the reference solution.

$$\int_\Omega B_j u_j d\Omega + \int_\Gamma T_j u_j d\Gamma = \int_\Omega b_j U_j d\Omega + \int_\Gamma t_j U_j d\Gamma \,. \tag{2.15}$$

Quantities indicated by capital letters in this formula and further on correspond to the reference problem. The displacement components along the boundary are denoted with u_j or U_j and the corresponding tractions with t_j or T_j. The body force components are termed with b_j or B_j. In the classical BEM formulation the fundamental **KELVIN** solution of a unit point load applied to an arbitrary source point P (ξ), with $\xi \in \Omega^*$, is considered in the unbounded space. The unit load acts in any of the three orthogonal directions, indicated by the unit vector e_i. This can be represented with a body force description

$$B_i = \delta(\xi,x)\, e_i \,, \tag{2.16}$$

where $\delta(\xi,x)$ represents the **DIRAC** delta function, ξ indicates the source point and $\mathbf{x} \in \Omega^*$ the current field point. Each component of the unit point load causes corresponding tractions and displacements. If each point load is taken independently, the tractions and displacements can be written in the form

$$\begin{aligned} T_j &= T_{ij}(\xi,x)\, e_i \,, \\ U_j &= U_{ij}(\xi,x)\, e_i \,, \end{aligned} \tag{2.17}$$

where $T_{ij}(\xi,x)$ and $U_{ij}(\xi,x)$ represent the tractions and displacements in the j direction at the point x corresponding to a unit force applied at the source point ξ and acting in the e_i direction.

The tensor expressions $U_{ij}(\xi,x)$ and $T_{ij}(\xi,x)$ defined in Eq. (2.17) are given by [12]

$$\begin{aligned} U_{ij}(\xi,x) &= \frac{1}{16\pi(1-\nu)\, G\, r}\{(3-4\nu)\delta_{ij} + r_{,i}\, r_{,j}\} \,, \\ T_{ij}(\xi,x) &= -\frac{1}{8\pi(1-\nu)r^2}\{[(1-2\nu)\delta_{ij} + 3\, r_{,i}\, r_{,j}]\frac{\partial r}{\partial n} - (1-2\nu)(r_{,i}\, n_j - r_{,j}\, n_i)\} \,, \end{aligned} \tag{2.18}$$

for the three-dimensional case and

$$\begin{aligned} U_{ij}(\xi,x) &= -\frac{1}{8\pi(1-\nu)G}\{(3-4\nu)\ln(r)\delta_{ij} - r_{,i}\, r_{,j}\} \,, \\ T_{ij}(\xi,x) &= -\frac{1}{4\pi(1-\nu)r}\{[(1-2\nu)\delta_{ij} + 2r_{,i}\, r_{,j}]\frac{\partial r}{\partial n} - (1-2\nu)(r_{,i}\, n_j - r_{,j}\, n_i)\} \,, \end{aligned} \tag{2.19}$$

for the two-dimensional plain strain problem. In these expressions

$$r(\xi,x) = \sqrt{r_i\, r_i} \qquad with\ r_i = x_i\,(x) - x_i\,(\xi) \tag{2.20}$$

represents the distance between the load point ξ and the field point x, and the derivatives of $r(\xi,x)$ with the respect to the coordinates of x are denoted with

$$r_{,i} = \frac{\partial r}{\partial x_i\,(x)} = \frac{r_i}{r} \,. \tag{2.21}$$

According to the well-known properties of the **DIRAC** delta function

$$\delta(\xi,x) = \begin{cases} 0 & \text{for } \xi \neq x\ , \\ \infty & \text{for } \xi = x\ , \end{cases} \tag{2.22}$$

and

$$\int_{\Omega} f(x)\,\delta(\xi,x)\,d\Omega(x) = f(\xi)\ , \tag{2.23}$$

for the case $\xi \in \Omega$, the first integral in Eq. (2.15) can be represented as

$$\int_{\Omega} B_i u_i\, d\Omega = u_i(\xi)\, e_i\ . \tag{2.24}$$

Together with the expression (2.17), Eq. (2.15) can be rewritten to represent the three separate components u_i of the displacement vector at ξ in the form

$$u_i(\xi) = \int_{\Gamma} U_{ij}(\xi,x)t_j(x)d\Gamma(x) - \int_{\Gamma} T_{ij}(\xi,x)u_j(x)d\Gamma(x) + \int_{\Omega} U_{ij}(\xi,x)b_j(x)d\Omega(x)\ . \tag{2.25}$$

Eq. (2.25) is known as **SOMIGLIANA**'s identity for displacements. If the body forces are prescribed and the boundary displacements and tractions are known throughout the boundary, **SOMIGLIANA**'s identity is a continuous representation of displacements at any point ξ .

2.2.2 STATE OF STRESS AT INTERNAL POINT

SOMIGLIANA's identity describes the components of the displacement vector at an internal point. The state of stress at this point can be obtained by combining the derivatives of Eq. (2.25) with respect to the coordinates of ξ to produce, corresponding to Eq. (2.5), the strain tensor ϵ_{ij} and then by substituting the result into **HOOKE**'s law. Following [12], the final formula is

$$\sigma_{ij}(\xi) = \int_{\Gamma} U_{ijk}(\xi,x)t_k(x)d\Gamma(x) - \int_{\Gamma} T_{ijk}(\xi,x)u_k(x)d\Gamma(x) + \int_{\Omega} U_{ijk}(\xi,x)b_k(x)d\Omega(x), \tag{2.26}$$

where the newly introduced tensors are written for the three- dimensional case as

$$U_{ijk}(\xi,x) = \frac{1}{8\pi(1-\nu)r^2}\{(1-2\nu)(r_{,k}\,\delta_{ij} + r_{,j}\,\delta_{ki} - r_{,i}\,\delta_{jk}) + 3\,r_{,i}\,r_{,j}\,r_{,k}\},$$

$$T_{ijk}(\xi,x) = \frac{G}{4\pi(1-\nu)r^3}\{3\frac{\partial r}{\partial n}[(1-2\nu)\delta_{ij}\,r_{,k} + \nu(\delta_{ik}\,r_{,j} + \delta_{jk}\,r_{,i}) - 5\,r_{,i}\,r_{,j}\,r_{,k}]$$

$$+ 3\nu(n_i\,r_{,j}\,r_{,k} + n_j\,r_{,i}\,r_{,k}) + (1-2\nu)(3n_k\,r_{,i}\,r_{,j}$$

$$+ n_j\,\delta_{ik} + n_i\,\delta_{jk}) - (1-4\nu)n_k\,\delta_{ij}\}, \tag{2.27}$$

and

$$U_{ijk}(\xi,x) = \frac{1}{4\pi(1-\nu)r}\{(1-2\nu)(r_{,k}\,\delta_{ij} + r_{,j}\,\delta_{ki} - r_{,i}\,\delta_{jk}) + 2\,r_{,i}\,r_{,j}\,r_{,k}\}.$$

$$T_{ijk}(\xi,x) = \frac{G}{2\pi(1-\nu)r^2}\{2\frac{\partial r}{\partial n}[(1-2\nu)\delta_{ij}\,r_{,k} + \nu(\delta_{ik}\,r_{,j} + \delta_{jk}\,r_{,i}) - 4r_{,i}\,r_{,j}\,r_{,k}]$$

$$+ 2\nu(n_i\,r_{,j}\,r_{,k} + n_j\,r_{,i}\,r_{,k}) + (1-2\nu)(2n_k\,r_{,i}\,r_{,j}$$

$$+ n_j\,\delta_{ik} + n_i\,\delta_{jk}) - (1-4\nu)n_k\,\delta_{ij}\}, \tag{2.28}$$

for the two-dimensional plain strain problem. Eqs. (2.28) are also valid for the state of plane stress, if the **POISSON** ratio ν has been formally replaced by $\nu^* = 1/(1+\nu)$. In order to derive expressions (2.27) and (2.28) the substitution

$$\frac{\partial r}{\partial x_i(\xi)} = -r_{,i} = -\frac{\partial r}{\partial x_i(x)} \tag{2.29}$$

was applied and the differentiation of Eq. (2.25) has to be carried out directly inside the integrals in an analytic manner.

2.2.3 BOUNDARY INTEGRAL EQUATION

SOMIGLIANA's identity, Eq. (2.25), describes the relations between the body forces and boundary tractions and displacements on one side and the displacement components of the internal source point ξ on the other side. In order to reduce it to a relation between the prescribed body forces and the boundary tractions and displacements only, we examine the limit transition of Eq. (2.25) for the case that the source point ξ tends to the boundary. Following [12], we assume, that the body can be represented as shown in Fig. 4. The source point ξ, now located on the boundary, is still considered as an internal point surrounded by a part of a spherical surface Γ_ϵ^* of radius ϵ.

The whole boundary can be divided into two parts

$$\Gamma = (\Gamma - \Gamma_\epsilon) + \Gamma_\epsilon^* \tag{2.30}$$

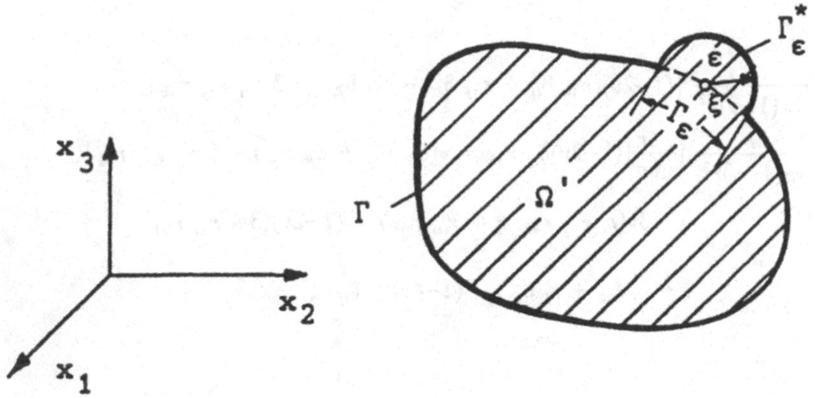

Fig. 4 Source point ξ located on the boundary surrounded by a spherical surface Γ_ε^*

and Eq. (2.25) can be written in the form

$$u_i(\xi) = \int_{\Gamma-\Gamma_\varepsilon+\Gamma_\varepsilon^*} U_{ij}(\xi,x)t_j(x)d\Gamma(x) - \int_{\Gamma-\Gamma_\varepsilon+\Gamma_\varepsilon^*} T_{ij}(\xi,x)u_j(x)d\Gamma(x) + \int_{\Omega'} U_{ij}(\xi,x)b_j(x)d\Omega(x) . \tag{2.31}$$

The first and the third integral on the right-hand side of Eq. (2.31) contain no singularities and can be interpreted in the normal sense of integration. The limit transition of the second integral can be carried out in the form [12].

$$\lim_{\varepsilon\to0} \int_{\Gamma-\Gamma_\varepsilon+\Gamma_\varepsilon^*} T_{ij}(\xi,x)u_j(x)d\Gamma(x) = \lim_{\varepsilon\to0} \int_{\Gamma_\varepsilon^*} T_{ij}(\xi,x)\,[u_j(x) - u_j(\xi)]d\Gamma(x) + \lim_{\varepsilon\to0} u_j(\xi) \int_{\Gamma_\varepsilon^*} T_{ij}(\xi,x)d\Gamma(x) + \lim_{\varepsilon\to0} \int_{\Gamma-\Gamma_\varepsilon} T_{ij}(\xi,x)u_j(x)d\Gamma(x) . \tag{2.32}$$

The first limit expression on the right-hand side of Eq. (2.32) vanishes from the condition of continuity of the displacements u_j and the third integral has to be taken in a CAUCHY principal value sense [13]. The second limit term can be put together with the left-hand side of Eq. (2.31) to form the expression

$$C_{ij}(\xi) = \delta_{ij} + \lim_{\varepsilon\to0} \int_{\Gamma_\varepsilon^*} T_{ij}(\xi,x)d\Gamma(x) . \tag{2.33}$$

Therefore, as $\epsilon \to 0$, Eq. (2.31) can be written in the form

$$C_{ij}(\xi)u_j(\xi) + \int_\Gamma T_{ij}(\xi,x)u_j(x)d\Gamma(x) = \int_\Gamma U_{ij}(\xi,x)t_j(x)d\Gamma(x) + \int_\Omega U_{ij}(\xi,x)b_j(x)d\Omega(x) , \tag{2.34}$$

where the singular integral on the left-hand side has to be interpreted in the **CAUCHY** principal value sense. Eq. (2.34) is valid for two and three dimensional problems. If the body forces are known, this equation represents a boundary integral equation for the unknown boundary data. This singular integral equation is the starting equation for the numerical boundary element solution technique.

The coefficient $C_{ij}(\xi)$, defined in Eq. (2.33), follows to $C_{ij}(\xi)=\delta_{ij}/2$, if the surface of the body is smooth and therefore the tangent plane at ξ is continuous. For more general boundaries with corners and edges closed form expressions of $C_{ij}(\xi)$ have been presented [14,15]. For practical applications, however, it is not necessary to know the explicit closed form expression of $C_{ij}(\xi)$. This coefficient, together with the corresponding principal value of the singular integral, can be computed indirectly by applying Eq. (2.34) to present a rigid body motion [16].

2.2.4 ALTERNATIVE WEIGHTED RESIDUAL FORMULATION

In this chapter, we shall come back for a moment to the beginning of the description of our problem in terms of partial differential equations. As an alternative to the way shown in the previous chapters, we follow now the theory described in [12] and transform the differential description into a corresponding integral formulation by means of weighted residual considerations.

The same problem as in chapter 2.2.1 is considered. We are seeking an approximate solution to the problem governed by **NAVIER**'s equation (2.12)

$$Gu_{j,ii} + \frac{G}{1-2\nu}u_{k,kj} + b_j = 0 \qquad in\ \Omega \tag{2.35}$$

with the boundary conditions

$$u_j(x) = \bar{u}_j(x) \qquad on\ \Gamma_u ,$$
$$t_j(x) = \bar{t}_j(x) \qquad on\ \Gamma_\sigma . \tag{2.36}$$

It is clear that an approximate solution gives rise to errors in each of these three expressions. These errors can be minimized by writing the following weighted residual statement

$$\int_\Omega [Gu_{j,ii} + \frac{G}{1-2\nu}u_{k,kj} + b_j]\, U_j d\Omega = \int_{\Gamma_u} (\bar{u}_j - u_j)T_j d\Gamma + \int_{\Gamma_\sigma} (t_j - \bar{t}_j)U_j d\Gamma , \tag{2.37}$$

where U_j and T_j are the displacements and tractions corresponding to a weighting field. Integrating the left-hand side integral of Eq. (2.37) twice by parts, we obtain

$$-\int_{\Omega} G[U_{j,ii} + \frac{1}{1-2\nu} U_{k,kj}]u_j d\Omega = \int_{\Omega} b_j U_j d\Omega$$
$$+ \int_{\Gamma} G[u_{j,i} U_j n_i + \frac{1}{1-2\nu} u_{k,k} U_j n_j] d\Gamma$$
$$- \int_{\Gamma} G[U_{j,i} u_j n_i + \frac{1}{1-2\nu} U_{k,k} u_j n_j] d\Gamma$$
$$+ \int_{\Gamma_u} (\bar{u}_j - u_j) T_j d\Gamma + \int_{\Gamma_\sigma} (t_j - \bar{t}_j) U_j d\Gamma . \tag{2.38}$$

Assuming that the strain-displacement relationship (2.5) and the constitutive equations (2.10) are valid for both the approximating and the weighting field, the second and the third integral on the right-hand side of Eq. (2.38) can be rewritten in the form

$$\int_{\Gamma} G[\cdots] d\Gamma = \int_{\Gamma} G[(u_{i,j} + u_{j,i}) + \frac{2\nu}{1-2\nu} \delta_{ij} u_{k,k}] n_i U_j d\Gamma$$
$$= \int_{\Gamma} t_j U_j d\Gamma ,$$
$$\int_{\Gamma} G[\cdots] d\Gamma = \int_{\Gamma} G[(U_{i,j} + U_{j,i}) + \frac{2\nu}{1-2\nu} \delta_{ij} U_{k,k}] n_i u_j d\Gamma$$
$$= \int_{\Gamma} T_j u_j d\Gamma , \tag{2.39}$$

and follows from Eq. (2.38)

$$-\int_{\Omega} G[U_{j,ii} + \frac{1}{1-2\nu} U_{k,kj}]u_j d\Omega = \int_{\Omega} b_j U_j d\Omega + \int_{\Gamma} t_j U_j d\Gamma$$
$$- \int_{\Gamma} T_j u_j d\Gamma . \tag{2.40}$$

For prescribed body forces b_j the domain integral on the right-hand side does not introduce any unknowns. In order to derive a boundary integral description for the unknown boundary data, we have to eliminate the domain integral on the left-hand side. This can be done by proposing either weighting field functions which satisfy the NAVIER's equation

$$U_{j,ii} + \frac{1}{1-2\nu} U_{k,kj} = 0 , \tag{2.41}$$

or by using fundamental solutions as weighted field functions. In the first case the domain integral vanishes, while in the latter the integrand contains the DIRAC delta function and allows, according to relation (2.23) an explicit evaluation of the domain integral. If we choose, for example, the fundamental KELVIN solution as weighted field function, Eq. (2.40) shall be transformed into SOMIGLIANA's identity. The further derivation of the basic boundary integral equation of the BEM can then be analogically completed as described in chapter 2.2.1. It is a decisive advantage of a weighted residual formulation that this technique permits a straightforward extension of the BEM to solve more complex partial differential equations, including e.g. nonlinearities.

2.2.5 EXTERNAL PROBLEMS WITH INFINITE REGIONS

Throughout the derivation presented so far, bounded bodies were considered. In this chapter we will extend the validity of the previous boundary integral description to infinite external regions. The possibility of this extension is an essential advantage of the BEM.

Following [12], we begin with a bounded external problem depicted in Fig. 5, where the inner boundary Γ precribes for example a cavity and the outer boundary Γ_R is assumed as the surface of a sphere with radius R and centered at point ξ .

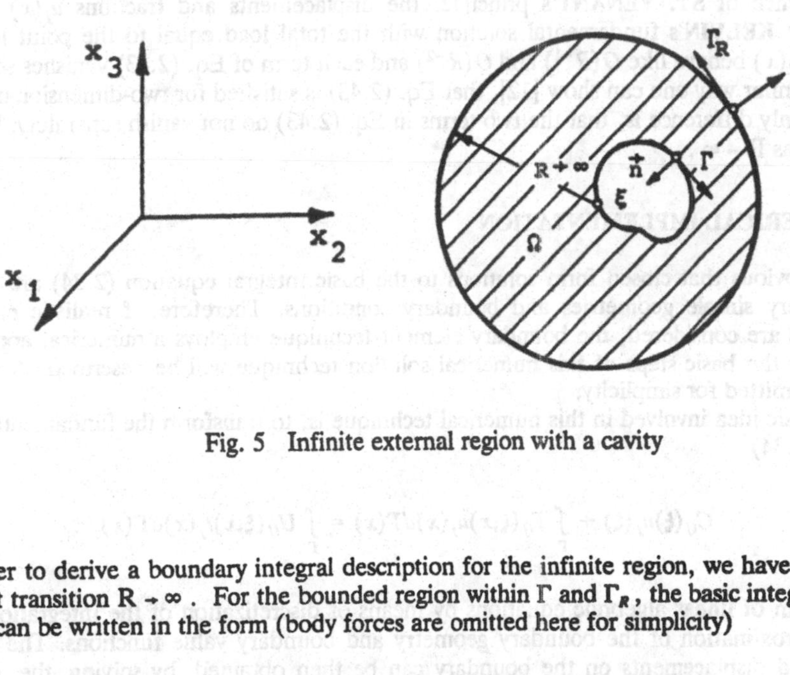

Fig. 5 Infinite external region with a cavity

In order to derive a boundary integral description for the infinite region, we have to examine the limit transition $R \to \infty$. For the bounded region within Γ and Γ_R, the basic integral equation (2.34) can be written in the form (body forces are omitted here for simplicity)

$$C_{ij}(\xi)u_j(\xi) + \int_{\Gamma+\Gamma_R} T_{ij}(\xi,x)u_j(x)d\Gamma(x) = \int_{\Gamma+\Gamma_R} U_{ij}(\xi,x)t_j(x)d\Gamma(x) . \tag{2.42}$$

For the limit transition $R \to \infty$, we have to examine the integrals over the outer boundary. If the limit condition

$$\lim_{R\to\infty} \int_{\Gamma_R} [T_{ij}(\xi,x)u_j(x) - U_{ij}(\xi,x)t_j(x)]d\Gamma(x) = 0 \tag{2.43}$$

is satisfied, Eq. (2.42) can be expressed in terms of boundary integrals over the inner boundary Γ

alone.

For three-dimensional problems, one finds the following estimates for the asymptotic behaviour at infinity

$$d\Gamma_R(x) = |f(R)|\, d\phi d\Theta\,, \qquad \textit{with } |f(R)| = O(R^2)\,,$$
$$U_{ij}(\xi,x) = O(R^{-1})\,,$$
$$T_{ij}(\xi,x) = O(R^{-2})\,, \tag{2.44}$$

where O(..) represents the asymptotic behaviour as $R \to \infty$. Therefore, condition (2.43) is satisfied , if the displacements and tractions $u_j(x)$ and $t_j(x)$ demonstrate the asymptotic behaviour $O(R^{-1})$ and $O(R^{-2})$ at infinity. In order to estimate the behaviour of $u_j(x)$ and $t_j(x)$ at infinity, we assume, that the total load applied over the inner boundary is not self-equilibrated. Then, as a consequence of **ST. VENANT**'s principle, the displacements and tractions $u_j(x)$ and $t_j(x)$ behave like **KELVIN**'s fundamental solution with the total load.equal to the point load.Thus, $u_j(x)$ and $t_j(x)$ behave like $O(R^{-1})$ and $O(R^{-2})$ and each term of Eq. (2.43) vanishes separately.

In a similar way one can show [12], that Eq. (2.43) is satisfied for two-dimensional problems also. The only difference is, that the two terms in Eq. (2.43) do not vanish separately, but cancel each other as $R \to \infty$.

2.3 NUMERICAL IMPLEMENTATION

It is obvious that closed form solutions to the basic integral equation (2.34) are attainable only for very simple geometries and boundary conditions. Therefore, if realistic engineering applications are considered, the boundary element technique employs a numerical approach. In this chapter the basic steps of this numerical solution technique will be described. Again, body forces are omitted for simplicity.

The basic idea involved in this numerical technique is, to transform the fundamental integral equation (2.34)

$$C_{ij}(\xi)u_j(\xi) + \int_\Gamma T_{ij}(\xi,x)u_j(x)d\Gamma(x) = \int_\Gamma U_{ij}(\xi,x)t_j(x)d\Gamma(x) \tag{2.45}$$

into a system of linear algebraic equations by means of discretization of the integration domain and by approximation of the boundary geometry and boundary value functions. The unknown tractions and displacements on the boundary can be then obtained by solving the system of equations including the boundary conditions. Then, in a second step, using **SOMIGLIANA**'s identity (2.25) for displacements or the single point evaluation (2.26) for stresses, one can determine the stresses and displacements within the elastic domain.

2.3.1 INTERPOLATION FUNCTIONS

In the plane case, the boundary geometry can be approximated by a combination of linear, circular and quadratic boundary elements. Although the special case of a circular element is very useful in practical applications, we refrain from an explicit description here. The linear element is defined by two and the quadratic element by three nodal points (Fig. 6). The Cartesian coordinates x_k, $k \in [1,2]$ of the current boundary point are expressed by the nodal point coordinates x_k^m, as well as by the shape functions $\Psi^m(\zeta)$. In the local coordinate ζ, $\zeta \in [-1,1]$ one obtains

$$x_k(\zeta) = \sum_{m=1}^{q} \Psi^m(\zeta) x_k^m, \tag{2.46}$$

where q corresponds to the grade of the approximation involved. The shape functions $\Psi^m(\zeta)$ are shown in Fig. 6.

In the spatial case, the surface geometry can be approximated by a combination of linear or curved triangular or quadrilateral elements, as shown in Fig. 7.

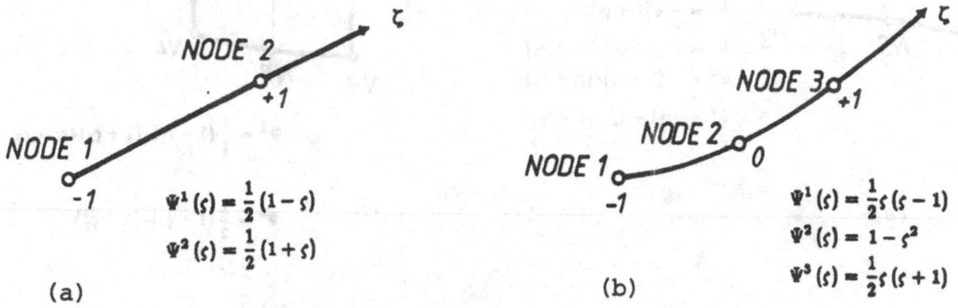

Fig. 6 Elements for geometic representation in plane problems.
(a) linear element; (b) quadratic element

The Cartesian coordinates x_k, k ∈ [1,2,3] of the current surface point are described analogous to Eq. (2.46)

$$x_k(\zeta_1,\zeta_2) = \sum_{m=1}^{q} \Psi^m(\zeta_1,\zeta_2) x_k^m, \tag{2.47}$$

with $\zeta_1,\zeta_2 \in [-1,1]$. The shape functions of these elements, which are similar to those used in the finite elements method, are shown in Fig. 7. Not all of them are specified in an explicit form. The remaining shape functions can be easily won out of them.

The boundary value functions u_j and t_j also have to be approximated in local coordinates by means of nodal point values u_j^m and t_j^m and the corresponding shape functions. For a current point on the element surface one has

$$u_j(\zeta) = \sum_{m=1}^{q} \Phi^m(\zeta) u_j^m,$$

$$t_j(\zeta) = \sum_{m=1}^{q} \Phi^m(\zeta) t_j^m, \tag{2.48}$$

$$\Psi^1(\zeta_1,\zeta_2) = -\frac{1}{2}\cdot(\zeta_1+\zeta_2)$$

$$\Psi^2(\zeta_1,\zeta_2) = \frac{1}{2}\cdot(1+\zeta_1)$$

$$\Psi^3(\zeta_1,\zeta_2) = \frac{1}{2}\cdot(1+\zeta_2)$$

$$\Psi^1 = \frac{1}{4}\cdot(1+\zeta_1)\cdot(1+\zeta_2)$$

$$\Psi^1 = \frac{1}{2}(\zeta_1+\zeta_2)\cdot(1+\zeta_1+\zeta_2)$$

$$\Psi^2 = \frac{1}{2}\zeta_1\cdot(1+\zeta_1)$$

$$\Psi^3 = \frac{1}{2}\zeta_2\cdot(1+\zeta_2)$$

$$\Psi^4 = -(1+\zeta_2)\cdot(\zeta_1+\zeta_2)$$

$$\Psi^5 = -(1+\zeta_1)\cdot(\zeta_1+\zeta_2)$$

$$\Psi^6 = (1+\zeta_1)\cdot(1+\zeta_2)$$

$$\Psi^1 = \frac{1}{4}\cdot(1+\zeta_1)\cdot(1+\zeta_2)(\zeta_1+\zeta_2-1)$$

$$\Psi^4 = \frac{1}{2}(1+\zeta_1)\cdot(1-\zeta_2^2)$$

(a)   (b)

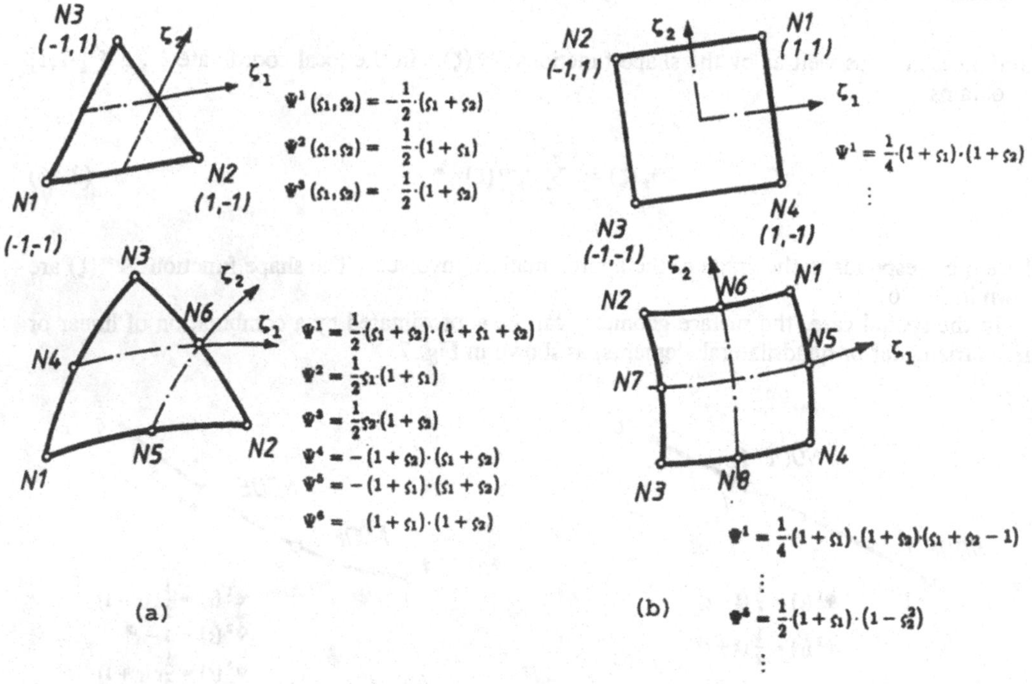

Fig. 7 Surface elements for geometry representation in spatial problems.
(a) linear and quadratic triangular surface element;
(b) linear and quadratic quadrilateral surface element

with $\zeta \in [-1,1]$ in the plane case and

$$u_j(\zeta_1,\zeta_2) = \sum_{m=1}^{q} \Phi^m(\zeta_1,\zeta_2)u_j^{\,m} \ ,$$

$$t_j(\zeta_1,\zeta_2) = \sum_{m=1}^{q} \Phi^m(\zeta_1,\zeta_2)t_j^{\,m} \ , \tag{2.49}$$

with $\zeta_1,\zeta_2 \in [-1,1]$ in the spatial case. The shape functions $\Phi^m(\zeta)$ for linear and quadratic approximations in the plane case are defined in Fig. 8. The corresponding shape functions $\Phi^m(\zeta_1,\zeta_2)$ for the spatial case are shown in Fig. 9.

2.3.2 TRANSFORMATION TO A SYSTEM OF LINEAR ALGEBRAIC EQUATIONS

In order to transform Eq. (2.45) into a set of linear algebraic equations, one subdivides the whole surface Γ into n surface elements Γ_l. Thus, the integrals over the whole boundary in Eq. (2.45) decompose into a sum of l, $l \in [1,n]$ integrals over single elements Γ_l.

$$\int_{\Gamma}[\ \cdots\ ]d\Gamma(x) = \sum_{l=1}^{n} \int_{\Gamma_l}[\ \cdots\ ]d\Gamma_l \ . \tag{2.50}$$

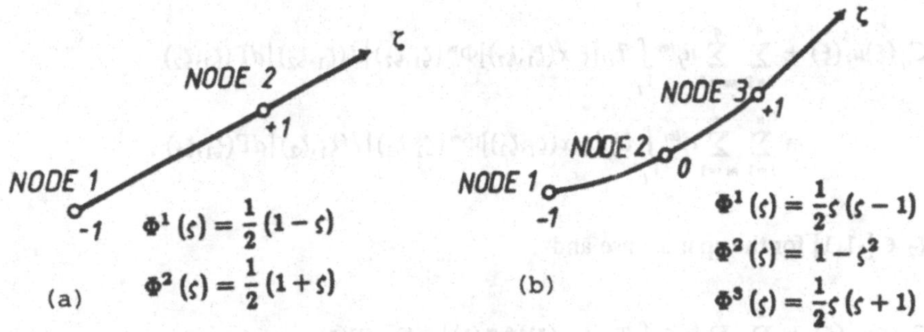

Fig. 8 Shape functions for boundary value representation in plane problems
(a) linear approximation; (b) quadratic approximation

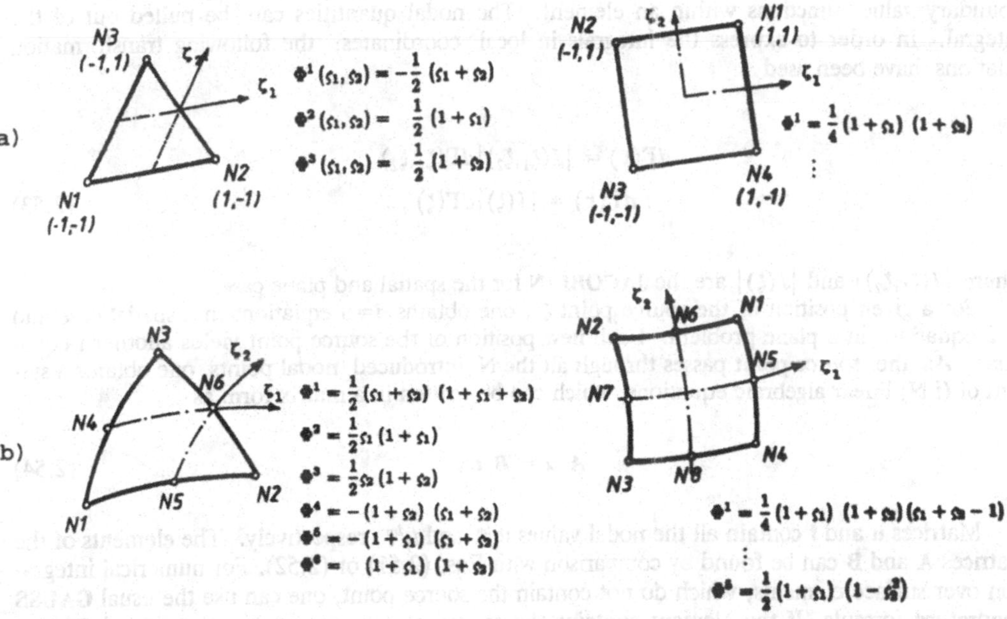

Fig. 9 Shape functions for boundary value representation
within triangular and quadrilateral surface elements
(a) linear approximation; (b) quadratic approximation

Then, according to Eqs. (2.48) and (2.49), the exact boundary value functions u_j and t_j within the l-th element Γ_l, $l \in [1,n]$, are represented approximately through q nodal values. Substituting the discretization and approximation of the boundary and approximation of the boundary value functions into the basic equation (2.45) one obtains the following discretized form

$$C_{ij}(\xi)u_j(\xi) + \sum_{l=1}^{n}\sum_{m=1}^{q} u_j^{lm} \int_{\Gamma_l} T_{ij}[\xi,x(\zeta_1,\zeta_2)]\Phi^m(\zeta_1,\zeta_2)\,|J(\zeta_1,\zeta_2)|\,d\Gamma(\zeta_1,\zeta_2)$$
$$= \sum_{l=1}^{n}\sum_{m=1}^{q} t_j^{lm} \int_{\Gamma_l} U_{ij}[\xi,x(\zeta_1,\zeta_2)]\Phi^m(\zeta_1,\zeta_2)\,|J(\zeta_1,\zeta_2)|\,d\Gamma(\zeta_1,\zeta_2)\,, \tag{2.51}$$

with $\zeta_1,\zeta_2 \in [-1,1]$ for the spatial case and

$$C_{ij}(\xi)u_j(\xi) + \sum_{l=1}^{n}\sum_{m=1}^{q} u_j^{lm} \int_{\Gamma_l} T_{ij}[\xi,x(\zeta)]\Phi^m(\zeta)\,|J(\zeta)|\,d\Gamma(\zeta)$$
$$= \sum_{l=1}^{n}\sum_{m=1}^{q} t_j^{lm} \int_{\Gamma_l} U_{ij}[\xi,x(\zeta)]\Phi^m(\zeta)\,|J(\zeta)|\,d\Gamma(\zeta)\,, \tag{2.52}$$

with $\zeta \in [-1,1]$ for the plane case. The first sum in each of these expressions corresponds to the summation over all the elements, while the second one stands for the approximation of the boundary value functions within an element. The nodal quantities can be pulled out of the integral. In order to express the integrals in local coordinates, the following transformation relations have been used

$$d\Gamma(x) = |J(\zeta_1,\zeta_2)|\,d\Gamma(\zeta_1,\zeta_2)\,,$$
$$d\Gamma(x) = |J(\zeta)|\,d\Gamma(\zeta)\,, \tag{2.53}$$

where $|J(\zeta_1,\zeta_2)|$ and $|J(\zeta)|$ are the **JACOBIAN** for the spatial and plane case.

For a given position of the source point ξ , one obtains i=3 equations in a spatial case and i=2 equations in a plane problem. Each new position of the source point yields another i equations. As the source point passes through all the N introduced nodal points, one obtains a system of (i N) linear algebraic equations, which can be written in a matrix form as

$$\boldsymbol{A}\,\boldsymbol{u} = \boldsymbol{B}\,\boldsymbol{t}\,. \tag{2.54}$$

Matrices **u** and **t** contain all the nodal values u_j^{lm} and t_j^{lm} , respectively. The elements of the matrices **A** and **B** can be found by comparison with Eqs. (2.51) or (2.52). For numerical integration over surface elements, which do not contain the source point, one can use the usual **GAUSS** quadrature formula. If the element contains the source point, one can use a weighted **GAUSS** quadrature formulation in the plane problem, whereas in the spatial case, when quadrilateral surface elements are considered, one has to divide the integral domain into triangular elements, to avoid the singularity. The diagonal elements of the matrix **A**, which contain the term $C_{ij}(\xi)$ and the principal value of the singular integral, can be determined implicitly, by considering the BEM formulation of a rigid body movement [16].

Matrix equation (2.54), together with the boundary conditions, is sufficient to determine the unknown nodal values of the boundary value functions $u_j(x)$ and $t_j(x)$. Depending on the boundary value problem, Eq. (2.54) has to be rearranged in the form

$$\boldsymbol{A}\,\boldsymbol{x} = \boldsymbol{b}\,, \tag{2.55}$$

where now the matrix **x** contains all the unknown nodal quantities. The right-hand side is known and can be computed. This system of linear algebraic equations has to be solved to obtain the unknown displacements and tractions over the entire surface Γ. Then, in a next step, the stresses and displacements within the body and the complete stress tensor at the boundary can be determined.

Matrix **A** in Eq. (2.55) is fully occupied and neither symmetric nor positive definite. Only these properties of the matrix **A** present a substantial disadvantage of BEM and may bring to naught all its other advantages, like the significant reductions of the set of equations, due to the discretization of the surface values only. However, the substructure technique, discussed in section 2.3.5, offers a way to compensate, at least partially, this disadvantage.

2.3.3 STRESSES AND DISPLACEMENTS AT INTERNAL POINTS

Knowing the nodal values of tractions and displacements at the boundary, one can apply SOMIGLIANA's identity (2.25) in the form

$$u_i(\xi) = \int_\Gamma U_{ij}(\xi,x)t_j(x)d\Gamma(x) - \int_\Gamma T_{ij}(\xi,x)u_j(x)d\Gamma(x)\,, \tag{2.56}$$

as well as the formula (2.26), derived in chapter 2.2.2,

$$\sigma_{ij}(\xi) = \int_\Gamma U_{ijk}(\xi,x)t_k(x)d\Gamma(x) - \int_\Gamma T_{ijk}(\xi,x)u_k(x)d\Gamma(x)\,, \tag{2.57}$$

in order to determine the stresses and displacements in an arbitrary point ξ of the interior.

Let us discretize the above equations, similarly as it has been done in the previous chapter. In result, we obtain the following matrix equations

$$\begin{aligned} \boldsymbol{u}(\xi) &= \boldsymbol{G}\,\boldsymbol{t} + \boldsymbol{H}\,\boldsymbol{u}\,, \\ \boldsymbol{\sigma}(\xi) &= \tilde{\boldsymbol{G}}\,\boldsymbol{t} + \tilde{\boldsymbol{H}}\,\boldsymbol{u}\,, \end{aligned} \tag{2.58}$$

where matrices $\boldsymbol{u}(\xi)$ and $\boldsymbol{\sigma}(\xi)$ contain, respectively, components of displacement and stress at the internal point under consideration. Matrices **G** and **H**, as well as $\tilde{\mathbf{G}}$ and $\tilde{\mathbf{H}}$, contain regular integrals of known functions and can be evaluated by means of the simple GAUSS quadrature formula. One encounters first difficulties when the internal point approaches the boundary. Here, the singularity of the source point shows increasingly its influence on the results of the computation of the integrals at regions neighbouring to the boundary. In consequence, one needs to increase the accuracy of the integration formulae.

2.3.4 STATE OF STRESS AT THE BOUNDARY

Originally, by means of BEM procedures one would obtain at the boundary the values of the tractions only. Still, an engineer needs also the knowledge of the whole state of stress in this area. Therefore, there have been developed two following ways of determination of the missing components of the stress tensor.

1. The first one consists of a limit transition, in the course of which the source point moves towards the boundary. One obtains in this procedure singular integrals in Eq. (2.57), which have to be evaluated according to their CAUCHY principal value. The method seems to be rather cumbersome, as quoted in the literature [17, 18]. It does not allow to expect for reasonable results.

2. There is a substantially simpler way, consisting of differentiation of the field of boundary displacements, which has been determined, although approximatively only, in the beginning of the procedure. This yields the local state of strain within the boundary and opens the way to the computation of the corresponding stress components. For this purpose, we introduce a local Cartesian coordinate system (x_i) at the stress point under consideration, see Fig. 10.

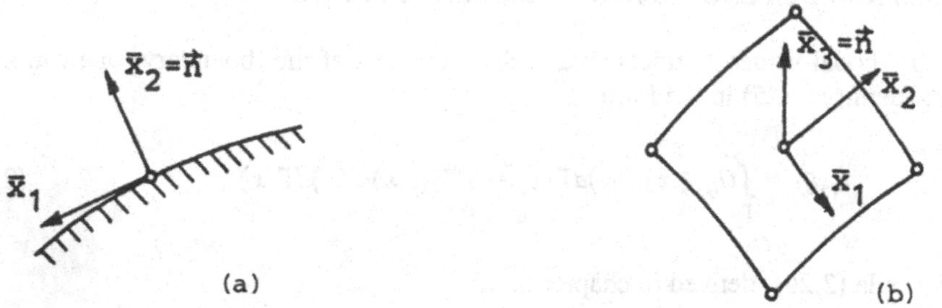

Fig. 10 Local Cartesian coordinate system defined at the stress point.
(a) two dimensional boundary element
(b) three dimensional boundary element.

It follows from Eq. (2.5), that in the general spatial case the strain is

$$\bar{\epsilon}_{ij} = \frac{1}{2}\left(\frac{\partial \bar{u}_i}{\partial \bar{x}_j} + \frac{\partial \bar{u}_j}{\partial \bar{x}_i}\right), \quad i,j = 1,2\,, \tag{2.59}$$

while in the plane case

$$\bar{\epsilon}_{11} = \frac{\partial \bar{u}_1}{\partial \bar{x}_1}\,. \tag{2.60}$$

Counting together with the known boundary tractions and including the relations, which follow from HOOKE's law, we obtain precisely as many known quantities or equations as needed to determine the missing components of the state of stress at the boundary. In the spatial case the known quantities are

$$\bar{\sigma}_{13} = \bar{t}_1 \quad ; \quad \bar{\sigma}_{23} = \bar{t}_2 \quad ; \quad \bar{\sigma}_{33} = \bar{t}_3 \, , \tag{2.61}$$

while expressions for the missing ones have the form

$$\bar{\sigma}_{11} = \frac{1}{1-\nu}\left[\nu \bar{t}_3 + 2G(\bar{\epsilon}_{11} + \nu \bar{\epsilon}_{22})\right] ,$$

$$\bar{\sigma}_{22} = \frac{1}{1-\nu}\left[\nu \bar{t}_3 + 2G(\bar{\epsilon}_{22} + \nu \bar{\epsilon}_{11})\right] . \tag{2.62}$$

In the plane case we need to distinguish between the plane state of stress and the plane state of strain. In the latter case we have

$$\bar{\sigma}_{12} = \bar{t}_1 \quad ; \quad \bar{\sigma}_{22} = \bar{t}_2 \, ,$$

$$\bar{\sigma}_{11} = \frac{1}{1-\nu}(\nu \bar{t}_2 + 2G\bar{\epsilon}_{11}) \, , \tag{2.63}$$

while in the former case one needs to substitute ν by $\nu^* = 1/(1+\nu)$ in the above formula. The second method is substantially simpler than the first one, since it does not contain any integration procedures. As its disadvantage, however, we have to consider the necessity to perform numerical differentiation of the boundary displacement, known only approximately. This leads to a decrease of the accuracy of the numerical evaluation of the state of stress at the boundary.

2.3.5 SUBSTRUCTURE TECHNIQUE

If the elastic region under consideration consists of a set of homogeneous subregions, the BEM procedure needs to be split in accordance with it. Let us consider the case of two subregions, illustrated on Fig. 11.

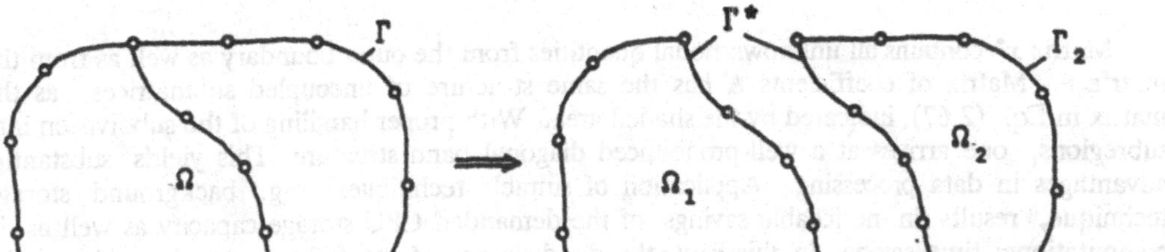

Fig. 11 Elastic domain Ω with boundary Γ divided in two subregions Ω_1 and Ω_2

According to Eq. (2.54) one can write a separate matrix equation

$$\begin{bmatrix} A_1 & A_1^* \end{bmatrix} \begin{bmatrix} u_1 \\ u_1^* \end{bmatrix} = \begin{bmatrix} B_1 & B_1^* \end{bmatrix} \begin{bmatrix} t_1 \\ t_1^* \end{bmatrix}, \quad \begin{bmatrix} A_2^* & A_2 \end{bmatrix} \begin{bmatrix} u_2^* \\ u_2 \end{bmatrix} = \begin{bmatrix} B_2^* & B_2 \end{bmatrix} \begin{bmatrix} t_2^* \\ t_2 \end{bmatrix}, \tag{2.64}$$

for each of the subregions. Quantities, denoted with the upper star, are related to the interface Γ^*. Continuity conditions at the interface have the form

$$\begin{aligned} u_1^* &= u_2^* = u^* , \\ t_1^* &= -t_2^* = t^* , \end{aligned} \tag{2.65}$$

and allow to couple the two matrix equations into a single one

$$\begin{bmatrix} A_1 & A_1^* & 0 \\ 0 & A_2^* & A_2 \end{bmatrix} \begin{bmatrix} u_1 \\ u^* \\ u_2 \end{bmatrix} = \begin{bmatrix} B_1 & B_1^* & 0 \\ 0 & -B_2^* & B_2 \end{bmatrix} \begin{bmatrix} t_1 \\ t^* \\ t_2 \end{bmatrix} . \tag{2.66}$$

Moving unknown interface tractions t^* to the left-hand side of the equations, one gets a linear system

$$\left[\begin{array}{c|ccc} A_1 & A_1^* & -B_1^* & 0 \\ \hline 0 & A_2^* & B_2^* & A_2 \end{array}\right] \begin{bmatrix} u_1 \\ u^* \\ t^* \\ u_2 \end{bmatrix} = \begin{bmatrix} B_1 & 0 \\ 0 & B_2 \end{bmatrix} \begin{bmatrix} t_1 \\ t_2 \end{bmatrix} . \tag{2.67}$$

Together with the boundary conditions, Eq. (2.67) makes possible to find the unknown quantities at the outer boundary and at the interface. Taking into account the boundary conditions one can represent Eq. (2.67) as

$$A^* \; x^* = b \; . \tag{2.68}$$

Matrix x^* contains all unknown nodal quantities from the outer boundary as well as from the interface. Matrix of coefficients A^* has the same structure of uncoupled submatrices, as the matrix in Eq. (2.67), indicated by the shaded area. With proper handling of the subdivision into subregions, one arrives at a well-pronounced diagonal band structure. This yields substantial advantages in data processing. Application of suitable techniques, e.g. background storage technique, results in noticeable savings of the demanded CPU storage capacity as well as in computational time saving. In this way, the disadvantage of the fully occupied matrix can be diminished, at least partially.

Application of the substructure technique is not limited to problems with inhomogeneities. Due to its particular properties it can be also applied successfully to homogeneous problems, as it will be shown further. Of particular use is the substructure technique in the case of complex structures, which otherwise would be practically unsolvable. Already in the first stage of the solution one obtains by means of this technique stresses and displacements at the interfaces. In some cases this allows to resign altogether of the evaluation of the solution in the points of the interior. In problems of crack analysis the substructure technique allows for a unique uncoupling of the crack surfaces, as it will be shown in chapter 3. This, in turn, leads to a very simple procedure of determination of the stress intensity factor at the crack tip.

2.3.6 DOUBLE NODE CONCEPT

The double node concept has been developed with the aim to improve the description of discontinuities in the boundary functions or boundary geometry. Its basic idea can be explained in a simplest way on the example of a plane problem, Fig. 12.

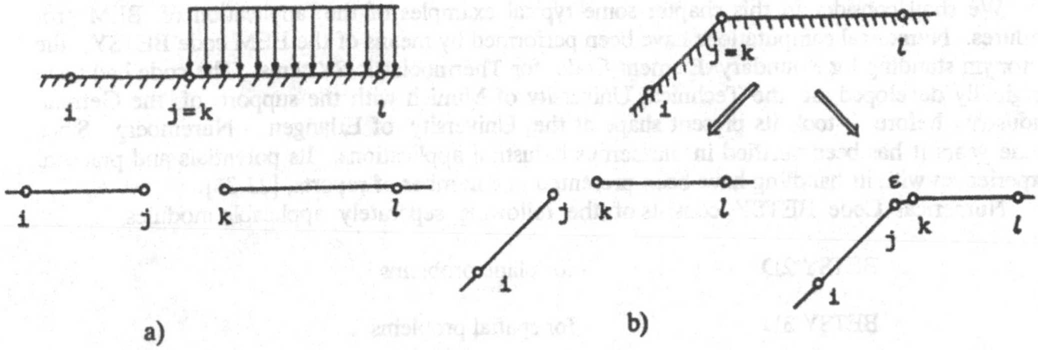

Fig. 12 Double node concept illustrated for
a) traction discontinuous, b) corners

At the site of the discontinuity one introduces formally two nodes, numbered separately j and k, although having identical coordinates. The nodes are originally uncoupled and considered as independent source points in Eq. (2.54). Since the displacement at both nodes is the same, in each direction there are two acceptable combinations of prescribed boundary conditions:

a) tractions at nodes j and k,
b) tractions at node j and displacements at node k (or vice versa).

The only difficulty connected with this form of the double node concept arises, when the boundary conditions for both nodes prescribe the same displacement component in at least one direction. This generates a singular matrix A. In this case the procedure proposed in [19] should be used. The condition of uniquenes of the stress tensor, as well as the condition of invariance of the trace of the strain tensor provide two extra equations, which can be used to remove the singularity of matrix A.

There is another way of handling the douple node concept. One can achieve the coupling of the two nodes by introduction of suitable connectivity conditions, which secure the compatibility and continuity of the solution. In the formulation of Eq. (2.54) one considers only one of the nodes as a source point. If the unknown nodal values for the other node have been already eliminated due to the connectivity condition, then the introduction of the double node concept does not bring any expansion of the system of equations. Otherwise, the nodal values have to be treated as additional unknowns, while the connectivity conditions need to be considered as additional equations. The double node concept can be extended formally to the spatial case, with the difference, however, that in this case there are 3 unknown nodal values at each node.

One can find also other analogous concepts in the literature. For instance, it has been proposed in [20] to introduce a fictitious element of length ϵ between the nodes j and k for boundaries containing corners, see Fig. 12 b. In fact, difficulties related to corners at the outer boundary or at the interfaces within the body, can be avoided by consideration of the so-called discontinuous elements, instead of the compatible ones, [21].

2.3.7 EXAMPLES

We shall consider in this chapter some typical examples of the application of BEM procedures. Numerical computations have been performed by means of the BEM code BETSY, the acronym standing for **B**oundary **E**lement Code for **T**hermoelastic **SY**stems. The code had been originally developed at the Technical University of Munich with the support of the German industry, before it took its present shape at the University of Erlangen - Nuremberg. Since some years it has been verified in numerous industrial applications. Its potentials and practical experiences with its handling have been presented in a number of reports, [22-26].

Numerical Code BETSY consists of the following separately applicable modules

BETSY 2D	- for plane problems
BETSY 3D	- for spatial problems
BETSY AXT, AX0, AX1	- for axisymmetric geometry

of which AXT has been developed for axially symmetric torsion, AX0 for axisymmetric external loads and AX1 for semi-analytical treatment of axially symmetric bodies under non-symmetric external loads. The external loading function is expanded into **FOURIER** series and the first term of this expansion finds its application in the AX1 module, as a model of bending and transversal forces. For further details see [26].

Example 2.1

As a first example for the plane problem we select the notch factor calculation for a brakeband of an automatic transmission [22]. Stress concentration problems of this type are usually typical examples, where BEM application promises advantages against other methods. In order to check it, a comparative calculation in FEM-code NASTRAN was made, using isoparametric formulations. To achieve a good comparison of the two methods with respect to computational time as well as convergence characteristic, different discretizations were chosen.

Fig. 13 shows the respective roughest and finest nets of grid and nodal points chosen for the calculation. Due to the symmetry, only one half of the unrolled brakeband was taken into account. The computation of stresses has been done under the assumption of a given constant displacement at the upper end of the band, with the other end being freely supported.

Fig. 14 shows the convergence characteristics of both methods (TRIM6- elements used for NASTRAN). When compared with the results shown in Fig. 15

it reveals a high demand on computation time for FEM (NASTRAN) against BEM, at comparable accuracy. For additional comparisons see [25].

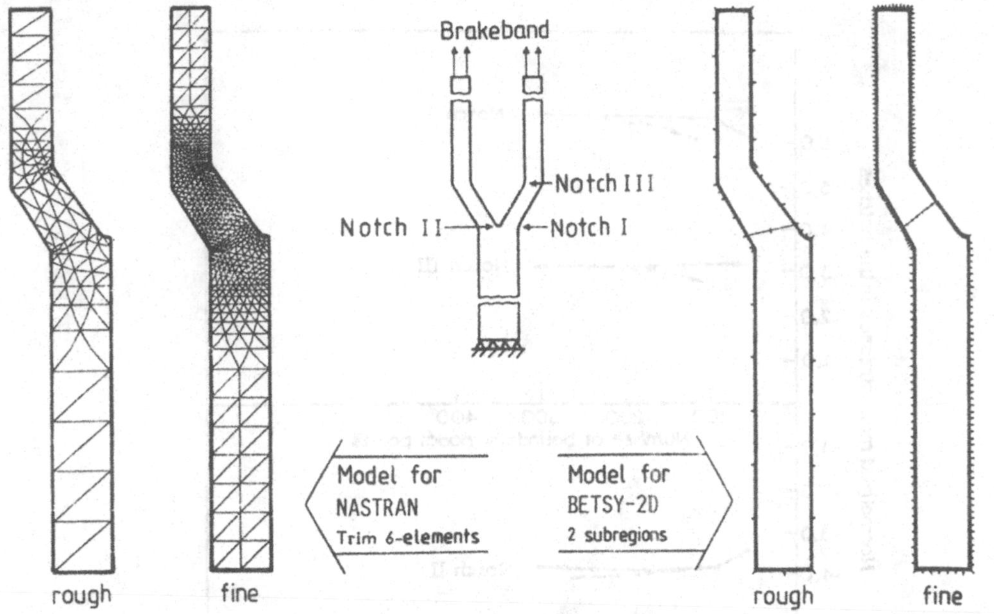

Fig. 13 Roughest and finest discretization of the unrolled brakeband

The computational model used for the BEM calculation consisted of two subregions only. Fig. 16 shows the influence of the further subregion partition on the computation time, at comparable accuracy.

The deformed and undeformed shapes of the brakeband are given in Fig. 16 for two cases: without any subdivision, and with eight subelements. The optimal computation time for a chosen number of subsections is arrived at, when all the subelements have approximately the same number of degrees of freedom. Under this condition, in the example shown here, one gets a minimum of CPU-time at seven to eight subelements. Naturally, the position of this minimum varies, depending upon the problem. Nevertheless, as shown in Fig. 16, the initial steep decrease of the computation time soon ceases with an increasing number of subsections. For practical purposes this means that even a rough assessment of this optimum saves sufficient time, when substructure technique is applied.

Another advantage of BEM, especially appreciable for practical applications in the industry, is a substantially shorter data generation time. In the example cited, the time required for data generation (before data input) as well as the time needed for the evaluation of the results were compared. Acknowledged specialists familiar with both methods were invited to participate in the comparison. The results show that the BEM model, depending upon the fineness of the meshes applied, takes about 8 to 24 times less time for the data preparation, and about 2.5 to 5 times less time for evaluation of the results, than the FEM model (NASTRAN).

Example 2.2

As an example for the three dimensional version we select the problem of a steering knuckle subjected to bending. Again a stress concentration problem and a notch factor computation is considered.

In this example, instead of a comparison with FEM, a comparison was made with the one-dimensional version BETSY-AX1. This program-module was specially developed for bending problems of geometrically axisymmetric structure elements.*) Therefore, it was

*)See chapter 2.6.2

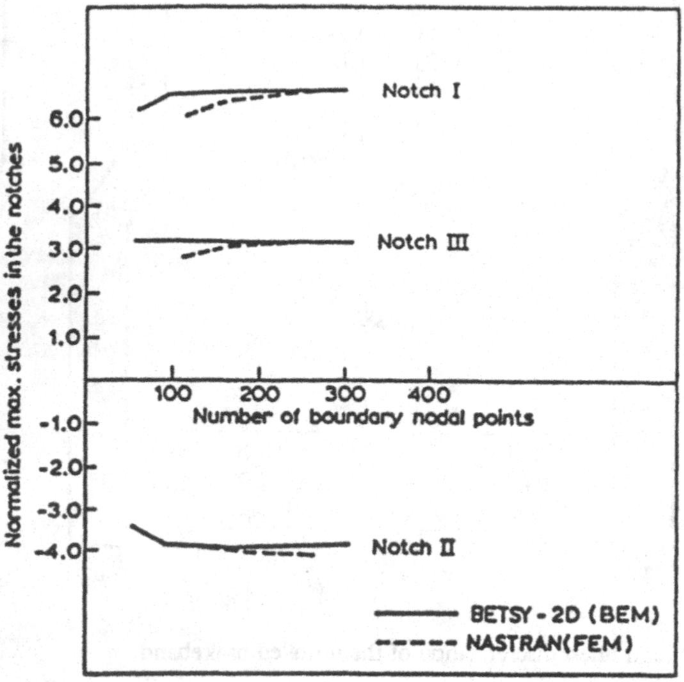

Fig. 14 Convergence characteristics for normalized maximum stresses in brakeband notches

necessary to re-model the steering knuckle in a rotational symmetric form, to approximate the transverse load with the first FOURIER cosine expansion term and to clamp the flange of the steering knuckle correspondingly.

Fig. 17 shows the modelled structure of the steering knuckle, halved due to its symmetry. The transverse load is applied at the top and the shaded area denotes the clamped surface. In order to reduce the computational expense, the steering knuckle was subdivided into four subelements. The dividing planes are indicated in the figure.

In the case of the finest discretization the surface was represented by 414 nodal points (that means 1242 degrees of freedom) altogether. The substructure division reduced the demand on the central storage capacity, since the maximal subelement contained 483 degrees of freedom only. The computations were made under application of quadratic shape functions and isoparametric surface elements.

Fig. 18 shows the values of the normalized maximal stresses in the three notches, for different discretizations. The results are compared with those obtained by BETSY-AX1. The rotational symmetric version used linear and quadratic shape functions, varying the fineness of the boundary division, in order to illustrate the convergence behaviour of BETSY-AX1 as well.

Also in BEM the overall data processing effort for the three-dimensional case increases in comparison with plane or axisymmetric cases. Nevertheless, due to the general property of BEM, which is a decrease by one dimension of the discretization volume, one can expect the same kind of advantages in the spatial case, as those shown in plane problems.

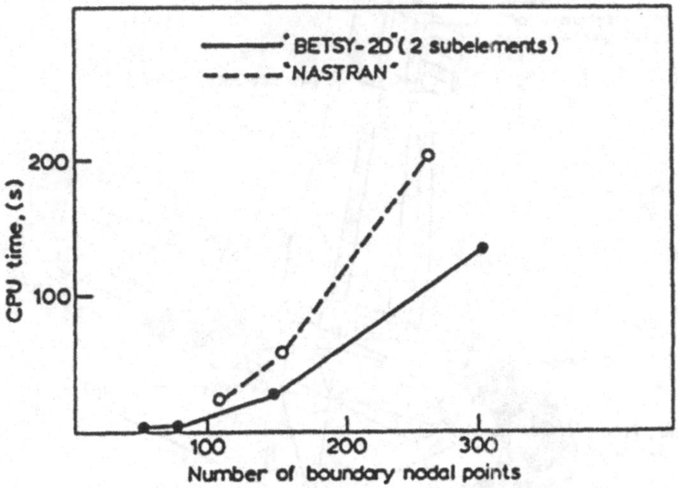

Fig. 15 Computation time characteristics

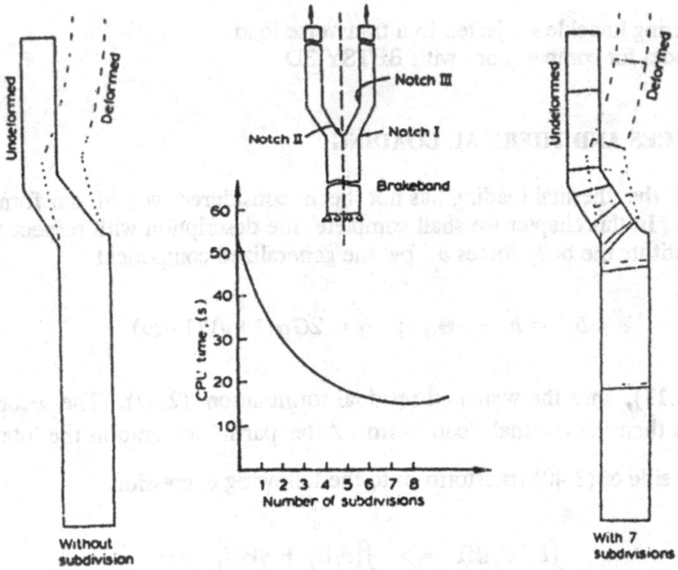

Fig. 16 Influence of substructuring on computation time for BETSY-2D

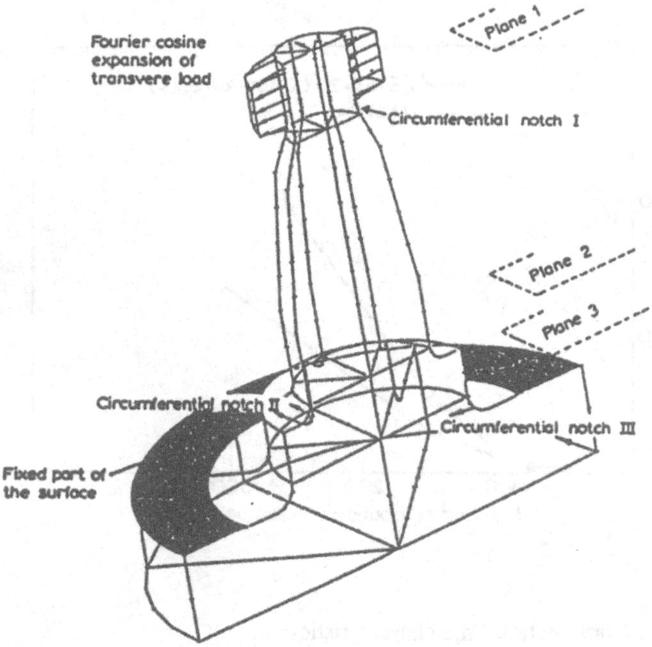

Fig. 17 Steering knuckle subjected to a transverse load.
Model for computations with BETSY-3D

2.4 BODY FORCES AND THERMAL LOADING

The case of the thermal loading has not been considered yet in our formulation of the BEM procedures. In this chapter we shall complete the description with respect to this. To this end, let us substitute the body forces b_j by the generalized component

$$b_j^* = b_j - \gamma\Theta,_j \;\; ; \;\; \gamma = 2G\alpha(1+\nu)/(1-2\nu)\, , \tag{2.69}$$

compare Eq. (2.13), into the weighted residual formulation (2.37). The second part of the equation contains then the thermal load term. After partial integration the integral $\int_\Omega b_j^* U_j d\Omega$ on the right-hand side of (2.40) transforms into the following expression

$$\int_\Omega b_j^* U_j d\Omega \;\; => \;\; \int_\Omega [b_j U_j + \gamma\Theta U_{j,j}] d\Omega\, . \tag{2.70}$$

We obtain now the basic boundary integral equation (2.34), valid for body forces and

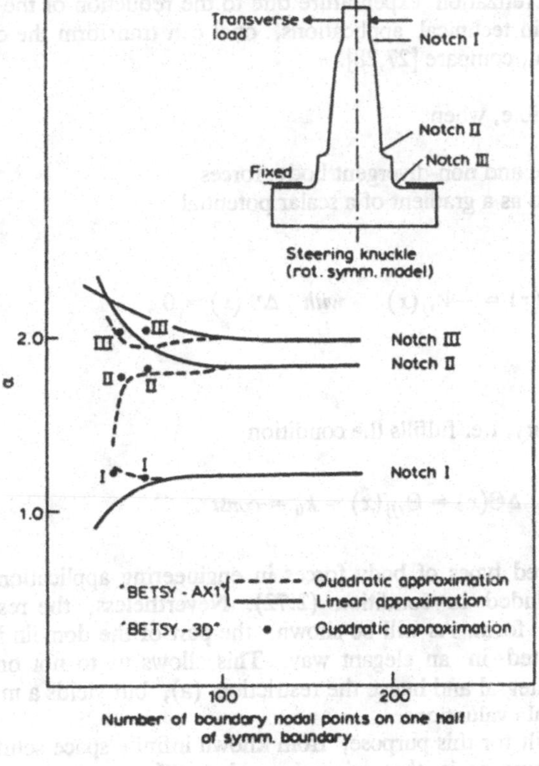

Fig. 18 Convergence characteristics for notch factors computation performed with BETSY-AX1, as compared with results of BETSY-3D. Notch factors are defined as

$$\alpha = \frac{\textit{max. stress in axial dir.}}{\textit{bending moment}/(\pi D^3/32)},$$

where D is the smallest notch diameter.

thermal loading in the form

$$C_{ij}(\xi)u_j(\xi) + \int_\Gamma [T_{ij}(\xi,x)u_j(x) - U_{ij}(\xi,x)t_j(x)]d\Gamma(x)$$
$$= \int_\Omega [b_j(x)U_j(\xi,x) + \gamma\Theta(x)U_{ij,j}(\xi,x)]d\Omega(x)\,. \tag{2.71}$$

Since we know the distribution of the body forces $b_j(x)$ and the field of temperature $\Theta(x)$ within the body, we can calculate the domain integral on the right-hand side. As a rule, the domain needs to be subdivided into a number of volume elements, in order to allow the computation of the integral. This leads, however, to the loss of the main advantage of the BEM procedure, which is lower discretization expenditure due to the reduction of the dimension. In some special cases, important in technical applications, one can transform the domain integral in (2.71) into a surface integral, compare [27,28].

The transformation is possible, when

(a) there are only conservative and non-divergent body forces, i.e. they can be represented as a gradient of a scalar potential harmonic function $\Psi(x)$

$$b_j(x) = -\Psi_{,j}(x) \qquad with \quad \Delta\Psi(x) = 0\,, \tag{2.72}$$

and

(b) temperature field is stationary, i.e. fulfills the condition

$$\Delta\Theta(x) = \Theta_{,jj}(x) = k_0 = const\,. \tag{2.73}$$

The most often encountered types of body forces in engineering applications like gravity and centrifugal forces are included in condition (2.72). Nevertheless, the restriction (2.72) need not be maintained in this form. As will be shown, the part of the domain integral due to volume forces can be eliminated in an elegant way. This allows us to not only avoid the transformation into a surface integral and hence the restriction (a), but yields a more favourable description for further numerical evaluations.

An auxiliary solution is built for this purpose, from known infinite space solution. Material constants G and ν are the same as in the original problem. The temperature difference is assumed to be zero.*) The only demand with respect to this auxiliary solution is that the body forces b_j should be equal to those of the actual problem. For gravity and centrifugal forces as well as for some combinations of point and line distributed body forces closed form analytical auxiliary solutions can be found.

If the components of the boundary displacements $\mathring{u}_j(x)$ and of the tractions $\mathring{t}_j(x)$ at the current point are known, one can introduce them in Eq. (2.71), instead of $u_j(x)$ and $t_j(x)$ respectively. Subtracting the new relation from Eq. (2.71) and introducing new quantities

$$\begin{aligned} u_j^*(x) &= u_j(x) - \mathring{u}_j(x)\,, \\ t_j^*(x) &= t_j(x) - \mathring{t}_j(x)\,, \end{aligned} \tag{2.74}$$

one obtains the basic integral equation in the form

*)In the code BETSY also the case of thermal loading with a linearly distributed temperature field can be treated in this way.

$$C_{ij}(\xi,x)u_j^*(\xi)+\int_\Gamma[T_{ij}(\xi,x)u_j^*(x)-U_{ij}(\xi,x)t_j^*(x)]d\Gamma(x)$$
$$=\gamma\int_\Omega\Theta(x)U_{ij,j}(\xi,x)d\Omega(x)\,, \tag{2.75}$$

in which the original domain integral due to volume forces has been eliminated. The remaining domain integral in Eq. (2.75) will now be transformed into a surface integral for the stationary case $\Delta\Theta = k_0$ and for simplicity we assume $k_0 = 0$. We shall present here another derivation than the one given in [22], which has the advantage of leading to a unified presentation for both spatial and plane problems.

To achieve this unified presentation it is convenient to express the reference solution through the stress function introduced by **PAPKOVICH** [29] and **NEUBER** [30].

According to:

$$F_i = \Phi_{i0} + [x_j(x) - x_j(\xi)]\,\Phi_{ij}\,, \tag{2.76}$$

with

$$\Delta\Phi_{i0} = 0;\quad \Delta\Phi_{ij} = 0 \ \text{ for } i,j = 1,2,3 \text{ and } \Delta\Delta F_i = 0\,, \tag{2.77}$$

the stress function F_i is set up of four harmonic functions.*) The subscript i indicates that the reference solution corresponds to a unit point load in e_i direction. Following [30] the divergence of displacement $U_{ij,j}(\xi,x)$ can be expressed in the form

$$U_{ij,j}(\xi,x) = \frac{1}{2G}(1-2\nu)F_{i,jj}(\xi,x)\,. \tag{2.78}$$

Exploying Eq. (2.78), as well as $\gamma = 2G\alpha(1+\nu)/(1-2\nu)$ in the remaining domain integral of Eq. (2.75), one obtains

$$\alpha(1+\nu)\int_\Omega\Theta(x)F_{i,jj}(\xi,x)d\Omega(x)\,. \tag{2.79}$$

Integrating expression (2.79) twice by parts and taking into account that $\Theta_{,jj}(x) = 0$, the domain integral can be transformed into a surface integral

---

*)NEUBER [30] showed that in the general case three harmonic functions are sufficient. Hence the name "three functions' rule".

$$\alpha(1+\nu)\int_\Omega \Theta(x)F_{i,jj}(\xi,x)d\Omega(x) = \alpha(1+\nu)\int_\Gamma [\Theta(x)\frac{\partial}{\partial n}F_i(\xi,x) - F_i(\xi,x)\frac{\partial}{\partial n}\Theta(x)]d\Gamma(x)\,, \tag{2.80}$$

where $\partial / \partial n$ denotes the outer normal derivative at the current boundary point x.

2.4.1 THERMAL LOADING IN THE THREE-DIMENSIONAL CASE

The reference solution for the point load F acting at the source point ξ in e_i direction can be expressed through the following harmonic functions:

$$\Phi_{i0} = 0\,; \quad \Phi_{ij} = \begin{cases} 0\,, & \text{for } i \neq j\,, \\ \dfrac{F}{8\pi(1-\nu)}\dfrac{1}{r(\xi,x)}\,, & \text{for } i = j\,, \end{cases} \tag{2.81}$$

where $r(\xi,x) = \sqrt{r_k r_k}$, with $r_k = x_k(x) - x_k(\xi)$, represents the distance between the current point x and the source point ξ. Substituting this into Eq. (2.76) one obtains the following stress function

$$F_i = \frac{F}{8\pi(1-\nu)} \cdot \frac{x_i(x) - x_i(\xi)}{r(\xi,x)}\,. \tag{2.82}$$

The arbitrary point load F has to be chosen as $F = 8\pi(1-\nu)$ in order to derive the same tensor expressions $U_{ij}(\xi,x)$ and $T_{ij}(\xi,x)$ as defined in Eq. (2.18). Substituting Eqs. (2.80) and (2.82) into (2.75) one obtains the final form of the basic integral equation in the general 3D case

$$C_{ij}(\xi)u_j^*(\xi) + \int_\Gamma [T_{ij}(\xi,x)u_j^*(x) - U_{ij}(\xi,x)t_j^*(x)]d\Gamma(x) =$$

$$= \alpha(1+\nu)\int_\Gamma [\Theta(x)\frac{\partial}{\partial n}(\frac{x_i(x)-x_i(\xi)}{r(\xi,x)}) - \frac{x_i(x)-x_i(\xi)}{r(\xi,x)}\frac{\partial}{\partial n}\Theta(x)]d\Gamma(x)\,. \tag{2.83}$$

2.4.2 THERMAL LOADING IN THE PLANE CASES

First we consider the special case of plane strain. The reference solution is built with a load distribution F constant along the x_3-axis and acting in e_i-direction. It will be described through the following harmonic functions:

$$\Phi_{i0} = 0\,; \quad \Phi_{ij} = \begin{cases} 0\,, & \text{for } i \neq j\,, \\ -\dfrac{F}{4\pi(1-\nu)}\ln r(\xi,x)\,, & \text{for } i = j\,. \end{cases} \tag{2.84}$$

with i,j $\in$ [1,2]. According to Eq. (2.76) the stress function becomes

$$F_i = -\frac{F}{4\pi(1-\nu)}[x_i(x)-x_i(\xi)]\ln r(\xi,x)\,, \tag{2.85}$$

or, if the assumption $F = 8\pi(1-\nu)$ is kept

$$F_i = -2\,[x_i(x)-x_i(\xi)]\ln r(\xi,x)\,. \tag{2.86}$$

By substitution of Eqs. (2.80) and (2.86) in Eq. (2.75) one obtains the basic equation, valid for the plane strain state.

$$C_{ij}(\xi)u_j^*(\xi) + \int_\Gamma [T_{ij}(\xi,x)u_j^*(x) - U_{ij}(\xi,x)t_j^*(x)]d\Gamma(x) =$$
$$= -2\alpha(1+\nu)\int_\Gamma \{\Theta(x)\frac{\partial}{\partial n}[(x_i(x)-x_i(\xi))\ln r(\xi,x)]$$
$$-[x_i(x)-x_i(\xi)]\ln r(\xi,x)\frac{\partial}{\partial n}\Theta(x)\}d\Gamma(x)\,. \tag{2.87}$$

The state of plane stress one can obtain through the following formal transition of the material constants:

$$G => G\,;\quad \nu => \frac{\nu}{1+\nu}\,;\qquad \alpha => \frac{1+\nu}{1-2\nu}\alpha\,. \tag{2.88}$$

The values of σ_{33} in the plane strain and ϵ_{33} in the plane stress can be calculated separately.

Similarly to the procedure described in the previous sections, Eqs. (2.83) and (2.87) can be transformed into a system of linear algebraic equations by means of discretization of the integration area and approximation of the boundary functions. Finally, one obtains in matrix form

$$\mathbf{A}\,\mathbf{u}^* = \mathbf{B}\,\mathbf{t}^* + \mathbf{C}\,\mathbf{\Theta} + \mathbf{D}\,\mathbf{\Theta}_{,n}\,, \tag{2.89}$$

where the matrices $\mathbf{\Theta}$ and $\mathbf{\Theta}_{,n}$ contain all the boundary nodal quantities Θ^{im} and $\Theta_{,n}{}^{im}$ of the prescribed thermal field, respectively. The elements of the matrices **C** and **D** contain integrals, which can handily be computed.

2.5 SYMMETRIC PROBLEMS

In the case, when both the geometry of the body and the boundary conditions are characterized by planes of symmetry, one can formulate within the BEM procedure two, entirely different, ways of solution. As an illustration let us consider the plane problem with a single axis of symmetry, presented on Fig. 19 a.

The first possibility, see Fig. 19 b, is to treat the conditions of symmetry at the symmetry axis as boundary conditions for one half of the body. Using this idea, which finds its

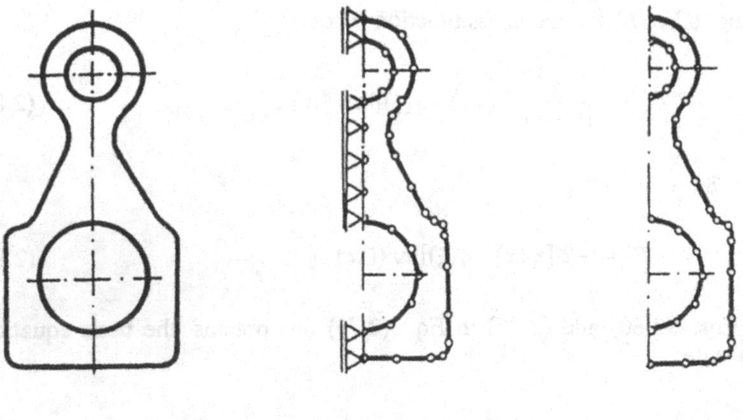

Fig. 19 Different calculation models for symmetric problems
(a) Symmetric plane problems
(b) Model 1, where half of the boundary and the symmetry lines are discretized
(c) Model 2, where only half of the boundary is discretized

application also in other numerical procedures, one obtains tractions at the symmetry line in the first step of solution already.

As an alternative, one can apply in the BEM procedure a modelling concept, Fig. 19 c, which allows to avoid the discretization of the symmetry line and, nevertheless, takes into consideration one of the symmetric halves only. In order to do that, one needs just to substitute quantities $T_{ij}(\xi,x)$ and $U_{ij}(\xi,x)$ in Eq. (2.34) with the following expressions

$$T_{ij}(\xi^{\pm},x) = T_{ij}(\xi^{+},x) + T_{ij}(\xi^{-},x)\,,$$
$$U_{ij}(\xi^{\pm},x) = U_{ij}(\xi^{+},x) + U_{ij}(\xi^{-},x)\,, \qquad (2.90)$$

which follow from the addition of the mirror reflections of the source points ξ^{+} and ξ^{-}. Application of this concept reduces the number of the degrees of freedom substantially. Stresses at the symmetry axis, if needed, can be computed in the second step of the procedure, when evaluating internal points.

2.6 AXISYMMETRIC PROBLEMS

In engineering applications one encounters quite often axially symmetric parts. It is important, therefore, to formulate a solving procedure, specific for this type of bodies. One begins with the introduction of a cylindrical system of coordinates (r,ϕ,z). When external loads are independent of the tangential coordinate ϕ, the problem is axisymmetric. In the particular case, when the axially symmetrical loading is oriented in the tangential direction ϕ, the problem turns into pure torsion. The latter case can be treated separately form the general axially symmetric loading in the (r,z)-plane.

In the BEM formulation the problem becomes one-dimensional. The range of integration extends over the boundary meridian only. In the case of simple, symmetrical body forces, like gravity or centrifugal forces, as well as for symmetrical stationary fields of temperature, one can transform the corresponding domain integral into a boundary integral, [23, 31]. General cases of non-symmetric external loads can be reduced into a series of one-dimensional problems of the above kind. To do that, one needs to develop the load function into a **FOURIER** series. Here, the zeroth component of the series corresponds to the axisymmetrical load. Each next component of the series allows analytical integration over the circumference and thus the reduction of the problem to a one-dimensional one. The solution of the original problem is represented then as a superposition of the partial solutions, which correspond to the components of the series [26, 32].

The BEM code BETSY contains procedure modules for symmetric torsion (module AXT), for axialsymmetric problems (module AX0), corresponding to the 0-th **FOURIER** series component, and for nonsymmetric loads, corresponding to the 1st **FOURIER** series component (module AX1), [22, 23]. As it has been shown in the example 2.2, procedure AX1 allows for treatment of problems with bending moments and transversal loads.

It is impossible to present within this short course all the details of the BEM procedures mentioned above. We should like, however, to point out some principal differences in the course of the derivation of separate procedures, omitting for simplicity the question of body forces and thermal loads. Details on torsional procedures, see [33], on axisymmetric problems, see [23] and on nonsymmetric loads, see [26, 32], while details of problems arising in connection with body forces and thermal loads can be found in [31].

2.6.1 FUNDAMENTAL RING LOAD SOLUTIONS

Within the BEM procedures for axially symmetric regions one can replace **KELVIN**'s fundamental solutions with the so-called ring load solutions. The term denotes spatial elastic solutions corresponding to circularly distributed linear loads. In the case of axisymmetric problems the load distribution with regard to the circumference is constant, Fig. 20.

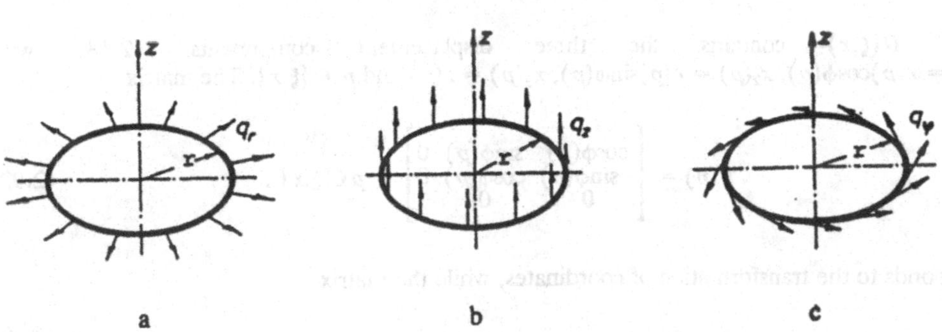

Fig. 20 Ring loads: (a) Radial load (b) axial load (c) tangential load

The derivation of the ring load solutions follows either directly from **NAVIER**'s equations, when body forces in a spatial problem are distributed along a given contour or from the integration of **KELVIN**'s solution along the circumference. Here we shall give a brief outline of the latter derivation [34].

Let us define cylindrical coordinates with the basis vectors (e_r, e_ϕ, e_z), see Fig. 21. We denote again the current point with x and the source point with ξ .

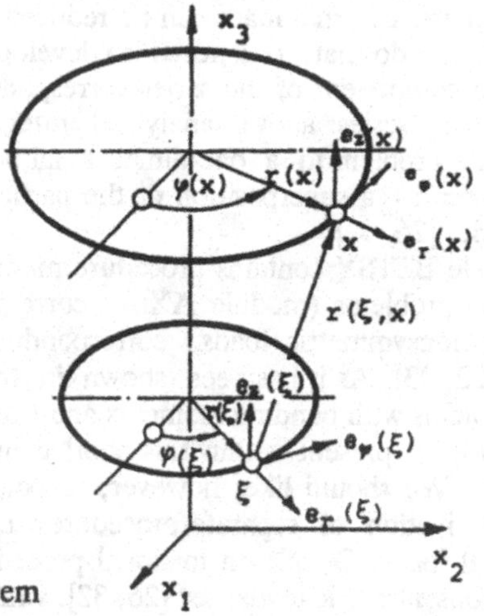

Fig. 21 Definition of the coordinate system

As a starting point we consider the fundamental displacements formula (2.18) of KELVIN's solution. By means of suitable transformations one expresses the displacements in the vector basis of the current point x, corresponding to a unit load applied in the source point ξ in the direction parallel to one of the vector basis directions of the source point ξ .

In the matrix form one can write it as the following transformation rule

$$U^*(\xi,x) = T^T(\xi)P(\xi)U(\xi,x)T(x)\ , \tag{2.91}$$

where $U(\xi,x)$ contains the three displacement components (2.18) with $x_1(p) = r(p)\cos\phi(p)$, $x_2(p) = r(p)\sin\phi(p)$, $x_3(p) = z(p)$ and $p \in \{\xi,x\}$. The matrix

$$T(p) = \begin{bmatrix} \cos\phi(p) & \sin\phi(p) & 0 \\ -\sin\phi(p) & \cos\phi(p) & 0 \\ 0 & 0 & 1 \end{bmatrix}\ , \ p\in\{\xi,x\}\ , \tag{2.92}$$

corresponds to the transformation of coordinates, while the matrix

$$P(\xi) = \begin{bmatrix} \dfrac{1}{\cos\phi(\xi)-\sin\phi(\xi)} & 0 & 0 \\ 0 & \dfrac{1}{\cos\phi(\xi)+\sin\phi(\xi)} & 0 \\ 0 & 0 & 1 \end{bmatrix} \tag{2.93}$$

reflects unit loads in the directions of the vector basis of the source point. The matrix $U^*(\xi,x)$ represents the displacements of the current point, expressed in the vector basis of x.

Integration of the displacement components (2.91) with regards to the circumferential coordinate ϕ

$$U_{ij}^*(\xi,x) = \frac{1}{2\pi}\int_0^{2\pi} U_{ij}^*(\xi,x)d\phi(\xi)\,, \quad i,j = r,\phi,z\,, \tag{2.94}$$

leads to the displacement field of the fundamental ring load solution, which is independent of the coordinate ϕ. The corresponding traction components $T_{ij}^*(\xi,x)$ can be calculated by means of the strain-displacements relation (2.5), **HOOKE**'s law (2.10) and external equilibrium conditions (2.4), after transforming them into cylindrical coordinates. For explicit expressions look in the literature [23].

2.6.2 BASIC INTEGRAL EQUATION

Substitution of the expressions for $U_{ij}^*(\xi,x)$ and $T_{ij}^*(\xi,x)$ into **SOMIGLIANA**'s identity transformed in the cylindrical coordinates allows for analytical integration along the circumference. With limit transition of the source point towards the boundary one comes to the basic integral equation in the form

$$\begin{aligned} C_{ij}(\xi)u_j(\xi) &+ 2\pi\int_\Gamma T_{ij}^*(\xi,x)u_j(x)r(x)d\Gamma(x) \\ &= 2\pi\int_\Gamma U_{ij}^*(\xi,x)t_j(x)r(x)d\Gamma(x)\,, \end{aligned} \tag{2.95}$$

where i,j correspond to r,z coordinates. The multiplier $2\pi\, r(x)$ results from the analytical integration with respect to ϕ. In the case of pure torsion one can apply only the tangential ring load solution. Index i in the above formula is here unnecessary.

2.6.3 HIGHER ORDER RING LOAD SOLUTIONS

Higher order ring load solutions can be applied in the case of non- symmetric load functions in order to allow for analytical integration with respect to the circumferential variable.

On Fig. 22 one can find examples of load distributions corresponding to the first component of the **FOURIER**'s series.

The derivation of the higher order fundamental solutions is analogous to the previous case. The expressions for U_{ij}^* and T_{ij}^* follow according to the order of the **FOURIER**'s component under consideration and are dependent on the variable ϕ. For instance, the first component yields

$$\begin{aligned} U_{ij}^*(\xi,x,\phi) &= \overline{U}_{ij}^*(\xi,x)\cos\phi + \overline{\overline{U}}_{ij}^*(\xi,x)\sin\phi\,, \\ T_{ij}^*(\xi,x,\phi) &= \overline{T}_{ij}^*(\xi,x)\cos\phi + \overline{\overline{T}}_{ij}^*(\xi,x)\sin\phi\,. \end{aligned} \tag{2.96}$$

Using **SOMIGLIANA**'s identity one can again conduct the analytical intregration with respect to the circumferential coordinate ϕ. For details of the derivation and of numerical implementation see [23,26].

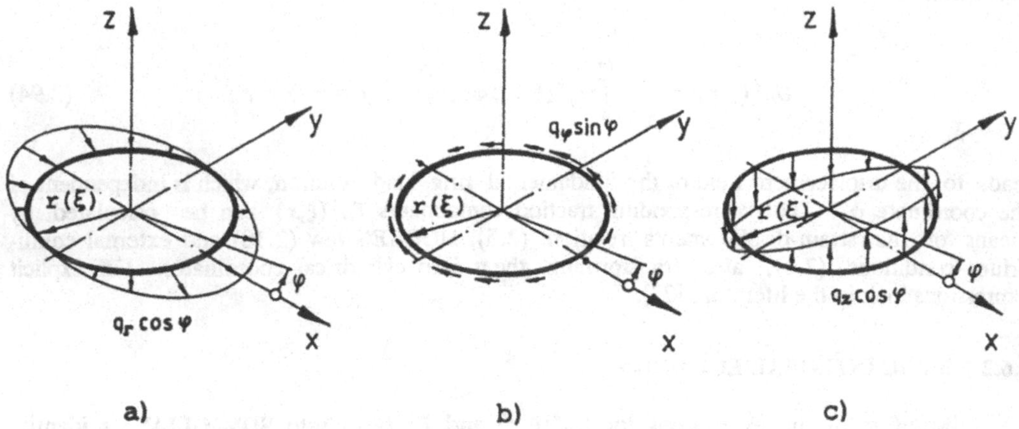

Fig. 22 (a) Radial, (b) circumferential and (c) axial ring sources

2.6.4 EXAMPLE

A simple notch problem should show the efficiency of the BEM procedure in application to axisymmetric problems.

Example 2.3

We have analysed a notched bar with the notch ratio $a/\rho = 0.868$ in tension, as shown on Fig. 23. For comparison, the result have been confronted with the analytical solution of **NEUBER** [35] as well as with the results of a FEM computation. In the latter case, the FEM procedure was based on problem-specific hybrid stress elements. The net of the FEM discretization and the boundary division in the BEM procedure are represented on the Fig. 23.

The result indicate a good coincidence regarding the notch factor α_K:

$\alpha_K = 1.353$ (BEM)

$\alpha_K = 1.356$ (FEM-hybrid elements)

$\alpha_K = 1.34$ (NEUBER)

There were 80 elements used in the FEM model and only 29 boundary elements needed in the BEM procedure.

Example 2.4

As an example of the treatment of non-symmetric loads on axi-symmetric parts we mentioned in chapter 2.3.7 the problem of a steering knuckle loaded with bending moments and transversal forces. The problem, which is substantially a spatial one, can be treated particularly efficiently as a series of one-dimensional problems, when the load function allows for representation with few FOURIER's components only. When this is the case, pre- and postprocessing expenditures, CPU-time and demands on the CPU storage capacity become incomparably lower than in the case of a general, three-dimensional approach.

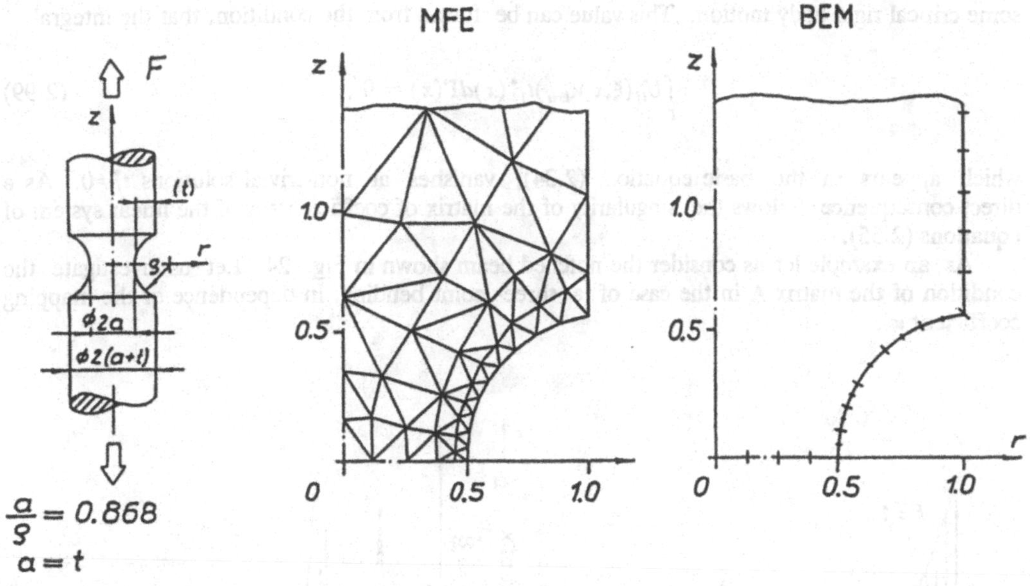

Fig. 23 Notch factor calculation

2.7 NON-UNIQUE SOLUTIONS IN PLANE PROBLEMS

The fundamental solution (2.19), applied in plane problems of elastostatics, contains a logarythmic component in the displacements

$$U_{ij}(\xi,x) = -\frac{1}{8\pi(1-\nu)G}\{(3-4\nu)\delta_{ij}\ln r - r_{,i}\, r_{,j}\}\,. \tag{2.97}$$

It is known, however, from the potential theory, see [36,37], that a logarithmic kernel function may lead to non-uniqueness of the solutions of the corresponding integral equation. This applies also to plane elastostatics, although, surprisingly, has not been noticed in the literature on engineering applications of the BEM. To point out the possibilities of the non-uniqueness let us introduce a central mapping $\bar{x}_i = \kappa x_i$, corresponding to the change of scale of the description of the problem. In view of the dependence $\bar{r} = \kappa r$ the logarythmic term, and with it the displacements in Eq. (2.97), depend on the mapping parameter κ as follows

$$U_{ij}(\xi,x,\kappa) = -\frac{1}{8\pi(1-\nu)G}\{(3-4\nu)\delta_{ij}(\ln r+\ln\kappa) - r_{,i}\, r_{,j}\}\,.. \tag{2.98}$$

It turns out that the central mapping is equivalent to the rigid body motion, corresponding to the displacements $U_{ij} = U_0\delta_{ij}\ln\kappa$. The non-uniqueness of the solutions of the basic equation of BEM arises at the critical value of the mapping coefficient κ_{kr}, corresponding to

some critical rigid body motion. This value can be found from the condition, that the integral

$$\int_{\Gamma} U_{ij}(\xi, x, \kappa_{krit}) t_j^*(x) d\Gamma(x) = 0\,, \tag{2.99}$$

which appears in the basic equation (2.34), vanishes at non-trivial solutions $t_j^* \neq 0$. As a direct consequence follows the singularity of the matrix of coefficients A of the linear system of equations (2.55).

As an example let us consider the notched beam shown in Fig. 24. Let us investigate the condition of the matrix A in the case of a three- point bending, in dependence of the mapping coefficient κ .

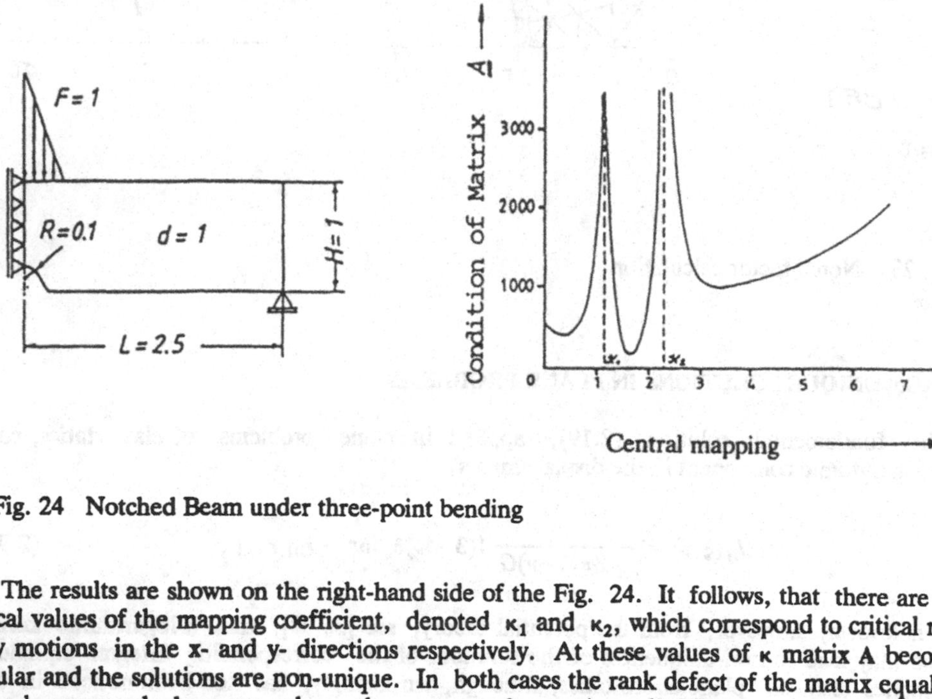

Fig. 24 Notched Beam under three-point bending

The results are shown on the right-hand side of the Fig. 24. It follows, that there are two critical values of the mapping coefficient, denoted κ_1 and κ_2, which correspond to critical rigid body motions in the x- and y- directions respectively. At these values of κ matrix A becomes singular and the solutions are non-unique. In both cases the rank defect of the matrix equals 1. It can be removed, however, when the system of equations (2.55) is completed with the global equilibrium condition in the direction of the corresponding coordinate axis.

In the following, two solving procedures are proposed, which allow for elimination of non-unique solutions [38].

1st proposal: Weighted global forces equilibrium condition.

Let us split away the rigid body motion term $U_{ij} = kU_0\delta_{ij}$ with an arbitrary weighting factor k from the first integral on the right-hand side of the Eq. (2.34)

$$C_{ij}(\xi)u_j(\xi) + \int_\Gamma [T_{ij}(\xi,x)u_j(x) - U_{ij}(\xi,x)t_j(x)]d\Gamma(x) - kU_0 \int_\Gamma t_j(x)\delta_{ij}d\Gamma(x) = 0 . \qquad (2.100)$$

Here, body forces have been omitted for simplicity. It is easy to see, that the global equilibrium condition for forces $\int_\Gamma t_i(x)d\Gamma(x) = 0$ enters into the equation with the weighting multiplier kU_0. The larger this multiplier is, the more exactly fulfills the approximative numerical solution the global forces equilibrium condition. This observation constitutes the foundation of our proposal, which has its origin in a known procedure of the potential theory, see [37]. Introduce formally the equilibrium terms denoted as ω_i for i = x,y

$$\omega_i = kU_0 \int_\Gamma t_i(x)d\Gamma(x) \qquad (2.101)$$

as new unknowns and include them into the solution vector $\boldsymbol{x}$. In order to obtain again a square coefficient matrix of the system of equations, the latter needs to be completed with two global equilibrium conditions for forces in the form

$$\int_\Gamma t_i(x)d\Gamma(x) = \frac{\omega_i}{kU_0} , \quad i = x,y . \qquad (2.102)$$

The larger is the weighting factor k chosen for the computations, the higher is the accuracy of satisfying the global forces equilibrium conditions. With k tending to infinity, the equilibrium conditions are fulfilled exactly within the frame of numerical approximation. The modified system of equations is presented schematically in (2.103).

$$\begin{bmatrix} & \begin{matrix}1 & 0\\ 0 & 1\\ 1 & 0\end{matrix} \\ \boldsymbol{A} & \\ & \\ \int t_x(x)d\Gamma_u & 0 \quad 0 \\ \int t_y(x)d\Gamma_u & 0 \quad 0 \end{bmatrix} \begin{bmatrix} \\ \boldsymbol{x} \\ \\ \omega_x \\ \omega_y \end{bmatrix} = \begin{bmatrix} \\ \boldsymbol{b} \\ \\ \int t_x(x)d\Gamma_\sigma \\ \int t_y(x)d\Gamma_\sigma \end{bmatrix} \qquad (2.103)$$

Also in the critical cases the solution fulfills the global equilibrium conditions for forces and guarantees thus the uniqueness. The quantities ω_x and ω_y represent some measures of the grade of the approximation.

2nd proposal: Least-squares solving procedure

Within the proposal 1 the solution complies to the global equilibrium of forces, but not to the equilibrium of moments. In problems containing elongated regions, as it is shown in the example illustrated on Fig. 25, this may have grave consequences. For these particular applications another procedure is proposed, taking into account on equal terms the global equilibrium of moments also.

We begin with the elimination of the influence of the quantities ω_x and ω_y on the matrix equation (2.103). To this purpose, one needs to compute the mean value of all evenly (or unevenly) numbered rows

$$\left[\frac{2}{N}\sum_{i=1}^{N/2} a_{2i-1,1};\ \frac{2}{N}\sum_{i=1}^{N/2} a_{2i-1,2};\ \ldots;\ \frac{2}{N}\sum_{i=1}^{N/2} a_{2i-1,N};\ 1;\ 0\right] = \frac{2}{N}\sum_{i=1}^{N/2} b_{2i-1}$$
$$\left[\frac{2}{N}\sum_{i=1}^{N/2} a_{2i\ ,1};\ \frac{2}{N}\sum_{i=1}^{N/2} a_{2i\ ,2};\ \ldots;\ \frac{2}{N}\sum_{i=1}^{N/2} a_{2i\ ,N};\ 0;\ 1\right] = \frac{2}{N}\sum_{i=1}^{N/2} b_{2i} \tag{2.104}$$

and then subtract them from the respective evenly (or unevenly) numbered rows. The rest forms the matrix $\boldsymbol{A}^*$, which does not contain any force equilibrium terms and its rank defect is equal two.

Now it is possible to complete the equation system with two global force and one global moment equilibrium conditions. The system is thus overdetermined and its solution takes the shape of a least-square procedure. The approximative solution satisfies all three global equilibrium conditions in the sense of the least-square procedure only. It guarantees also the uniqueness of the solution.

As it is shown in the example, Fig. 25, where due to L/H = 5 an elongated region yields higher errors in the evaluation of the equilibrium of moments, the solving procedure based on the 2nd proposal has obvious advantages over the proposal 1.

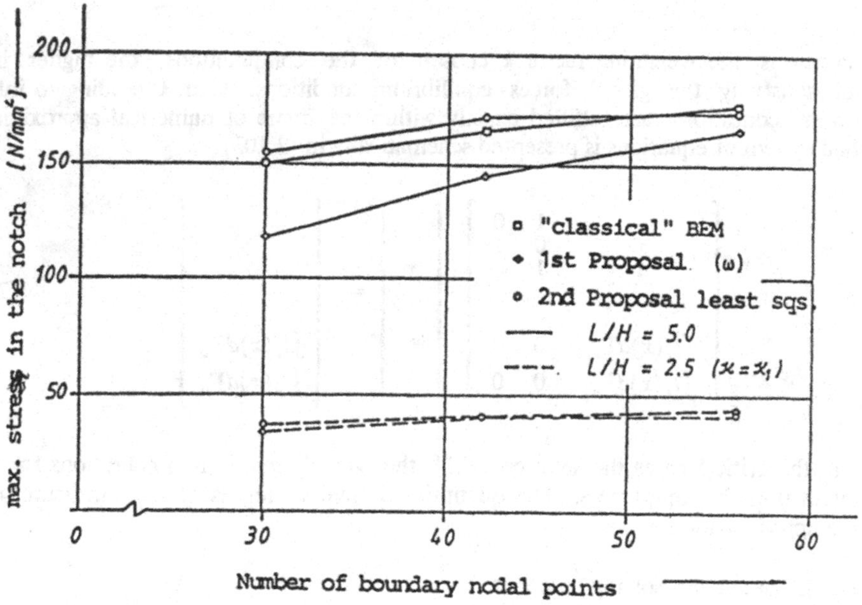

Fig. 25 Notch factor calculation of a notched bar unter three point bending

Example 2.5

Using one of the above proposals we can solve now uniquely the problem of the notched beam, Fig. 24 left, for critical values κ_1 and κ_2 of the mapping coefficient. The maximal notch stresses in dependence of the number of the boundary nodes, computed by means of each of the proposals, are represented on Fig. 25 with dashed lines for the case of the critical value of the mapping coefficient κ_1. The differences between the two procedures are in this case not so obvious. To point out the slower convergence for elongated regions at the 1st proposal a slender notched beam with the ratio L/H=5 has been considered, see continuous lines. Since at this ratio the problem is not critical any more, the results could be compared with the "classical" BEM procedure. Obviously, taking into account the global equilibrium of moments yields faster convergence in problems containing elongated regions.

3. APPLICATION IN LINEAR FRACTURE MECHANICS

3.1 INTRODUCTION

Linear fracture mechanics represents one of the most interesting application areas for numerical methods. In the following chapters we should like to point out a variety of applications of the BEM in this area. The main goal of the procedures will be here the computation of the stress intensity factors (SIF). It is well-known, that the knowledge of the stress intensity factor is necessary e.g. for the forecast of the crack growth of a fatigue crack or for the determination of the residual strength of a cracked (brittle) structure. The factor characterizes the intensity of the elastic singular stress field in the neighbourhood of the crack tip and depends substantially on the geometry and external loads.

Stress intensity factors have been analytically computed for a wide range of different geometries and loads. The results are presented in handbooks e.g. **TADA** [39], **SIH** [40], **ROOKE** and **CARTWRIGHT** [41], for cases with idealized simple geometry and loads. For more complex structures, not included in handbooks, the stress intensity factor has to be computed separately for each specific crack problem. A variety of methods have been developed for these numerical solutions. As far as the FEM is considered, special elements, including the right singularity of the asymptotic stress-strain-field in the vicinity of the crack tip, have been developed as well as modifications of ordinary elements to improve the accuracy of the results. But still a fine mesh has to be used around the crack tip. As a consequence, the number of unknowns increases and a large amount of man-power is needed for the preparation of the input data. Practically, all numerical techniques, developed within the FEM with the goal of computation of the stress intensity factor, can be transferred without difficulties into the BEM, with the obvious advantages following from the diminished dimension of the formulation of the problem. A particularly interesting version has been proposed by **SNYDER** and **CRUSE** [42]. Applying suitable **GREEN**'s functions as fundamental solutions, they managed to get rid of the discretization of the crack surface in the case of a tractionfree crack. This technique has been generalized by **KUHN** [43] for the case of a field of cracks.

After a short introduction into the fundamental concepts of linear fracture mechanics we should like to mention briefly the method of extrapolation. This ralatively simple procedure is applicable to any standard code of the BEM, provided it contains the substructure technique. By introduction of singular boundary elements [44] one can then improve the results further. To complete the discussion, we shall present the technique of application of **GREEN**'s functions as fundamental solutions [42].

3.2 FUNDAMENTAL CONCEPTS OF LINEAR FRACTURE MECHANICS

Let us discuss briefly the so-called K-concept of the linear fracture mechanics. Detailed descriptions can be found in the literature, see [45,46].

3.2.1 DISPLACEMENTS AND STRESSES NEAR THE CRACK TIP

One distinguishes in the fracture mechanics the following three basic crack opening (or crack loading) modes, Fig. 26.

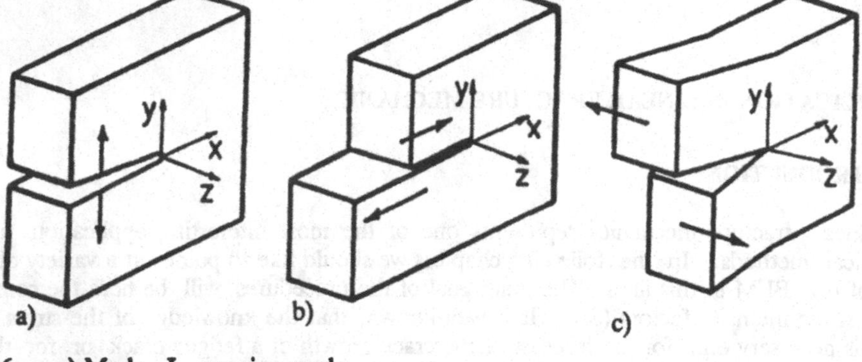

Fig. 26 a) Mode I, opening mode
b) Mode II, sliding mode
c) Mode III, tearing mode

An independent elastic field in the vicinity of the crack tip corresponds to each of these modes of loading. The superposition of the three modes describes the general case of cracking. In a plane problem only mode I and mode II can be present. Mode III corresponds to an anti-plane state of strain problem.

In Fig. 27 **GRIFFITH**'s problem of a finite crack of length 2a in an infinite plate under tension is considered. Due to **WESTERGAARD** [47], the asymptotic terms of the displacement components in the vicinity of the crack tip are

$$\begin{bmatrix} u_x \\ u_y \end{bmatrix} = \frac{K_I}{4G}\sqrt{\frac{r}{2\pi}}\begin{bmatrix} (2\kappa-1)\cos\frac{\Theta}{2} + \cos\frac{3\Theta}{2} \\ (2\kappa+1)\sin\frac{\Theta}{2} - \sin\frac{3\Theta}{2} \end{bmatrix} + \frac{K_{II}}{4G}\sqrt{\frac{r}{2\pi}}\begin{bmatrix} (2\kappa+3)\sin\frac{\Theta}{2} + \sin\frac{3\Theta}{2} \\ -(2\kappa-3)\cos\frac{\Theta}{2} - \cos\frac{3\Theta}{2} \end{bmatrix}, \tag{3.1}$$

where G is the shear modulus and κ stands for

$$\kappa = 3-4\nu \qquad \textit{in the plane strain case} ,$$
$$\kappa = \frac{3-\nu}{1+\nu} \qquad \textit{in the plane stress case} . \tag{3.2}$$

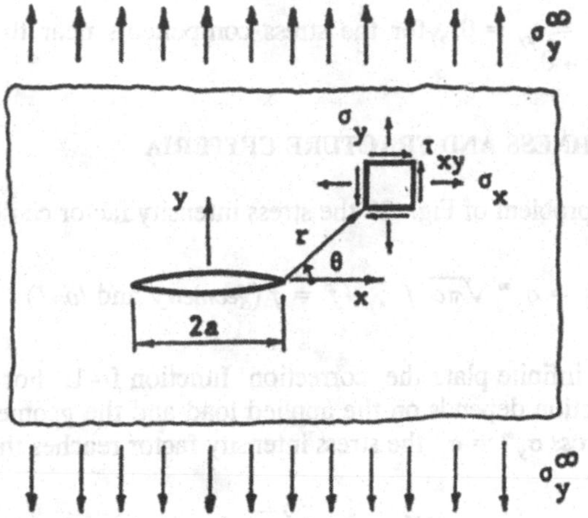

Fig. 27 **GRIFFITH crack in an infinite plate**

The corresponding asymptotic terms of the stress field near the crack tip are found to be

$$\begin{bmatrix}\sigma_x\\ \sigma_y\\ \tau_{xy}\end{bmatrix} = \frac{K_I \cos\frac{\Theta}{2}}{\sqrt{2\pi r}}\begin{bmatrix}1-\sin\frac{\Theta}{2}\sin\frac{3\Theta}{2}\\ 1+\sin\frac{\Theta}{2}\sin\frac{3\Theta}{2}\\ \sin\frac{\Theta}{2}\cos\frac{3\Theta}{2}\end{bmatrix} + \frac{K_{II}\sin\frac{\Theta}{2}}{\sqrt{2\pi r}}\begin{bmatrix}2+\cos\frac{\Theta}{2}\cos\frac{3\Theta}{2}\\ \cos\frac{\Theta}{2}\cos\frac{3\Theta}{2}\\ \cos\frac{\Theta}{2}-\sin\frac{3\Theta}{2}\end{bmatrix} \tag{3.3}$$

with

$$\sigma_z = \nu\,(\sigma_x+\sigma_y) \qquad \textit{in plane strain} ,$$
$$\sigma_z = 0 \qquad \textit{in plane stress} . \tag{3.4}$$

The indices I and II for the stress intensity factors K_I and K_{II} refer to the principal modes of Fig. 26, respectively. As can be seen, the stress field is singular as $r \to 0$. If a crack problem of a finite plate is considered and the crack tip is close to a boundary, terms of order $r^{1/2}$ and $r^{3/2}$ should be added.

For the mode III case one obtains

$$u_x = u_y = 0 \; ; \quad u_z = \frac{2K_{III}}{G}\sqrt{\frac{r}{2\pi}}\sin\frac{\Theta}{2} \; ; \tag{3.5}$$

for the displacements and

$$\tau_{xz} = -\frac{K_{III}}{\sqrt{2\pi r}} \sin\frac{\Theta}{2} ; \quad \tau_{yz} = \frac{K_{III}}{\sqrt{2\pi r}} \cos\frac{\Theta}{2} , \tag{3.6}$$

as well as $\sigma_x = \sigma_y = \sigma_z = \tau_{xy} = 0$, for the stress components near the crack tip. Again the stress field is singular as $r \to 0$.

3.2.2 FRACTURE TOUGHNESS AND FRACTURE CRITERIA

For the **GRIFFITH** problem of Fig. 27 the stress intensity factor can be written in the form

$$K_I = \sigma_y{}^\infty \sqrt{\pi a}\ f \ ; \quad f = f\,(geometry \text{ and } load) \ , \tag{3.7}$$

where in the case of an infinite plate the correction function f=1. For the more general case of a finite plate this function depends on the applied load and the geometry. If the crack starts growing for an applied stress $\sigma_y{}^\infty = \sigma_c$ the stress intensity factor reaches the critical value

$$K_{Ic} = \sigma_c \sqrt{\pi a}\ f \ . \tag{3.8}$$

The fracture criterion is thus written in the form

$$K_I = K_{Ic} \ . \tag{3.9}$$

K_{Ic} is the so-called "fracture toughness" and can usually be considered as a material constant. For the two other opening modes analogous fracture criteria

$$K_{II} = K_{IIc} \ ; \quad K_{III} = K_{IIIc} \tag{3.10}$$

are valid. In the case of mixed-mode problems one formulates within the K-concept an equivalent stress intensity factor

$$K_v = K_v(K_I , K_{II} , K_{III}) \ , \tag{3.11}$$

which needs to be compared in a mixed-mode fracture criterion

$$K_v(K_I , K_{II} , K_{III}) = K_{vc}(K_{Ic} , K_{IIc} , K_{IIIc}) \tag{3.12}$$

with the experimental value of the mixed-mode fracture toughness, considered as a material characteristic. Knowing the values of the stress intensity factors and the fracture toughness one can determine by means of the fracture criteria, whether the crack is going to grow.

Besides the simple K-concept one finds within the linear elastic fracture mechanics a variety of other fracture criteria. We should like to mention here only two of them. The first one, going back to **GRIFFITH**'s "energy criterion" [48], demands that the growth of the crack starts only then, when the so-called "energy release rate" arrives at a critical value, characteristic for the material. The second one is based on the concept of the path-independent J-integral [49]. This integral represents a generalized expression for the energy release rate due to crack propagation, which may also be valid if there is appreciable crack tip plasticity.

3.3 STRESS INTENSITY FACTOR CALCULATION

3.3.1 K-FACTOR CALCULATION FROM NODAL VALUES

The simplest way to determine the stress intensity factor is to compute numerically the solution in the vicinity of the crack tip and to compare it with the asymptotic solution near to the crack tip. One can apply it either to the stresses or to the displacements in this area. In order to apply the BEM one needs to start with the suitable modelling of the crack problem. The simplest case corresponds to a crack located on a symmetry axis, Fig. 28. In the numerical model, which is usually applied for this case, the symmetry conditions lead to the uncoupling separation of the crack edges. In the case of an arbitrary orientation of the crack, Fig. 29, the uncoupling separation of the crack edges can be achieved by means of the introduction of a corresponding substructure. In both cases we obtain with the BEM technique the approximate stress distribution in front of the crack tip as well as the crack opening displacement.

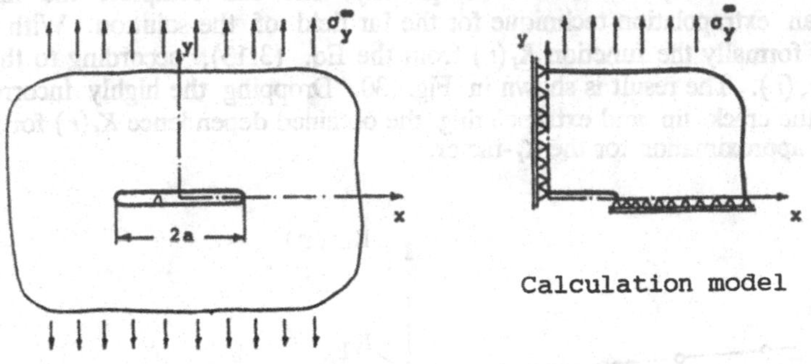

Fig. 28 Crack located in a symmetric plane

The singularity of the near field at the crack tip is presented also approximately only. One can improve the result by applying the so-called "singular boundary elements" [44]. These are linear elements with a quadratic function approximation, which can simulate the $1/\sqrt{r}$-singularity of the near field by means of a displacement of the middle node by 1/4 of the length of the element.

The approximate values of the K-factors, determined from the nodal values of the solution, can be illustrated on the example of the **GRIFFITH**'s crack shown on Fig. 28. First, one computes the approximate stress distribution $\sigma_y(r)$ in front of the crack. Comparing it with the asymptotic stress field near to the crack tip, Eq. (3.3), one obtains at $\Theta = 0$ the stress intensity factor.

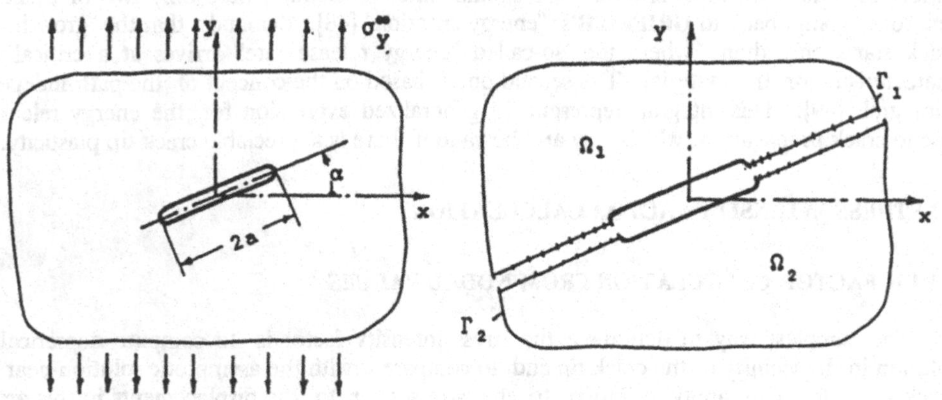

Fig. 29 Crack separation through substructure technique

$$K_I = \lim_{r \to 0} K_I(r) = \lim_{r \to 0} \sqrt{2\pi r}\ \sigma_y(r) \tag{3.13}$$

Its value for r tending to zero is well defined here and constant. Since the numerical procedure can not fully simulate the singularity, one can complete the limit transition by involving an extrapolation technique for the far field of the solution. With this purpose, one determines formally the function $K_I(r)$ from the Eq. (3.13), according to the nodal values of the stress $\sigma_y(r)$. The result is shown in Fig. 30. Dropping the highly incorrect values in the vicinity of the crack tip and extrapolating the obtained dependence $K_I(r)$ for $r \to 0$, one obtains the desired approximation for the K_I-factor.

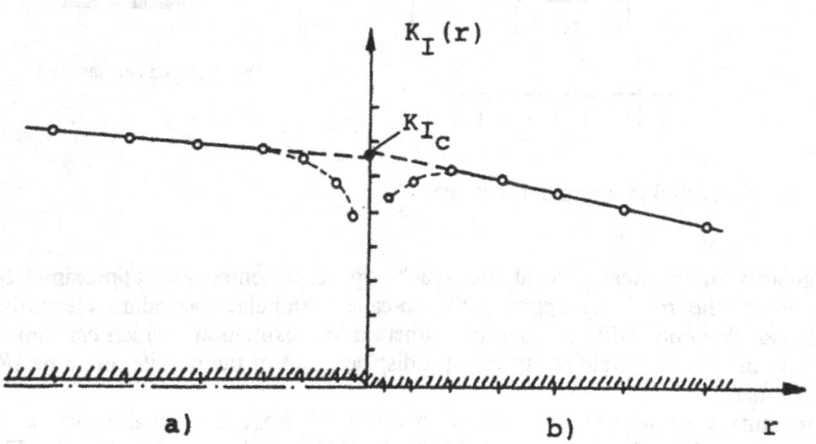

Fig. 30 Extrapolation of $K_I(r)$ as $r \to 0$
a) from nodal displacements
b) from nodal stresses

Similarly, one can determine the K_I-factor from the displacements of the crack edges (crack opening). From Eq. (3.1) one obtains for $\Theta = \pm\pi$

$$K_I = \lim_{r\to 0} K_I(r) = \lim_{r\to 0} \frac{2G}{(1+\kappa)} \sqrt{\frac{\pi}{2r}} [u_y(r,\pi) - u_y(r,-\pi)] , \tag{3.14}$$

where κ corresponds to formula (3.2). Also in this case one gets formally a dependence $K_I(r)$. The extrapolation procedure for $r \to 0$ is shown on Fig. 30 and can be considered as a complementary and checking procedure. The same procedure of the computation of the K-factor can be applied for other load (opening) modes and can be carried out within any BEM code. Still, one should expect errors up to 10 percent.

We shall not enter here into the details of other BEM calculations of the K-factor, based on other fracture criteria, e.g. criteria for

- o the release of potential energy
- o the release of elastic energy or for
- o J-integral.

Instead, we shall present now the elegant alternative procedure proposed by **SNYDER** and **CRUSE** [42], in which by application of special **GREEN's** functions one can altogether avoid the discretization of the crack boundary.

3.3.2 K-FACTOR CALCULATION WITHOUT CRACK DISCRETIZATION

We shall illustrate the procedure on the example of a **GRIFFITH's** crack in an uniaxially loaded plate, shown in Fig. 31. The following assumptions

- o the crack is straight
- o the surface of the crack is tractionfree
- o the crack opens under the loads considered

are valid. When considering the basic integral equation (2.34) of the BEM and the corresponding boundary value problem for this example, we have to realize, that the boundary consists here of the outer part Γ and the crack surface Γ_R.

We have thus

$$C_{ij}(\xi)u_j(\xi) + \int_{\Gamma} [T_{ij}(\xi,x)u_j(x) - U_{ij}(\xi,x)t_j(x)]d\Gamma$$
$$+ \int_{\Gamma_R} [T_{ij}(\xi,x)u_j(x) - U_{ij}(\xi,x)t_j(x)]d\Gamma_R = 0 , \tag{3.15}$$

where body forces have been omitted. **SNYDER** and **CRUSE** [42] have proposed for this type of problems to apply a **GREEN's** solution, illustrated on the left-hand side of the Fig. 32, instead of relying on the common BEM technique of unit load fundamental solutions. One can combine such a solution as a superposition of the partial solutions a) and b), shown on the right-hand side of Fig. 32. The **GREEN's** functions for these problems have been given by **ERDOGAN** [50] and **MEWS** [51] in the isotropic case and by **SNYDER** and **CRUSE** [42] for an anisotropic case.

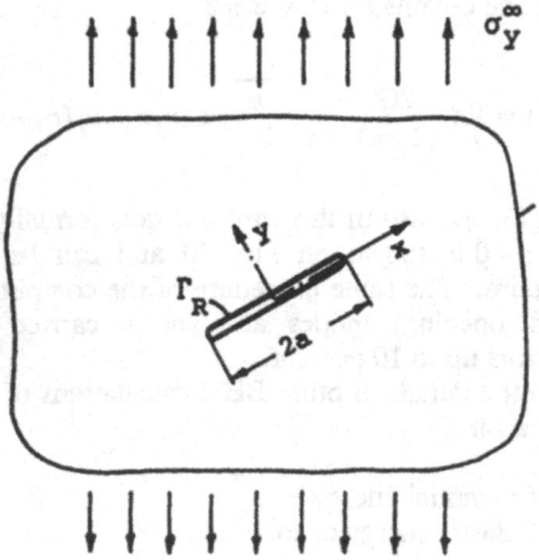

Fig. 31 **GRIFFITH-crack under tension**

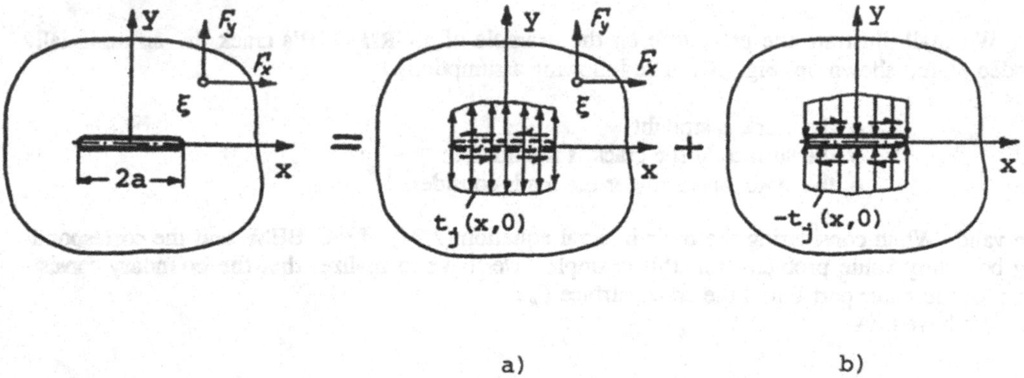

Fig. 32 **GREEN's solution**

The unit load is applied in an arbitrary source point ξ of the infinite plate, containing a tractionsfree crack of the same geometry as the crack in the finite disc under consideration. The derivation and inplementation of the GREEN's functions is conducted in complex variables. We shall restrain here from giving explicit expressions of the formulae, indicating formally the kernel functions and referring to the corresponding literature, see [51]. We obtain thus

$$T_{ij}^*(\xi,x) = T_{ij}^*[\xi(z,\bar{z}), x(z,\bar{z})] \; ; \quad U_{ij}^*(\xi,x) = U_{ij}^*[\xi(z,\bar{z}), x(z,\bar{z})] \; , \tag{3.16}$$

where

$$T_{ij}^*(\xi,x) = 0 \quad on \; \Gamma_R \; . \tag{3.17}$$

In view of the boundary condition (3.17) and the tractionsless crack with $t_j = 0$ on Γ_R, the whole crack boundary integral expression in Eq. (3.15) vanishes. Thus, without any explicit discretization of the crack boundary it becomes possible to determine the whole state of stress and strain.

The stress intensity factor can be computed now by including the procedure of internal point evaluation. We need, however, to investigate the limit transition of the internal source point to the tip of the crack. There are two ways to perform this evaluation. **SNYDER** and **CRUSE** [42] have determined the stress intensity factors from the stress field. **MEWS** [51] has obtained more simple expressions by computing the stress intensity factor from the displacement field. Comparative computations yield approximately the same degree of accuracy for the two methods.

Similarly as in chapter 2.3.3, Eq. (2.57), we get the components of the stress state in the internal point from the following equation

$$\sigma_{ij}(\xi) = \int_\Gamma [U_{ijk}^*(\xi,x) t_k(x) - T_{ijk}^*(\xi,x) u_k(x)] d\Gamma(x) \; , \tag{3.18}$$

with new kernel functions U_{ijk}^* and T_{ijk}^*, due to the application of the **GREEN**'s functions. In the vicinity of the crack tip one can present these kernels in the form

$$U_{ijk}^* = \frac{1}{\sqrt{r}} U_{ijk}^{**} \; ; \quad T_{ijk}^* = \frac{1}{\sqrt{r}} T_{ijk}^{**} \; , \tag{3.19}$$

so that the $1/\sqrt{r}$-singularity can be analytically presented as a multiplier. We can write now the Eq. (3.18) for the near field at the crack tip as

$$\tilde{\sigma}_{ij}(\xi_0) = \lim_{r\to 0} \sqrt{r}\,\sigma_{ij}(\xi) = \lim_{r\to 0} \int_\Gamma [U_{ijk}^{**}(\xi,x) t_k(x) - T_{ijk}^{**}(\xi,x) u_k(x)] d\Gamma(x) \; . \tag{3.20}$$

The stress intensity factors follow from the formula (3.3) for $\Theta = 0$ and at $r \to 0$, i.e. at $\xi \to \xi_0$, in the following form

$$K_I = \lim_{r\to 0} (\sqrt{2\pi r}\; \sigma_{yy}(\xi)) = \sqrt{2\pi}\, \tilde{\sigma}_{yy}(\xi_0) \; ,$$
$$K_{II} = \lim_{r\to 0} (\sqrt{2\pi r}\; \sigma_{xy}(\xi)) = \sqrt{2\pi}\, \tilde{\sigma}_{xy}(\xi_0) \; . \tag{3.21}$$

Substituting correspondingly the expressions (3.20), we can write the formulae for the K-factors as

$$K_I = \sqrt{2\pi} \int_\Gamma [U_{yyk}{}^{**}(\xi_0,x)t_k(x) - T_{yyk}{}^{**}(\xi_0,x)u_k(x)]d\Gamma(x) \,,$$

$$K_{II} = \sqrt{2\pi} \int_\Gamma [U_{xyk}{}^{**}(\xi_0,x)t_k(x) - T_{xyk}{}^{**}(\xi_0,x)u_k(x)]d\Gamma(x) \,, \tag{3.21}$$

They express the stress intensities as simple closed-contour integrals of known functions determined on the external boundary. In the procedure proposed by **MEWS** [51] one begins with the displacement field in the vicinity of the crack. After substracting the displacement of the crack tip itself, one obtains the field of the displacements towards the crack tip. It serves further for the separation of the multiplier $\sqrt{r}$. The formulae for the K-factors follow now from the transition $r \to 0$ in the form of simple closed contour integrals over known boundary functions.

3.3.3 MULTIPLE-CRACK PROBLEMS

The procedure proposed by **SNYDER** and **CRUSE** has been generalized by **KUHN** [43] for the case of a plate with multiple cracks, see Fig. 33.

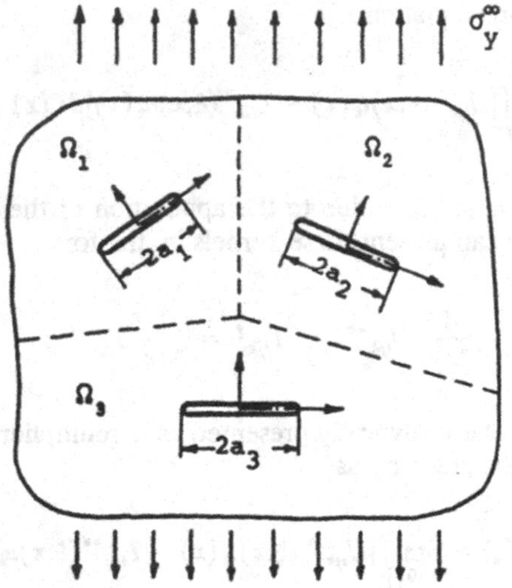

Fig. 33 Plate with multiple cracks

With respect to the substructure technique one can introduce namely in each of the subregions a corresponding **GREEN**'s function. Decomposing the problem into the subelements containing single cracks, one can apply **SNYDER** and **CRUSE**'s procedure within each of the subregions. In the first step of the computation one needs then to discretize the outer boundary and the division lines of the subelements only. The computation of the K-factors from the formulae (3.22) is performed then in the following steps. Here, the integration contour consists each time of the respective elements boundary. Still, the condition of the opening of all the cracks under the given external load must be observed, otherwise the assumption of a tractionfree crack surface would be violated. This point of view has been considered by **MEWS** [51]. In his procedure, the crack opening is being checked at each step of the computation by applying the internal point evaluation technique. If, at some point, calculations show, that there is contact between the opposite crack surfaces, an iterativ procedure is included in order to take into account the changing boundary conditions.

3.4 EXAMPLES

The following examples have been computed by means of the **SIFCA** code, [51].

Example 3.1

As the first numerical test let us consider the infinite plate with a crack oriented under the angle α to the direction of the external tensile load, as shown on Fig. 34. The numerical model is indicated here with the dashed line, while the complete data set is given on the right-hand side of the figure. The computation has been performed for 40 boundary elements.

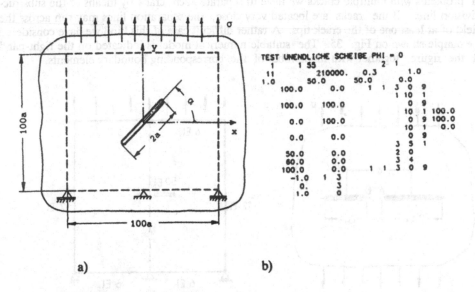

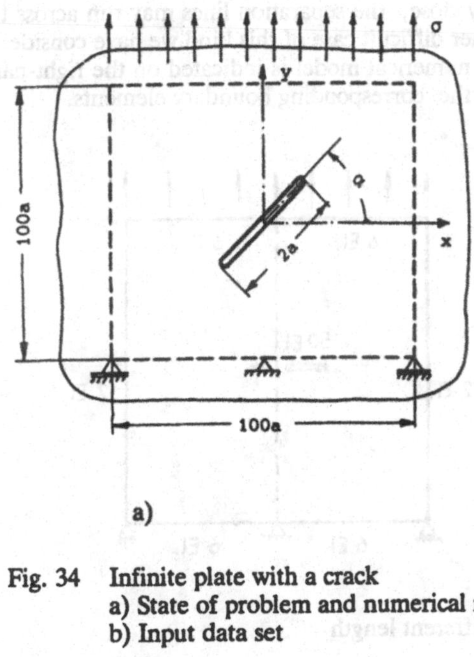

a)

```
TEST UNENDLICHE SCHEIBE PHI = 0
  1      1 55                       2  1
 11          210000.   0.3            1.0
1.0         50.0       50.0         0.0
    100.0        0.0       1   1   3   0   9
                                   1  10   1
    100.0     100.0                    0   9
                                      10   1  100.0
      0.0     100.0                    0   9  100.0
                                      10   1    0.0
      0.0       0.0                    0   9
                                   3   5   1
     50.0       0.0                2   0
     60.0       0.0                3   4
    100.0       0.0        1   1   3   0   9
     -1.0    1   3
       0.        3
      1.0        0
```

b)

Fig. 34 Infinite plate with a crack
a) State of problem and numerical model
b) Input data set

In Table 3.1 the results of the computations are given for two different values of the angle α. The comparative values of the stress intensity factors have been taken from [40].

It is of particular interest to point out that in the procedure considered here we need to change only one parameter in the data set in order to take into account the change of the orientation angle.

$K_I/(\sqrt{\pi a}\,\sigma)$ α	comparative values		calculated values			
	$K_I/(\sqrt{\pi a}\,\sigma)$	$K_{II}/(\sqrt{\pi a}\,\sigma)$	$K_I/(\sqrt{\pi a}\,\sigma)$	%	$K_{II}/(\sqrt{\pi a}\,\sigma)$	%
0°	1.0	-	1.0002	0.02	-	-
45°	0.5	0.5	0.5002	0.04	0.5001	0.02

Table 3.1 Stress intensity factors

Example 3.2

In problems with multiple cracks we have to separate each crack by means of the substructure division lines. If the cracks are located very close, the separation lines may run across the near field of at least one of the crack tips. A rather difficult case of this kind we have considered in the example shown on Fig. 35. The suitable numerical model is indicated on the right-hand side of the figure, together with the number of the corresponding boundary elements.

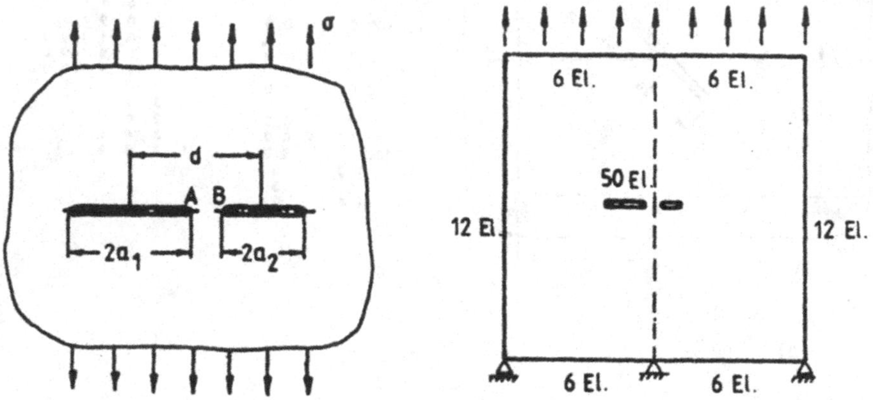

Fig. 35 Infinite plate with two cracks of different length

Three different values of the distance between the cracks have been considered. The K-factors corresponding to the crack tips A and B have been related to the analytical results, known from the literature [40]. The results have been summarized in Table 3.2. The largest error of the computations is equal to 1.7 percent.

$\frac{a_1 + a_2}{d}$	comparative values		calculated values			
	K_{IA}	K_{IB}	K_{IA}	%	K_{IB}	%
0.7	269.46	212.7	269.3	0.06	213.17	0.22
0.8	284.0	235.56	284.32	0.11	236.1	0.23
0.9	318.59	280.58	320.6	0.64	285.19	1.64

Table 3.2 Stress intensity factors

4. REFERENCES

[1] Fredholm, I., Sur une classe d'equations fonctionelles, Acta Math. 27, 365-390 (1903).

[2] Muskhelishvili, N.I., Some Basic Problems of the Mathematical Theory of Elasticity, Noordhoff, Groningen, 1953.

[3] Mikhlin, S.G., Integral Equations, Pergamon, New York, 1957.

[4] Smirnov, V.J., Integral equations and partial differential equations, in A Course in Higher Mathematics, Vol. IV, Addison-Wesley, London, 1964.

[5] Kupradze, O.D., Potential Methods in the Theory of Elasticity, Daniel Davey & Co., New York, 1965.

[6] Kellog, O.D., Foundations of Potential Theory, Dover, New York, 1953.

[7] Jaswon, M.A., Integral equation methods in potential theory, I, Proc. Roy. Soc. Ser. A 275, 23-32 (1963).

[8] Symm, G.T., Integral equation methods in potential theory, II, Proc. Roy. Soc. Ser. A 275, 33-46 (1963).

[9] Massonnet, C.E., Numerical Use of Integral Procedures, in Stress Analysis (O.C. Zienkiewicz and G.S. Holister, Eds.), Wiley, London, 1966.

[10] Rizzo, R.J., An integral equation approach to boundary value problems of classical elastostatics, Q. appl. Math. 25, 83-95 (1967).

[11] Cruse, T.A., and Rizzo, F.J., A direct formulation and numerical solution of the general transient elasto-dynamic problem, I, J. Math. Anal. Appl. 22, 244-259 (1968).

[12] Brebbia, C.A., Telles, J.C.F. and Wrobel, L.C., Boundary Element Techniques - Theory and Applications in Engineering, Springer Verlag, Berlin, Heidelberg, New York, Tokyo, 1984.

[13] Zabreyko, P.P., et al., Integral Equations - A Reference Text, Noordhoff, Amsterdam, 1975.

[14] Riccardella, P.C., An implementation of the boundary integral technique for planar problems in elasticity and elasto-plasticity, Report No. SM-73-10, Dept. Mech. Engng., Carnegie Mellon Univ., Pittsburg, 1973.

[15] Hartmann, F., Computing the C-matrix in non-smooth boundary points, in New Developments in Boundary Element Methods (C.A. Brebbia, Ed.), pp. 367-379, Butterworths, London, 1980, CML Southampton, 1983.

[16] Lachat, J.C., A further development of the boundary integral technique for elastostatics, Ph.D. Thesis, University of Southampton, 1975.

[17] Cruse, T.A., Mathematical foundations of the boundary integral equation method in solid mechanics, Report No. AFOSR-TR-77-1002, Pratt and Whitney Aircraft Group, 1977.

[18] Cruse, T.A., and Vanburen, W., Three-dimensional elastic stress analysis of a fracture specimen with an edge crack, Int. J. Fracture Mech. 7, 1-15 (1971).

[19] Chaudonneret, M., On the discontinuity of the stress vector in the boundary integral equation method for elastic analysis, in Recent Advances in Boundary Element Methods (C.A. Brebbia, Ed.), pp. 185-194, Pentech Press, London, 1986.

[20] Brebbia, C.A., The Boundary Element Method for Engineers, Pentech Press, London, Halstead Press, New York, 1978 (second edition, 1980).

[21] Danson, D., BEASY (Boundary Element Analysis System), Computational Mechanics Centre, Manual: Southampton, U.K., 1982.

[22] Kuhn, G., and Möhrmann, W., Boundary element method in elastostatics: theory and application, Appl. Math. Modelling, 1983, Vol. 7.

[23] Drexler, W., Ein Beitrag zur Lösung rotationssymmetrischer, thermoelastischer Kerbprobleme mittels der Randintegralgleichungsmethode, Diss. Techn. Universität München, 1982.

[24] Neureiter, W., Boundary-Element-Programmrealisierung zur Lösung von zwei- und dreidimensionalen thermoelastischen Problemen mit Volumenkräften, Diss. Techn. Universität München, 1982.

[25] Radaj, D., Möhrmann, W. and Schilberth, G., Economy Convergence of Notch Stress Analysis Using Boundary and Finite Element Methods. Int. J. Numerical Methods Engng. 20, 565-572, 1984.

[26] Mayr, M., Drexler, W. and Kuhn, G., A semianalytical boundary integral approach for axisymmetric elastic bodies with arbitrary boundary conditions, Int. J. Solids Struct., 1980, 16, 863.

[27] Stippes, M., and Rizzo, F.J., A Note on the body force integral of classical elastostatics, Z. Angew. Math. Phys. 28, 339-341 (1977).

[28] Rizzo, F.J., and Shippy, D.J., An advanced boundary integral equation method for three dimensional thermoelasticity, Int. J. Numerical Methods Engng. 11, 1753-1768, 1977.

[29] Papkovitch, P.F., Solution generale des equations d'elasticite, C.R. Acad. Sci. Paris 1932, 195.

[30] Neuber, H., Ein neuer Ansatz zur Lösung räumlicher Probleme der Elastizitätstheorie, Der Hohlkegel unter Einzellast als Beispiel, ZAMM 1934, 14, 203.

[31] Cruse, T.A., Snow, D.W., and Wilson, R.B., Numerical solutions in axisymmetric elasticity, Comput. Structures 7, 445-451, (1977).

[32] Shippy, D.J., Rizzo, F.J., and Nigan, R.K., A boundary integral equation method for axisymmetric elastostatic bodies under arbitrary surface loads, in Proc. 2nd Int. Symp. on Innovative Numerical Analysis in Appl. Engng. Sci. (R.P. Snow et al., Eds.), University of Virginia Press, Charlottesville, 1980.

[33] Mayr, M. und Neureiter, W., Ein numerisches Verfahren zur Lösung des axialsymmetrischen Torsionsproblems, Ingenieur-Archiv 46, 137-142 (1977).

[34] Mayr, M., On the numerical solution of axisymmetric elasticity problems using an integral equation approach, Mech. Res. Com. 3, 393-398 (1976).

[35] Neuber, H., Kerbspannungslehre, Grundlagen für genaue Festigkeitsberechnung mit Berücksichtigung von Konstruktionsform und Werkstoff, 2. Auflage, Springer Verlag, Berlin 1957.

[36] Jaswon, M.A. and Symm, G.T., Integral Equation Methods in Potential Theory and Elastostatics, Academic Press, London, 1977.

[37] Christiansen, S., Numerical Treatment of an Integral Equation originating from a two-dimensional Dirichlet Boundary Value Problem, in: Abrecht, J. and Collatz, L., (Eds.), Numerische Behandlung von Integralgleichungen, Birkhäuser Verlag, 1979.

[38] Kuhn, G., Löbel, G., Potrc, I., Kritisches Lösungsverhalten der direkten Randelementmethode bei logarithmischem Kern, ZAMM 67, 1987.

[39] Tada, H., Paris, P. and Irwin, G.R., The Stress Analysis of Crack Handbook, DEL Research Corporation, Hellertown, Pennsylvania, 1973.

[40] Sih, G.C., Handbook of Stress Intensity Factors, Institut of Fracture and Solid Mechanics, Lehigh University, 1973.

[41] Rooke, D.P. and Cartwright, D.J., Compendium of Stress Intensity Factors, Her Majesty's Stationary Office, London, 1975.

[42] Snyder, M.D. and Cruse, T.A., Boundary integral equation analysis of cracked anisotropic plates, Int. Journal of Fracture, 11, 315-328, 1975.

[43] Kuhn, G., Numerische Behandlung von Mehrfachrissen in ebenen Scheiben, ZAMM 61, 1981.

[44] Blandfort, G.E., Ingraffea, A.R. and Liggett, J.A., Two-dimensional stress intensity factor computations using the boundary element method, Int. Journ. Num. Meth. Eng., 17, 387-404, 1981.

[45] Sih, G.C. and Liebowitz, H., Mathematical theories of brittle fracture, in Fracture: An Advanced Treatise, 2, (Ed. H. Liebowitz), Academic Press, N.Y., 67-190, 1968.

[46] Broek, D., Elementary Engineering Fracture Mechanics, Sijthoff and Noordhoff, 1978.

[47] Westergaard, H.W., Bearing Pressures and Cracks, J. Appl. Mech., 6, 1939.

[48] Griffith, A.A., The phenomena of rupture and flow in solids, Phil. Trans. Royal Soc., London, 163-198, 1921.

[49] Rice, J.R., A Path Independent Integral and the Approximate Analysis of Strain Concentration by Notches and Cracks, J. Appl. Mech., 35, 17-26, 1968.

[50] Erdogan, F., On the stress distribution in plates with colinear cuts under arbitrary loads, Proceedings of the 4. U.S. National Congress of Appl. Mech., 1962.

[51] Mews, H., Berechnung von K-Faktoren mittels der Boundary Element Methode, Diss. Universität Erlangen-Nürnberg (erscheint).

NUMERICAL TREATMENT
OF CORNER AND CRACK SINGULARITIES

H. Blum
Universität des Saarlandes, Saarbrücken, F.R.G.

ABSTRACT

The accuracy of standard finite element and boundary element methods for elliptic boundary value problems on polygonal or polyhedral domains is significantly reduced due to the influence of corner singularities. Based on the validity of a singular decomposition of the solution, this pollution effect is described, with respect to different norms. Several procedures are proposed and analyzed to recover the full order of convergence for the Galerkin solution, as well as for the stress intensity factors.

1. INTRODUCTION

Optimal order error estimates for finite element or boundary element approximations for elliptic boundary value problems essentially depend on the differentiability properties of the solution and on the validity of optimal a priori estimates. These assumptions are typically satisfied for elliptic problems

$$Lu = f \ , \text{ in } \Omega \ , \quad Bu = b \ , \text{ on } \partial\Omega \ , \tag{1.1}$$

on smoothly bounded domains Ω if the coefficients of the differential operators L and B are sufficiently regular. However, in the presence of irregular boundary points z_n , $n = 1,\dots N$, of the domain Ω such as corners or cracks and/or in points of changing boundary conditions, the solution u of (1.1) contains local singularities. In linear problems, u admits a decomposition of the form

$$u = \sum_{n=1}^{N} \sum_{i=1}^{I_n} K_{n;i} s_{n;i} + U \ , \tag{1.2}$$

where $s_{n;i} \sim r_n^{\alpha_{n;i}}$ are certain "singular functions" , r_n denotes the distance to the corner z_n , and U is a regular remainder. The number I_n of singularities which may be split from u , at z_n , depends on the smoothness of the data. The coefficients $K_{n;i}$ which are sometimes called *stress intensity factors* are nontrivial functionals of f and b . For the following, let us assume that there exists only one critical point , at $z = 0$, and let us omit the index n .

The presence of the corner singularity s_1 naturally leads to bad approximations near the irregular point z , if α_1 is small. Further, the error is "polluted" by the singular function all over the domain, even if it is measured in arbitrary weak norms. Therefore, it is also impossible to extract accurate information from the approximate solutions by means of a suitable postprocessing, e.g. for the coefficients $K_{n;i}$ in (1.2).

In order to overcome these difficulties several modifications of the standard discretization methods have been proposed in the literature. The best known technique is the method of mesh refinements which is generally applicable as soon as estimates are known for the exponent α_1 of the leading singularity near a critical point. However, in many applications, including some nonlinear equations, more detailed information about the structure of the solution is available, for instance,

- Singular decomposition of the solution of the type (1.2), explicit or approximate knowledge of the singular functions.

- Representation formulas for the stress intensity factors, $K_i = K_i(f,b,u)$.
- Information about the asymptotic behavior of the Green function or special fundamental solutions.

These theoretical results can efficiently be used for the construction of improved approximation procedures, and for the numerical calculation of the stress intensity factors. The analysis and comparison of some of these modifications is the main purpose of the following sections. All results are presented here only for the two-dimensional case although most of them carry over to three space dimensions as soon as a singular expansion of the type (1.2) and formulas for coefficients can be made available for them, see also [23].
The paper is organized as follows.

In Section 2 some results by Kondrat'ev [1] and Maz'ja-Plamenevskij [2] are reviewed who have shown singular expansions of the type (1.2) near conical boundary points for general linear elliptic systems. Moreover, there has been shown an integral representation formula $K_i = K_i(f,b)$ which is useful for the numerical approximations as well as for the error analysis of discretization procedures.

In Section 3, we describe the asymptotic error orders and the pollution effect for the standard finite element method. For the usual Ritz projection $\mathcal{R}_h u$ of u for second order problems the rate of convergence is reduced to

$$\|u-\mathcal{R}_h u\|_{L_2} + h^{\alpha_1}\|\nabla(u-\mathcal{R}_h u)\|_{L_2} + h^{\alpha_1}\|u-\mathcal{R}_h u\|_{L_\infty} = O(h^{2\alpha_1}) , \quad (1.3)$$

if $\alpha_1 < 1$. Away from the irregular points, however, the pointwise error behaves asymptotically like the error with respect to the L_2-norm. The situation is even worse if nonstandard methods like mixed finite element methods for fourth order problems are used. Here, the presence of the singular functions may completely destroy the convergence of the method or leads to an approximation of a wrong limit function.

In Section 4 several modifications of the finite element method for linear problems are presented to recover the best possible order of approximation. First, we shall briefly discuss some results on mesh refinements. Our main subject, however, is the description of techniques employing augmented trial and test spaces.

1. The singular function method (Fix method, SFM) [3].
The SFM makes use of trial and test spaces which are augmented by several singular functions. This leads to optimal mean square convergence rates for the solution. The convergence of the stress intensity factors, however, is sometimes very poor. They usually must be corrected by a suitable postprocessing.

2. The dual singular function method (DSFM) [4].
Here again the trial space is augmented. In addition, the exact representation $K_i = K_i(f,b)$ is used which involves the singular functions of the dual problem (L_*,B_*) . This method gives optimal pointwise results since it behaves like the usual finite element approximation of the smooth remainder and it approximates the coefficients K_i with the best possible accuracy.
3. A general iterative correction method.
It may be considered as a slight disadvantage of the DSFM that the evaluation of $K_i(f,b)$ is sometimes rather complicated, in particular for nonlinear problems. A variant of the method, recently proposed by Dobrowolski [5], makes use of approximate representation formulas $K_i^j(f,b)$, $j=1,2,\dots$, to define an iterative correction procedure for the solution u . These formulas only need the values of the approximate solution and the explicit form of the singular functions. Because of its simple structure the method seems to be suitable also for the calculation of stress intensity factors near edges in three dimensional problems where the SFM and the DSFM appear to be too complicated. It should be noticed that all the modifications above can be implemented such that only linear systems for the original stiffness matrix have to be solved.

Finally, we discuss the error behavior on piecewise uniform grids. Here, asymptotic expansions with respect to fractional powers of the discretization parameter can be shown for the error of the singular functions. This makes it possible to remove the pollution effect by means of Richardson extrapolation or related defect correction procedures. This technique is most efficient if several singularities are present with the same exponents as is the case in problems with many cracks or along edges in three space dimensions. Here, the influence of all singularities can be removed by only one step of extrapolation. Further, it is possible to handle the effects from several corners simultaneously.

The final Section 5 contains some remarks on the numerical treatment of corner singularites in the boundary element method. We concentrate on Galerkin approximations for which the error analysis in mean square norms is similar to that for the finite element method, see, e.g., [24]. The key to the error analysis is a Gårding inequality for the integral operators which must hold independent of the smoothness of the boundary. It turns out that integral equations of the first kind should be prefered as the basis for the boundary element method since for these such results are available on general Lipschitz domains [31].

For an improvement of convergence, up to now, mostly mesh refinements have been used. Moreover, the SFM has been analyzed and implemented, e.g., for several problems in 2D- and 3D- elasticity and mixed boundary value problems for the Laplacian, see, e.g., [6],[7],[30]. It is shown that also the DSFM can be applied in connection with boundary element methods. Here, a representation formula for the K_i can be taken which only involves the Cauchy data on the boundary and, therefore, is well-suited for the application in direct methods.

As a further modification of the boundary element method we mention the application of special fundamental solutions which satisfy homogeneous boundary conditions near the corner point. By this, the Cauchy data must be calculated only away from the singularity and are thus obtained with optimal accuracy. Clearly, such special fundamental solutions are available only in very specific situations. For more details and numerical calculations for crack problems in 2D- elasticity based on this approach we refer to [33].

Most of the techniques described in Section 4 only for the linear case carry over to nonlinear problems as soon as the validity of the theoretical basis (1.2) is ensured. In semilinear equations, containing only lower order nonlinear terms, the leading singularities are determined only by the linear principal part and also the stress intensity factors can be calculated by similar formulas [18],[34]. For quasilinear problems, however, the results are much less complete. For equations of the p-harmonic type,

$$-\operatorname{div}((1+|\nabla u|^2)^{p/2-1}\nabla u) = f , \quad 1 < p < \infty ,$$

an expansion of the form

$$u = Ks + U$$

has been found in [35], see also [8]. The representation formulas for K, which are now given by nonlinear functions $K = F(f,u,U,K)$, can again be used for an iterative correction of the solution, but the evaluation is sometimes quite complicated [9].

2. SINGULAR DECOMPOSITION OF THE SOLUTION

As a first model problem we consider the mixed boundary value problem for the Laplacian in a plane polygonal domain $\Omega \subset \mathbb{R}^2$ with boundary $\partial\Omega$,

$$\begin{aligned} -\Delta u &= f , \text{ in } \Omega , \\ u &= 0 , \text{ on } \Gamma_u , \\ \frac{\partial u}{\partial n} &= 0 , \text{ on } \Gamma_\sigma . \end{aligned} \tag{2.1}$$

Here, $\Gamma_u \cup \Gamma_\sigma = \partial\Omega$, $\Delta u = u_{xx}+u_{yy}$, and $\frac{\partial}{\partial n}$ is the directional derivative with respect to the outer normal vector. In the following, $L_p(\Omega)$ denotes the usual Lebesgue spaces and $H^k(\Omega) = H^{k,2}(\Omega)$ are the Sobolev spaces of square integrable functions with (generalized) derivatives up to the order k in $L_2(\Omega)$.

It is well known that the weak solution $u \in V$ of

$$a(u,\varphi) = \int_\Omega \nabla u \nabla\varphi \, dx = \int_\Omega f\varphi \, dx \ , \quad \forall \ \varphi \in V \ , \qquad (2.2)$$

has bounded derivatives of arbitrary order in the interior of Ω and along smooth parts of the boundary as long as f is sufficiently smooth. Here, $V = \{v \in H^1(\Omega) | \ v=0 \text{ on } \Gamma_u\}$ and $\int u dx = \int f dx = 0$ is required for $\Gamma_u = \emptyset$. Near corner points, or points where the boundary conditions change the derivatives of u may become singular.

We introduce polar coordinates (r,φ) around an irregular point z and denote by $\tau = \tau(r)$ a smooth cut-off function, $\tau = 1$ $(r \le r_0)$, $\tau = 0 \ (r \ge r_1)$, $0 < r_0 < r_1$. Naturally, the singular behavior of u near z coincides with that of τu on an infinite sector C , with the same interior angle ω and vertex z .

We have to distinguish between two classes of singular components. First, there are singularities which are completely determined by local properties of the data near z . As a simple example we consider the Dirichlet problem, $\Gamma_\sigma = \emptyset$, for the right hand side $f = 1$. Then, for $\omega \neq \pi/2$, the polynomial $s_0 = x(y-\tan\omega \ x)/(2\tan\omega)$ solves (2.1) on the infinite cone C . For $\omega = \pi/2$, however, the solution is

$$s_0 = \pi^{-1} r^2 (\ln(r)\sin 2\varphi + \varphi\cos 2\varphi + (\pi/2)\sin^2\varphi) \ , \qquad (2.3)$$

the second order derivatives of which become singular at $r = 0$. We see that this function will occur in the weak solution u , for smooth data f , if and only if $f(z) \neq 0$. In general, singularities of this first type can be removed by local modification of the data, i.e. by subtracting particular solutions.

The second type of singular functions is much more important as they usually influence the global behavior of the discretization error. The functions are multiples $K\bar{s}$ of the functions

$$\bar{s} = r^\alpha t_\alpha(\varphi) \ , \qquad (2.4)$$

which solve the homogeneous equation on the infinite cone C with homogeneous boundary data. The coefficients K are determined as functions of the data, see below. These singularities may be part of the solution u even if f is smooth and vanishes identically near z . Since

$$\Delta(r^\alpha t_\alpha(\varphi)) = r^{\alpha-2}(t''_\alpha(\varphi) + \alpha^2 t_\alpha(\varphi)) \ , \qquad (2.5)$$

the possible values α and the corresponding angular parts are determined as the solutions of the one dimensional eigenvalue problem

$$L(\alpha,\partial_\varphi) = t''_\alpha(\varphi) + \alpha^2 t_\alpha(\varphi) = 0 \ , \text{ in } (0,\omega) \ , \qquad (2.6)$$

subject to the boundary conditions

$$\begin{aligned} t_\alpha(0) &= t_\alpha(\omega) = 0 \text{ , in the case (u/u),} \\ t_\alpha(0) &= t'_\alpha(\omega) = 0 \text{ , in the case } (u/\sigma), \\ t'_\alpha(0) &= t'_\alpha(\omega) = 0 \text{ , in the case } (\sigma,\sigma). \end{aligned} \tag{2.7}$$

The three cases (u/u), (u/σ), and (σ/σ) refer to the combination of boundary conditions on the adjacent edges of the corner. The corresponding eigensolutions are

$$\begin{aligned} &\alpha = m\pi/\omega &&, t_\alpha(\varphi) = \sin(m\pi\varphi/\omega) &&\text{, for (u/u),} \\ &\alpha = (m-1/2)\pi/\omega &&, t_\alpha(\varphi) = \sin((m-1/2)\pi\varphi/\omega) &&\text{, for } (u/\sigma), \\ &\alpha = m\pi/\omega &&, t_\alpha(\varphi) = \cos(m\pi\varphi/\omega) &&\text{, for } (\sigma/\sigma), \end{aligned} \tag{2.8}$$

for $m \in \mathbb{Z}$. Clearly, only for $\alpha > 0$, the functions $\bar{s}$ have bounded energy, $\|\nabla s\|_\Omega < \infty$. This excludes negative values of m for possible singular contributions of the weak solution u .

As a second example, we consider the biharmonic boundary value problem arising from the Kirchhoff plate theory,

$$\begin{aligned} \Delta^2 u &= p/D \quad (= f) \text{ ,} && \text{in } \Omega \text{ ,} \\ u &= \frac{\partial u}{\partial n} && = 0 \text{ , on } \Gamma_c \text{ ,} \\ u &= n\cdot M\cdot n && = 0 \text{ , on } \Gamma_s \text{ ,} \\ n\cdot M\cdot n &= \partial_s(n\cdot M\cdot t) + \nabla\cdot M\cdot n && = 0 \text{ , on } \Gamma_f \text{ .} \end{aligned} \tag{2.9}$$

Here, u describes the deflection of a thin elastic plate which is clamped on Γ_c , simply supported on Γ_s , and free on Γ_f . M is the tensor of bending moments and p denotes the vertical force density. The singular functions for problem (2.9) are of the form

$$\bar{s} = r^{1+\alpha} t_\alpha(\varphi) \text{ ,} \tag{2.10}$$

satisfying $\Delta^2 \bar{s} = 0$, and homogeneous boundary conditions. Again, the equation $\Delta^2(r^{1+\alpha} t_\alpha(\varphi)) = 0$ results in an eigenvalue problem for the angular part

$$L(\alpha,\partial_\varphi)t_\alpha = \partial^4_\varphi t_\alpha + [(1+\alpha)^2 + (1-\alpha)^2]\partial^2_\varphi t_\alpha + (1+\alpha)^2(1-\alpha)^2 t_\alpha = 0 \text{ ,} \tag{2.11}$$

and the corresponding boundary conditions $B(\alpha,\partial_\varphi)t_\alpha = 0$. Besides the singular functions of type (2.10), also solutions of the form

$$\bar{s} = r^{1+\alpha}(\ln(r)t_{\alpha 1} + ct_{\alpha 2}) \tag{2.12}$$

are possible for exceptional inner angles ω. For a dicussion of the possible exponents α for all combinations of boundary conditions, we refer to [10].

The theoretical justification for the restriction to the two types of singular functions mentioned above has been given by Kondrat'ev [1]. His results can be summarized as follows.

If L and B are *homogeneous* differential operators with constant coefficients of order $2m$ and m_j, $j=1,\dots,m$, respectively, the weak solution u of the problem

$$Lu = f, \text{ in } \Omega, \quad Bu = b, \text{ on } \partial\Omega, \tag{2.13}$$

around an irregular boundary point z admits the singular decomposition

$$u = \sum_{i=1}^{I} \sum_{l=1}^{M(\alpha_i)} K_{il}s_{il} + U, \tag{2.14}$$

provided that the data (f,b) are sufficiently smooth in the interior and, at the corner, satisfy certain growth conditions specified below. Here, $s_{il} = \tau\bar{s}_{il}$, and $\bar{s}_{il}$ are the solutions of the homogeneous problem of the form

$$\bar{s}_{il} = r^{m-1+\alpha_i} \sum_{j=0}^{l-1} c_{ij}^{l}(\ln r)^j t_{i,l-j}, \quad l = 1,\dots,M(\alpha_i), \tag{2.15}$$

where c_{ij}^{l} are certain constants and $M(\alpha_i) \geq 1$. For most applications we find $M(\alpha_i) = 1$ such that no logarithmic terms appear in (2.15), and (2.14) reduces to

$$u = \sum_{i=1}^{I} K_i s_i + U. \tag{2.16}$$

The differentiabilty properties of the "smooth" remainder U, around z, depend on the choice of I, i.e. on the number of singularities removed from u. It is implicitly understood that the regularity of s_i is increasing for larger values of i, $\alpha_i \leq \alpha_j$, for $i \leq j$.

For problems with variable coefficients and/or lower order derivatives in (L,B) the leading singularities are the same, i.e., they are completely determined by the principal part. As an example we consider the equation

$$-\Delta u + u = f, \text{ in } \Omega,$$

subject to homogeneous Dirichlet conditions. By treating the lower order term as part of the right hand side, we see that each singularity Ks, in u, induces a higher order singular function $\tilde{s}$, namely the solution of

$$-\Delta\tilde{s} = -Ks \text{ , in } \Omega \text{ ,}$$

independent of the smoothness of f. For example, on a slit domain, $\omega = 2\pi$, we have $\tilde{s} = -1/6\, r^{5/2}\sin(\varphi/2)$ if we take leading the singular function $s_1 = r^{1/2}\sin(\varphi/2)$ as right hand side. These additional singular parts of the solution can often be determined analytically, once the expansion (2.16) is known, see [1]. Therefore, in the following, we shall restrict ourselves to homogeneous differential operators.

The results of Kondrat'ev have been generalized to linear elliptic systems by Maz'ja and Plamenevskij [2]. In addition, they have derived the following representation formula for the determination of the stress intensity factors K_i.

Let (L,B) satisfy a Green identity of the form

$$(Lu,v)_\Omega + (Bu,Sv)_{\partial\Omega} = (u,L_* v)_\Omega + (S_* u,B_* v)_{\partial\Omega} \text{ .} \tag{2.17}$$

Further, let $\bar{s}_{-i}$ denote the singular functions of the dual boundary value problem $L_*\bar{s}_{-i} = B_*\bar{s}_{-i} = 0$, on the infinite cone C, with exponent $m-1-\alpha_i$, normalized by the condition (2.19), below. Then, for the coefficients K_i in (2.16) there holds

$$K_i = (L(\tau u),\bar{s}_{-i})_C + [B(\tau u),S\bar{s}_{-i}]_{\partial C} \text{ .} \tag{2.18}$$

This representation is easily proved using the Green identity (2.17), separately on a finite cone C_1 and its complement $C_2 = C\backslash C_1$. The integrals on the right hand side exist if

$$(f,r^{m-1-\alpha_i})_\Omega < \infty \quad \text{and} \quad (b,Sr^{m-1-\alpha_i})_{\partial\Omega} < \infty \text{ .}$$

This specifies the growth conditions on the data, mentioned above, which guarantee the validity of (2.14) or (2.16), respectively. Below, we shall make use of the representation (2.18) in one of the following equivalent forms, on the bounded domain Ω.

<u>Theorem 2.1</u>: *The coefficients* K_i *in (2.16) satisfy the relations*

(i) $K_i = (f,\bar{s}_{-i})_\Omega + (b,S\bar{s}_{-i})_{\partial\Omega} - (S_* u,B_*\bar{s}_{-i})_{\partial\Omega}$.

(ii) $K_i = (f,s_{-i})_\Omega - (u,L_* s_{-i})_\Omega + (b,Ss_{-i})_{\partial\Omega} - (S_* u,B_* s_{-i})_{\partial\Omega}$.

(iii) $K_i = (f,s'_{-i})_\Omega + (b,Ss'_{-i})_{\partial\Omega}$.

Here, $s_{-i} = \tau\bar{s}_{-i}$ *and* $s'_{-i} = s_{-i} - v_i$ *, where* v_i *is the (weak) solution of the problem* $L_* v_i = L_* s_{-i}$ *, in* Ω *and* $B_* v_i = B_* s_{-i}$ *, on* $\partial\Omega$.

Notice that $L_* s_{-i}$ and $B_* s_{-i}$ are smooth data although s_{-i} is singular near z . Therefore, $v_i \neq s_{-i}$, and s'_{-i} are solutions of the homogeneous problem (2.13) with the same singular behavior near z as s_{-i} . All three representation formulas contain unknown functions, the solution u itself, in (i) and (ii), and the function v_i , in (iii). Nevertheless, they are of great importance for numerical applications, see Section 4, below. As a particular consequence of Theorem 2.1 we have the normalizing condition

$$(Ls_j,\bar{s}_{-i})_\Omega + (Bs_j,S\bar{s}_{-i})_{\partial\Omega} - (S_* s_j,B_*\bar{s}_{-i})_{\partial\Omega} = \delta_{ij} . \qquad (2.19)$$

Example: For selfadjoint problems, $L = L_*$, $B = B_*$, the functions s_{-i} are obtained from s_i , up to a normalizing factor, by reflecting the exponent at $m-1$ and taking the same angular part.

$L=-\Delta$, B=id: $\bar{s}_i = r^{i\pi/\omega}\sin(i\pi\varphi/\omega)$, $\bar{s}_{-i} = (i\pi)^{-1} r^{-i\pi/\omega}\sin(i\pi\varphi/\omega)$,

$L=\Delta^2$: $\bar{s}_i = r^{1+\alpha_i} t_i(\varphi)$, $\bar{s}_{-i} = c_i r^{1-\alpha_i} t_i(\varphi)$.

As a further consequence of Theorem 2.1 (iii) we have the following.

<u>Corollary 2.2</u>: *The functions* $\bar{s}_{-i}(x)$ *and* $s'_{-i}(x)$ *describe the i-th stress intensity factor of the Green function on* C *, and* Ω *, respectively, with source point at* x .

For the proof we simply observe that the Green function is the solution of problem (2.13), corresponding to homogeneous boundary conditions, $b = 0$, and right hand side $f = \delta_x$, the Dirac distribution. Choosing a sequence approximating δ_x in Theorem 2.1 (i), with Ω replaced by C , and (iii), respectively, we immediately obtain the desired result.

3. CONVERGENCE OF THE USUAL FINITE ELEMENT METHOD

In this section, we want to study the influence of corner singularities on the convergence behavior of standard finite element approximations. For simplicity, we assume that Ω is a polygonal domain with only one significant corner z, for which the singular parts in (2.16) destroy the validity of the desired a priori bounds.

By $T_h = \{K\}$ we denote a subdivision of Ω into elements of size h, $0 < h \leq 1$, which is regular in the usual sense, i.e. each element is contained in a circle of diameter $C_1 h$ and contains a circle of diameter $C_2 h$ for some constants C_1, C_2. In the following, c stands for a generic constant which may vary with the context and is independent of h, and of the solution u.

We first consider conformal approximations for problem (2.1). To this end, we define the finite element spaces

$$V_h = \{v_h \in C^0(\overline{\Omega}) \mid v_h|_K \in P \supset P_{m-1},\ v_h = 0 \text{ on } \Gamma_u,\ m \geq 2\}, \tag{3.1}$$

where P_{m-1} denotes the space of polynomials of degree less or equal $m-1$. For the natural interpolant $\mathcal{I}_h v \in V_h$ of a given function $v \in V$ there holds

$$\|\nabla^l (v - \mathcal{I}_h v)\|_{L_p(K)} \leq c h^{k-l} \|\nabla^k v\|_{L_p(K)}, \quad \begin{array}{l} 0 \leq l \leq k \leq m, \\ 1 \leq p \leq \infty, \end{array} \tag{3.2}$$

where ∇^l denotes the tensor of all derivatives of order l. Now, we define the Ritz projection $\mathcal{R}_h u \in V_h$ of the weak solution u of problem (2.2), by the relation

$$a(\mathcal{R}_h u, \varphi_h) = (\nabla \mathcal{R}_h u, \nabla \varphi_h) = (f, \varphi_h), \quad \forall\, \varphi_h \in V_h. \tag{3.3}$$

Using the projection property of $\mathcal{R}_h$, we immediately get the energy estimate

$$\|\nabla(u - \mathcal{R}_h u)\|_{L_2} = \inf_{\varphi_h \in V_h} \|\nabla(u - \varphi_h)\|_{L_2}. \tag{3.4}$$

The regularity of u is in general determined by the first singular function s_1. Therefore, in view of (3.4), we cannot expect a better estimate than

$$\|\nabla(u - \mathcal{R}_h u)\|_{L_2} \leq c K_1 \|\nabla(s_1 - \mathcal{R}_h s_1)\|_{L_2} \leq c K_1 \|\nabla(s_1 - \mathcal{I}_h s_1)\|_{L_2}. \tag{3.5}$$

In view of this estimate the error behavior of the usual finite element method is described by the following result. Here, and below, we use the abbreviation $\alpha = \alpha_1$.

Theorem 3.1: *For the leading singular function* $s_1 = r^\alpha t(\varphi)$, *with* $\alpha < 1$, *there hold the mean square estimates*

$$\|s_1-\mathcal{R}_h s_1\|_{L_2} + h^\alpha \|\nabla(s_1-\mathcal{R}_h s_1)\|_{L_2} \le ch^{2\alpha} ,$$

and the pointwise estimate

$$\|s_1-\mathcal{R}_h s_1\|_{L_\infty} \le ch^\alpha .$$

All estimates cannot be improved.

Thus, the order of convergence is significantly reduced compared, e.g., to the best possible order $O(h^{m-1})$ for the error in the energy norm for smooth solutions, and is independent of the approximation order m of the finite element spaces.

Proof. We give the proof only for the mean square estimates. The pointwise result is obtained, e.g., by means of the techniques in [29], see also Theorem 3.2, below. By the required regularity of the mesh, there exist $O(i)$ elements K_{ij} , $j=1,\dots,O(i)$, of the triangulation T_h such that $ih \le \mathrm{dist}(K_{ij},z) \le (i+1)h$, $i=1,\dots,O(1/h)$. For fixed i , we denote the union of those elements by A_i . Since $\nabla^2 s_1 \sim r^{\alpha-2}$, we immediately obtain the following estimate in view of (3.4) and the approximation property (3.2),

$$\begin{aligned}
\|\nabla(s_1-\mathcal{R}_h s_1)\|^2_{L_2(\Omega)} &\le \sum_{i=1}^{O(1/h)} \sum_{j=1}^{O(i)} \|\nabla(s_1-\mathcal{P}_h s_1)\|^2_{L_2(K_{ij})} + O(h^{2\alpha}) \\
&\le ch^2 \sum_{i=1}^{O(1/h)} \|\nabla^2 s_1\|^2_{L_2(A_i)} + O(h^{2\alpha}) \\
&\le ch^2 h^{2\alpha-2} \sum_{i=1}^{O(1/h)} i^{2\alpha-3} + O(h^{2\alpha}) .
\end{aligned}$$

The sum on the right is bounded uniformly for $h \to 0$, since $\alpha < 1$, which proves the desired energy estimate. The L_2-estimate is obtained via a standard duality argument. Let $v \in V$ denote the solution of the auxiliary problem

$$a(v,\varphi) = (s_1-\mathcal{R}_h s_1,\varphi) , \quad \forall \varphi \in V . \tag{3.6}$$

Then, by choosing $\varphi = s_1-\mathcal{R}_h s_1$, and using again the projection property of $\mathcal{R}_h$, we get

$$\|s_1-\mathcal{R}_h s_1\|_{L_2}^2 \lesssim \|\nabla(s_1-\mathcal{R}_h s_1)\|_{L_2}\|\nabla(v-\mathcal{I}_h v)\|_{L_2} . \tag{3.7}$$

Since the solution v will also contain the singular function s_1, this only gives us an additional power $O(h^\alpha)$ for the L_2-error. There remains to show the sharpness of the estimates. In view of $\nabla s_1 \sim r^{\alpha-1}$, the error for the gradient can not be of better order than $O(h^\alpha)$, in radial direction, for any degree $m-1$ of approximating polynomials. The sharpness of the L_2-estimate follows from the inequalities

$$ch^{2\alpha} \le \|\nabla(s_1-\mathcal{R}_h s_1)\|_{L_2}^2 = (-\Delta s_1, s_1-\mathcal{R}_h s_1) \le \|\Delta s_1\|_{L_2(\Omega_0)}\|s_1-\mathcal{R}_h s_1\|_{L_2(\Omega_0)} ,$$

where Ω_0 is the support of Δs_1. □

The last inequality indicates that the L_2-error is *polluted* all over the domain. Even in the subdomain Ω_0, away from the corner point z, the order of convergence is not of optimal order, although here s_1 is arbitrarily smooth. By arguments of the same type it is seen that $O(h^{2\alpha})$ is the best possible order of convergence even if the error is measured in arbitrary weak norms.

The strongest singularity for problem (2.1) occurs for slit domains, $\omega = 2\pi$, in corners of type (u,σ). Here, $\alpha = 1/4$, and the best possible error order is $O(h^{1/2})$. For the approximation of $-\Delta u = 1$ by piecewise linear elements on a slit unit square, relative errors in the interior of about 8% have been found in [17], even for a mesh size $h=1/80$. This indicates that usually bad asymptotic orders result in large absolute values for the error, for any reasonable mesh size h.

The situation described above is typical for all conformal finite element approximations. Let us, e.g., consider the Kirchhoff plate bending problem (2.9). Here, of course, we have to assume that the elements of V_h are in $C^1(\Omega)$, satisfying the essential boundary conditions. Then, Theorem 3.1 carries over, for $\alpha < 2$, if the energy estimate is replaced by

$$a(s_1-\mathcal{R}_h s_1, s_1-\mathcal{R}_h s_1) \le c\|\nabla^2(s_1-\mathcal{I}_h s_1)\|_{L_2}^2 \le ch^{2\alpha} . \tag{3.8}$$

The lowest values of α for the Kirchhoff plate occur in corners where the plate is simply supported on both adjacent edges. Here, $s_1 = r^{1+\alpha}t_\alpha(\varphi)$ is of the form $r^{\pi/\omega}\sin(\pi\varphi/\omega)$, for $\omega \le \pi$, $r^{2-\pi/\omega}\sin(\pi\varphi/\omega)$, for $\pi \le \omega \le 3\pi/2$, and $r^{2\pi/\omega}\sin(2\pi\varphi/\omega)$, for $3\pi/2 \le \omega \le 2\pi$. Hence, $\alpha \to 0$ as $\omega \to \pi$ and $\omega \to 2\pi$, and the convergence will be arbitrary slow in these cases.

In Section 4, below, we shall discuss several modifications of the standard method to recover the full order of convergence. The analysis will essentially depend on the approximation properties for the smooth remainder U of $u^{\cdot}$ in (2.16). Let us, therefore, briefly report some convergence results for *smooth* solutions U, not containing the leading singular functions.

We restrict ourselves to problem (2.2), using piecewise linear finite elements, and we assume that $\alpha < 1$. For $U \in H^2(\Omega)$, clearly the energy estimate is of optimal order. In view of (3.4) and the approximation property (3.2) we have $\|\nabla(U-\mathcal{R}_h U)\| \le ch\|U\|_{2,2}$. For the L_2-error, however, we can not guarantee a better estimate than

$$\|U-\mathcal{R}_h U\|_{L_2} \le ch^{1+\alpha}\|U\|_{H^{2,2}}, \tag{3.9}$$

since the solution v of the auxiliary problem will in general contain the leading singularity s_1, although U is smooth. The mean square estimates can be improved if U satisfies additional growth conditions. The key results in this context are estimates in weighted Sobolev norms ([17],[21]) which can also be used to control the pointwise error. For some positive κ, we set $\sigma^2 = (r^2+\kappa^2 h^2)$ and define a weighted L_2-norm by

$$\|u\|^2_{(\gamma)} = \sum_{K\in T_h} \int_K \sigma^\gamma |u|^2 dx . \tag{3.10}$$

With respect to these norms, no pollution takes place if the weight γ is suitably chosen.

<u>Theorem 3.2</u>: *For any* $U \in H^{2,2}(\Omega)$ *and any* γ *satisfying* $-2 < \gamma < -2+2\alpha$, *there hold the optimal order estimates*

$$\|U-\mathcal{R}_h U\|_{(\gamma)} + h\|\nabla(U-\mathcal{R}_h U)\|_{(\gamma)} \le ch^2\|\nabla^2 U\|_{(\gamma)} . \tag{3.11}$$

if the constant κ *is chosen sufficiently large.*

Since $\gamma < 0$, the estimate is uniform for $h \to 0$ if

$$\int_\Omega r^\gamma |\nabla^2 U|^2 dx < \infty , \tag{3.12}$$

i.e., if the smooth weight function σ may be replaced by r on the right hand side of (3.11). In general, this seems to be the weakest condition on U to remove completely the pollution effect. Clearly, also the L_2-error is of optimal order if (3.12) is satisfied.

Combining (3.11) with the results from [13], one can also derive pointwise error estimates. It turns out that the Ritz projection is nearly stable in the L_∞-norm,

$$\|U-\mathscr{R}_h U\|_{L_\infty} \leq ch^{-\epsilon}\|U-\mathscr{I}_h U\|_{L_\infty}, \quad \epsilon > 0, \tag{3.13}$$

where the factor $h^{-\epsilon}$ may be replaced by $|\log(h)|$ for most interior angles, see also [28]. If U has bounded second derivatives this gives us the nearly optimal pointwise result $O(h^{2-\epsilon})$.

A further information which we shall need in Section 4 and which is also of interest in itself is the description of the pointwise error away from the corner point z, where the solution is smooth. In view of Theorem 3.1, the best possible order we can expect is $O(h^{2\alpha})$. In fact it can be shown that, up to a factor $|\log(h)|$, the pointwise error in the interior behaves like the L_2-error. Let us describe the results in more detail only for the singular functions s_i. From Corollary 2.2 we have the expansion

$$g_x = \sum_{j=1}^{I} s'_{-j}(x)s_j + G_x, \tag{3.14}$$

where g_x denotes the Green function with source point $x \in \Omega$. The remainder G_x contains the logarithmic singularity, at x, but satisfies the growth condition (3.12), near z, where γ is chosen as in Theorem 3.2. Using the definition of the Green function and the usual orthogonality relation, this gives us the error identity

$$(s_i-\mathscr{R}_h s_i)(x) = (\nabla(s_i-\mathscr{R}_h s_i), \nabla g_x) \tag{3.15}$$

$$= \sum_{j=1}^{I} s'_{-j}(x)(\nabla(s_i-\mathscr{R}_h s_i), \nabla(s_j-\mathscr{R}_h s_j)) + (\nabla(s_i-\mathscr{R}_h s_i), \nabla(G_x-\mathscr{R}_h G_x)).$$

Since the singular behavior of the functions s'_{-j} at z is explicitly known, $s'_{-j}(x) \sim r^{-\alpha_j}$, this can be used to prove the following estimate from [13]; see also [21] for a more detailed discussion.

<u>Theorem 3.3</u>: *For the leading singular functions* s_i, $i=1,\ldots,I$, *such that* $\alpha_1+\alpha_i < 2$, *there hold the pointwise estimates*

$$|(s_i-\mathscr{R}_h s_i)(x)| \leq cr^{-\alpha_1} h^{\alpha_i+\alpha_1}|\log(h)|, \text{ for } r \geq h.$$

By the Cauchy-Schwarz inequality and estimate (3.11) we obtain that

$$(\nabla(s_i-\mathcal{R}_h s_i),\nabla(s_j-\mathcal{R}_h s_j)) \lesssim ch^{\alpha_i+\alpha_j}\ , \text{ for } \alpha_i+\alpha_j < 2\ . \qquad (3.16)$$

The remainder in (3.15) is of the order $O(r^{\alpha-2}h^2|\log(h)|)$ as is proved in [21], using the methods of [29]. This shows the desired result in the interior and describes how the error increases if x approaches the corner point z. Notice that all terms on the right of (3.15) are of the same order for $r \approx h$, reflecting the global pointwise estimate (3.13). The fact that the pollution terms in (3.15) are multiples of the dual singular functions has already been observed in [36].

The preceding results seem to be typical for conformal (and nonconformal) finite element approximations and most of them carry over to fourth order problems, see also [18]. Let us, at the end of this section discuss an example which shows that the effect of corner singularities may be even worse if *nonstandard* finite element methods are used. Here, instead of a reduced order of approximation, we shall see convergence to a *wrong* limit function. A similar example based on a least squares approach is given in [22].

We again consider a simply supported Kirchhoff plate. Then, on a polygonal domain Ω, the equations (2.9) can be rewritten in the equivalent form

$$-\Delta v = f\ , \text{ in } \Omega\ , \quad -\Delta u = v\ , \text{ in } \Omega\ , \quad u = \Delta u = 0\ , \text{ on } \partial\Omega\ . \qquad (3.17)$$

This system has the same solution u as (2.9) as long as we require bounded elastic energy, $u \in H^{2,2}(\Omega)$. On the other hand, (3.17) may be interpreted as a sequence of two Dirichlet problems for the Laplacian.

Find $(\bar{u},\bar{v}) \in \mathring{H}^{1,2}(\Omega)\times\mathring{H}^{1,2}(\Omega)$ such that

$$(\nabla\bar{v},\nabla\varphi) = (f,\varphi)\ , \quad (\nabla\bar{u},\nabla\psi) = (\bar{v},\psi)\ , \quad \forall\ (\varphi,\psi) \in (\mathring{H}^{1,2}(\Omega))^2 \qquad (3.18)$$

Here, $\mathring{H}^{1,2}(\Omega)$ is the space of functions in $H^{1,2}(\Omega)$, satisfying homogeneous boundary conditions. From (2.8) we know that, in the presence of a nonconvex corner, the unique solution $\bar{u}$ from (3.18) does not belong to $H^{2,2}(\Omega)$, except for certain f. Therefore, the pair $(\bar{u},\bar{v})$ does not coincide with the Kirchhoff plate solution.

Now as an approximation of the system (3.17) we choose the Ciarlet-Raviart method.

Find $(u_h,v_h) \in V_h\times V_h$ such that

$$(\nabla v_h,\nabla\varphi_h) = (f,\varphi_h)\ , \quad (\nabla u_h,\nabla\psi_h) = (v_h,\psi_h)\ , \quad (\varphi_h,\psi_h) \in V_h\times V_h\ . \qquad (3.19)$$

Clearly, (3.19) coincides with the usual Ritz projection of (3.18). The pair (u_h,v_h) converges to $(\bar{u},\bar{v})$ and, hence, does *not* converge to the plate bending solution in the presence of a nonconvex corner.

This situation does not change if the formulation (3.17) is replaced by introducing the full tensor of bending moments, M, instead of the scalar quantity v,

$$\underline{C}^{-1}[M] = \nabla^2 u \ , \ \text{in } \Omega \ , \quad \nabla^2 : M = f \ , \ \text{in } \Omega \ . \qquad (3.20)$$

Here, $\underline{C}$ denotes the elasticity tensor. Again, the pair (u_h, M_h) of approximate solutions of (3.20) converges to a wrong limit, for simply supported plates on nonconvex polygons, if the mixed problem (3.20) is discretized using continuous finite elements as in Herrmann's second method [11].

4. IMPROVED FINITE ELEMENT METHODS AND NUMERICAL CALCULATION OF STRESS INTENSITY FACTORS

The bad convergence results of the preceding section show that the standard finite element method is not well-suited for the approximation of corner problems. In this section we describe several modifications of the usual method in order to avoid or reduce the pollution effect.

4.1 Optimal mesh refinements

The best known modification of the standard finite element method which is most widely used in practice is the employment of refined meshes. Let us again consider the model problem (2.1). First, we discuss the error in L_2-norms. In virtue of the estimate (3.5) it is sufficient to use a mesh for which the first singular function s_1 is approximated with optimal order with respect to the energy norm. Since the derivatives of s_1 only grow in radial direction, this is achieved by subdivisions which asymptotically, for $h \to 0$, meet the following assumptions.

(i) A unit distance away from the corner point the triangulation is regular with meshsize h.

(ii) Around z there are $O(1/h)$ concentric circles, with radii $R_i = i^Z h^Z$ and width $d_i = R_{i+1} - R_i = O(i^{Z-1} h^Z)$, for some fixed $Z \geq 1$, such that the i-th annulus A_i is divided regularly into elements K_{ij}, $j=1,\ldots,O(i)$.

Such a mesh is called of *refinement type* Z. Notice that the asymptotic number of elements is still $O(h^{-2})$. The size of the smallest elements immediately at the corner is h^Z. Triangulations of this type can easily be constructed for triangular elements, for quadrilaterals one usually introduces blind nodes.

Let us now assume that $\alpha = \alpha_1 < 1$, such that a refinement is needed even for the approximation with piecewise linear elements. Proceeding as in the proof of Theorem 3.1 and observing that now $|\nabla^2 s_1|_{K_{ij}}| \le cR_i^{\alpha-2}$ we obtain the following, for $\alpha Z > 1$,

$$\begin{aligned} \|\nabla(s_1-\mathcal{R}_h s_1)\|_{L_2}^2 &\le \sum_{i=1}^{O(1/h)} d_i^2 \|\nabla^2 s_1\|_{L_2(A_i)}^2 + O(h^2) \\ &\le ch^{2\alpha Z} \sum_{i=1}^{O(1/h)} i^{2\alpha Z-3} + O(h^2) . \end{aligned} \tag{4.1}$$

The sum on the right in (4.1) is not bounded uniformly for $h \to 0$, but can be estimated by $O(h^{2-2\alpha Z})$. Combined with the leading factor $h^{2\alpha Z}$ we have the desired optimal energy estimate. Once this is established, the standard duality argument of the form (3.6) gives us a full additional power of h , for the L_2-error. Thus, we have shown the following result.

<u>Theorem 4.1</u>: *Let the triangulation* T_h *be of refinement type* Z *such that* $\alpha Z > 1$. *Then, the Ritz projection onto piecewise linear elements satisfies the optimal mean square estimates*

$$\|u-\mathcal{R}_h u\|_{L_2} + h\|\nabla(u-\mathcal{R}_h u)\|_{L_2} = O(h^2) . \tag{4.2}$$

Clearly, the estimate (4.1) can also be used for the determination of the order of convergence for higher order elements , $m > 2$. The optimal order $O(h^{m-1})$, now is obtained if Z is chosen according to $\alpha Z > m-1$. This, however leads to very small elements near the corner point z . From the practical point of view, a value of $Z \approx 2$ seems reasonable. For linear elements, $m = 2$, this corresponds to a singularity with $\alpha = 1/2$ as in the pure Dirichlet or Neumann problem for the Laplacian or in crack problems in linear elasticity. For further properties of finite element spaces on refined meshes we refer to [12].

The question whether an even stronger refinement will give also optimal pointwise error estimates has been treated in [13]. Since the interpolation error satisfies

$$\|s_1-\mathcal{I}_h s_1\|_{L_\infty(K_{ij})} \le ci^{\alpha Z-2}h^{\alpha Z} + O(h^2) , \tag{4.3}$$

for piecewise linear elements, optimal pointwise convergence can be expected only for $\alpha Z > 2$. In fact, such a strong refinement turns out to be sufficient to guarantee

$$\|u-\mathcal{R}_h u\|_{L_\infty} = O(h^{2-\epsilon}) . \tag{4.4}$$

Analogous conditions hold for higher order elements.
For $\alpha = 1/2$ this means $Z > 4$ which is not realistic. However, in

most applications one is only interested in the pointwise error away from the corner. By the interior L_∞-estimates in Section 3, this can be expected to behave like the L_2-error such that already a milder refinement leads to a significant improvement.

In practice, the exact growth of the solution u and its derivatives is not always known a priori. In such situations the optimal layout for refined meshes is frequently controlled by adaptive procedures using *a posteriori* estimates of the error. Here, local indicators of the (energy-)error of a discrete solution control where the triangulation should be modified for further calculations [14]. Although the asymptotic reliability of these procedures is rigorously justified only for very specific problems (in particular in one space dimension) it seems to work very well in practice, see, e.g., [15], for second order problems, and [16], for the Mindlin plate.

4.2 The singular function method (SFM)

In the following two subsections we discuss two modified finite element schemes which both seem very natural in view of the representation (2.16) in combination with Theorem 2.1. Both methods construct approximate solutions in the augmented spline spaces

$$V_h^I = V_h \oplus \{s_1,\dots,s_I\} , \tag{4.5}$$

where $s_1,\dots,s_I$ are the singular functions in (2.16). Clearly, the number I of functions added to the approximating subspaces will determine the accuracy of the procedures.

First, we study the singular function method (SFM) presented, e.g., in [3]. A more detailed analysis, following ideas of Schatz [27] can be found in [17] and [18].

The SFM makes use of the spaces V_h^I as trial and test spaces, i.e., a modified Ritz projection $\mathcal{R}_h^I u \in V_h^I$ is determined by the relation

$$a(\mathcal{R}_h^I u,\varphi_h) = (f,\varphi_h) , \quad \forall\ \varphi_h \in V_h^I . \tag{4.6}$$

By definition, the solution $\mathcal{R}_h^I u$ is of the form

$$\mathcal{R}_h^I u = \sum_{i=1}^{I} K_i^I s_i + v_h , \tag{4.7}$$

for some $v_h \in V_h$. The coefficients K_i^I can be taken as approximate stress intensity factors but they sometimes converge very badly, see Theorem 4.3, below. From (4.6) we immediately obtain the improved orthogonality relation

$$a(u-\mathcal{R}_h^I u,\varphi_h) = 0 , \quad \forall\ \varphi_h \in V_h^I . \tag{4.8}$$

This implies the identity

$$a(u-\mathcal{R}_h^I u, u-\mathcal{R}_h^I u) = a(u-\mathcal{R}_h^I u, \sum_{i=1}^{I} K_i s_i + U-\mathcal{R}_h U) = a(u-\mathcal{R}_h^I u, U-\mathcal{R}_h U) \ . \qquad (4.9)$$

We shall in the following assume that the regularity of the smooth remainder U on the right is determined by the subsequent singular function, s_{I+1}, with exponent α_{I+1}. Then, (4.9) gives us optimal convergence results in mean square norms which are stated here for the approximation of problem (2.2).

Theorem 4.2: *For the SFM using* I *singular functions there holds*

$$\|\nabla(u-\mathcal{R}_h^I u)\|_{L_2} = \mathcal{O}(h^{\min\{m-1,\alpha_{I+1}\}}) \ ,$$

$$\|u-\mathcal{R}_h^I u\|_{L_2} = \mathcal{O}(h^{\min\{m,\alpha_{I+1}+1,2\alpha_{I+1}\}}) \ .$$

Proof. The energy estimate immediately follows from (4.9) by means of the Cauchy-Schwarz inequality and the approximation property (3.2). For the second estimate, we again apply the standard duality argument (3.6), for the right hand side $u-\mathcal{R}_h^I u$. In view of (4.8), we get

$$\|u-\mathcal{R}_h^I u\|_{L_2}^2 = a(u-\mathcal{R}_h^I u, V-\mathcal{R}_h V) \ ,$$

where V is the smooth remainder of the solution v of the auxiliary problem. Since V again behaves like s_{I+1}, this gives us the desired additional power of h. □

In order study the error for the approximate stress intensity factors K_i^I we have to look at the identity (4.7) more carefully. As a consequence of the definition (4.6), used for $\varphi_h \in V_h$, we see that

$$v_h = \mathcal{R}_h u - \sum_{i=1}^{I} K_i^I \mathcal{R}_h s_i \ , \qquad (4.10)$$

where again $\mathcal{R}_h u$, $\mathcal{R}_h s_i$ are the usual Ritz projections. In virtue of the expansion (2.16) and (4.7) this gives us the following error identity which is typical for *all* discretization methods employing the augmented spline spaces V_h^I,

$$u-\mathcal{R}_h^I u = \sum_{i=1}^{I} (K_i-K_i^I)(s_i-\mathcal{R}_h s_i) + (U-\mathcal{R}_h U) \ . \qquad (4.11)$$

This relation contains all information about the error for the stress intensity factors.

As an example for the analysis we study the simplest case $I = 1$, i.e. V_h is augmented by only one singular function. Here, (4.11) takes the form

$$u-\mathcal{R}_h^1 u = (K_1-K_1^1)(s_1-\mathcal{R}_h s_1) + (U-\mathcal{R}_h U) . \tag{4.12}$$

Using again the relation (4.8), for $\varphi_h = s_1-\mathcal{R}_h s_1$, this gives us

$$\begin{aligned} 0 &= a(u-\mathcal{R}_h^1 u, s_1-\mathcal{R}_h s_1) = \\ &= (K_1-K_1^1)\, a(s_1-\mathcal{R}_h s_1, s_1-\mathcal{R}_h s_1) + a(U-\mathcal{R}_h U, s_1-\mathcal{R}_h s_1) . \end{aligned} \tag{4.13}$$

From this we deduce the identity

$$K_1-K_1^1 = -a(U-\mathcal{R}_h U, s_1-\mathcal{R}_h s_1) \;/\; a(s_1-\mathcal{R}_h s_1, s_1-\mathcal{R}_h s_1) . \tag{4.14}$$

Now, we can state the following estimate for the approximation of (2.2) by means of piecewise linear elements.

Theorem 4.3: *In the SFM using one singular function, the error for the leading stress intensity factor satisfies the estimate*

$$|K_1-K_1^1| = \mathcal{O}(h^{\min\{2,\alpha_1+\alpha_2\}-\min\{2,2\alpha_1\}}) .$$

Proof. Since we have assumed that $U \approx s_2$, we can apply (3.16) to control the nominator. The denominator is estimated from below by $\mathcal{O}(h^{2\alpha})$, see Theorem 3.1. Clearly, for both estimates, the upper bound is 2, corresponding to the degree of the approximating elements. □

Notice that the convergence for K_1 is very bad if $\alpha_1 \approx \alpha_2$, and no convergence can be shown in the more regular case $\alpha_1 \geq 1$.

By similar arguments, using $s_2-\mathcal{R}_h^1 s_2$ and $s_1-\mathcal{R}_h s_1$ as test functions in (4.8), error identities for the K_1 and K_2 are obtained in the case $I = 2$. Here, again using (3.16), the following result is derived for the leading stress intensity factor.

$$|K_1-K_1^2| = \begin{cases} \mathcal{O}(h^{\min\{2,\alpha_1+\alpha_2\}-\min\{2,2\alpha_1\}}) , & \text{if } \alpha_2 \geq 1 \\ \mathcal{O}(h^{\min\{2,\alpha_2+\alpha_3\}-\alpha_2-\alpha_1}) , & \text{if } \alpha_2 < 1 \end{cases} . \tag{4.15}$$

This estimate and Theorem 3.1 already indicate the general *rules* for the error behavior of the SFM.

(i) It is essentially the difference of the exponents of subsequent singular functions which determines the convergence rate of stress intensity factors, not the regularity of the remainder U.

(ii) The convergence is not improved adding an additional singular function s_I, if s_I is already approximated with optimal order with respect to the energy norm.

Theorem 4.3 immediately carries over to the approximation of the plate bending problem (2.9). Here, the difference of two singular exponents may become arbitrarily small. For an L-shaped clamped plate, e.g., there holds $\alpha_1 \approx 0.54$ and $\alpha_2 \approx 0.91$ and the difference decreases for $\omega \to 2\pi$. The approximation of (2.9) by a mixed method using piecewise linear elements resulted in a typical relative error for $|K_1 - K_1^1|$ of about 40%, for $h=1/70$ (about 4000 unknowns) although the deflection converged very well, see [18]. This shows that in many cases the calculated stress intensity factors in the SFM should be corrected by means of a suitable postprocessing, for example by employing one of formulas discussed in the next subsection.

Let us conclude the discussion with some remarks on the numerical implementation of the SFM. It is widely considered as a great disadvantage that the stiffness matrix in (4.6) contains I full rows and columns, since for the original sparse matrix there often exist very fast solution procedures. This can be avoided by implementing the SFM in the following equivalent form, where only the usual Ritz projections are needed.

Step 0. Determine $\mathcal{R}_h s_i$ and calculate the matrix

$$a_{ij}^I = a(s_i - \mathcal{R}_h s_i, s_j), \quad i,j=1,\dots,I .$$

Step 1. Calculate the usual Ritz projection $\mathcal{R}_h u$.

Step 2. Determine the approximate stress intensity factors K_i^I by solving the I×I-system (4.16)

$$\sum_{i=1}^{I} a_{ij}^I K_i^I = (f, s_j) - a(\mathcal{R}_h u, s_j), \quad j=1,\dots,I .$$

Step 3. Set $\mathcal{R}_h^I u = \mathcal{R}_h u + \sum_{i=1}^{I} K_i^I (s_i - \mathcal{R}_h s_i)$.

If only the matrix (a_{ij}^I) from Step 0 is stored, but not the whole solution vectors $\mathcal{R}_h s_i$, Step 3 must be replaced by

Step 3'. Solve $a(U_h, \varphi_h) = (f, \varphi_h) - \sum_{i=1}^{I} K_i^I a(s_i, \varphi_h), \quad \forall\, \varphi_h \in V_h^I$,

and define $\mathcal{R}_h^I u = U_h + \sum_{i=1}^{I} K_i^I s_i$.

Notice that Step 0 must be performed only *once* for a given domain. Thus, formulation (4.16) shows that the computational work is only slightly increased, compared to the usual Ritz projection, at least if the problem must be solved for many right hand sides. Moreover, one can stop the procedure after Step 2 if only the approximations of the stress intensity factors are desired.

A second remark concerns the definition of the spaces V_h^I. Instead of including the singular functions s_i one sometimes should prefer to use $\bar{s}_i$, not involving a cut-off function τ. This avoids oscillations in the calculation of $\mathcal{R}_h s_i$ which are caused by the differentiation of τ, on coarse meshes. Otherwise, in particular for fourth order problems, the asymptotic error orders become significant only for relatively small discretization parameters, see [18].

4.3 The dual singular function method (DSFM)

In this subsection we present several procedures which, in addition to the explicit knowledge of the singular functions s_i, also employ the integral representation formulas for the stress intensity factors given in Theorem 2.1. We concentrate on the formulas (ii) and (iii) whereas an application of (i) is studied in Section 5 in connection with boundary element approximations.

4.3.1 Direct evaluation of the stress intensity factors

First, we consider an application of formula (iii). For our model problem (2.2) it reads as

$$K_i = (f, s'_{-i}) = (f, s_{-i}) - (f, v_i) . \tag{4.17}$$

where v_i is the weak solution of

$$(\nabla v_i, \nabla\varphi) = (-\Delta s_{-i}, \varphi) , \quad \forall \varphi \in V . \tag{4.18}$$

Clearly, the solution of (4.18) is not known analytically except for very specific problems. However, formula (4.17) can be used for the calculation of K_i if a sufficiently good approximation of v_i has been determined. This leads to the following simple algorithm.

Step 0. Calculate an approximate solution $\tilde{v}_i$ of equation (4.18). (4.19)

Step 1. For given f, set $\tilde{K}_i = (f, s_{-i}) - (f, \tilde{v}_i)$.

Step 2. Calculate $U_h = \mathcal{R}_h u - \sum_{i=1}^{I} \tilde{K}_i \mathcal{R}_h s_i$ and set

$$u_h = U_h + \sum_{i=1}^{I} \tilde{K}_i s_i \ .$$

Usually, Step 2 is performed by calculating the Ritz projection for the right hand side $f + \sum \tilde{K}_i \Delta s_i$. Again, Step 0 is necessary only once for each domain Ω . Then, the evaluation of only *one* integral in Step 1 gives us the information on the stress intensity factors for any given data. Since

$$K_i - \tilde{K}_i = -(f, v_i - \tilde{v}_i) \ , \tag{4.20}$$

the relative error for the coefficients is bounded by ϵ if the approximation $\tilde{v}_i$ satisfies $\|v_i - \tilde{v}_i\| \le \epsilon$. To approximate (4.18), one may choose the usual Ritz projection, as in [19]. However, the error may be large since v_i will, in general, contain the first singular function s_1 . It seems preferable to employ one of the improved methods described below.

4.3.2. An iterative procedure

The following variant of the finite element methods make use of the second representation formula in Theorem 2.1. For problem (2.1) it takes the form

$$K_i = (f, s_{-i}) - (u, -\Delta s_{-i}) \ , \tag{4.21}$$

containing the unknown solution itself. However, by replacing u in the last integral by some approximate solution we get the coefficients with the same accuracy. In turn, this information can be used for the improvement of the solution.

The following iterative scheme based on this idea is found in [4].

Step 0. Choose some starting values $\tilde{K}_i^0$, $i=1,\dots,I$.

Step 1. For $j \ge 1$, determine an approximation for the regular remainder by

$$U_h^j = \mathcal{R}_h u - \sum_{i=1}^{I} \tilde{K}_i^{j-1} \mathcal{R}_h s_i = \mathcal{R}_h U + \sum_{i=1}^{I} (K_i - \tilde{K}_i^{j-1}) \mathcal{R}_h s_i \ . \tag{4.22}$$

Step 2. Set $u_h^j = U_h^j + \sum_{i=1}^{I} \tilde{K}_i^{j-1} s_i$.

Step 3. Define corrected stress intensity factors by

$$\tilde{K}_i^j = (f, s_{-i}) - (u_h^j, -\Delta s_{-i}) \ .$$

The analysis for this procedure is very simple because of the following basic error identities.

Theorem 4.4: *For* $j \geq 1$, *the solution of the iterative scheme* (4.22) *satisfies the relations*

$$u - u_h^j = \sum_{i=1}^{I} (K_i - \tilde{K}_i^{j-1})(s_i - \mathcal{R}_h s_i) + (U - \mathcal{R}_h U) \ , \qquad (4.23)$$

$$K_i - \tilde{K}_i^j = -(u - u_h^j, -\Delta s_{-i}) \ , \quad i = 1, \ldots, I \ . \qquad (4.24)$$

Usually, the initial guess for the stress intensity factors is 0 and, therefore, $u_h^1 = \mathcal{R}_h u$. Since by Theorem 3.1 we have $\|u - \mathcal{R}_h u\| = O(h^{2\alpha})$, for the L_2-error, the same holds true for $|K_i - \tilde{K}_i^1|$ by (4.24). Now inserting this into (4.23), we see that the rate of convergence is improved in the next step by the order of $\|s_1 - \mathcal{R}_h s_1\|$, i.e., again by $O(h^{2\alpha})$. After a (small) finite number of steps the error order of the remainder $\|U - \mathcal{R}_h U\|$ is reached which is of course the best possible for this method. Finally, we observe that Δs_{-i} are smooth functions with support bounded away from z . Therefore, by (4.24), the error for *all* stress intensity factors is determined by that of $U - \mathcal{R}_h U$ in an arbitrarily weak interior norm.

Of course, the iterative correction method (4.22) can be applied to more general elliptic problems as long as Theorem 2.1 remains valid for the coefficients K_i . Then, the right hand side in identity (4.24) is replaced by $-[(u - u_h^j, L_* s_{-i}) + (S_*(u - u_h^j), B_* s_{-i})]$ such that again the approximation error must be estimated only away from the singular point.

4.3.3. The dual singular function method [4]

Forming the difference of two subsequent iterates in (4.23) we obtain the following contraction estimate, for $j \geq 1$,

$$|\tilde{K}_i^{j+1} - \tilde{K}_i^j| \leq \max\{ |\tilde{K}_l^j - \tilde{K}_l^{j-1}| \|s_l - \mathcal{R}_h s_l\|, \ l = 1, \ldots, I\} \ . \qquad (4.25)$$

Since $\|s_l - \mathcal{R}_h s_l\| \to 0$ for $h \to 0$, this shows that the iterate stress intensity factors converge, as $j \to \infty$, if h is sufficiently small.

This in turn implies also the convergence of the functions u_h^j. Let us denote the limit values by

$$\mathcal{R}_h^{-I}u \in V_h^I \quad \text{and} \quad K_i^{-I}\ , \ i=1,\dots,I\ ,$$

where the notation $-I$ indicates the use of I singular functions of the dual problem in the representation formulas. By a simple calculation, using in particular the normalizing identity (2.19), it is seen that the limit can be obtained directly by the following procedure.

Step 0. Determine $\mathcal{R}_h s_i$ and calculate the matrix

$$a_{ij}^{-I} = (-\Delta s_i, s_{-j}) - (s_i, -\Delta s_{-j})\ , \quad i,j = 1,\dots I\ .$$

Step 1. Calculate the usual Ritz projection $\mathcal{R}_h u$.

Step 2. Determine approximate stress intensity factors (4.26)
K_i^{-I} by solving the I×I-system

$$\sum_{i=1}^{I} K_i^{-I} a_{ij}^{-I} = (f, s_{-j}) - (\mathcal{R}_h u, -\Delta s_{-j})\ .$$

Step 3. Set $\mathcal{R}_h^{-I}u = \mathcal{R}_h u + \sum_{i=1}^{I} K_i^{-I}(s_i - \mathcal{R}_h s_i)$.

Thus, the idea of iteration (4.22) may be neglected in this context where the coefficients K_i are evaluated by the exact representation formula, see also the following subsection. Notice that

$$a_{ij}^{-I} \to \delta_{ij}\ , \quad \text{as} \quad h \to 0\ ,$$

as a consequence of (2.19), such that the matrix in Step 2 is regular for sufficiently small h.

By the formal identity with the scheme (4.16), the limit functions of the DSFM are obtained with the same amount of computational work as in the SFM. An additional difficulty may arise from the evaluation of the integrals (f, s_{-i}) because of the singularity of the dual singular function at z. On the other hand, the error behavior is significantly improved, especially for the coefficients K_i, such that this additional effort seems to be justified.

We collect the error identities for the DSFM in the main theorem of this section.

Theorem 4.5: *The approximations* $\mathcal{R}_h^{-I}u \in V_h^I$ *and* K_i^{-I}, $i=1,\ldots,I$, *defined by (4.26), satisfy the relations*

$$\text{(i)} \qquad u-\mathcal{R}_h^{-I}u = \sum_{i=1}^{I}(K_i-K_i^{-I})(s_i-\mathcal{R}_h s_i) + (U-\mathcal{R}_h U),$$

$$\text{(ii)} \qquad K_i-K_i^{-I} = -(u-\mathcal{R}_h^{-I}u, -\Delta s_{-i}),$$

$$\text{(iii)} \qquad \sum_{i=1}^{I} a_{ij}^{-I}(K_i-K_i^{-I}) = -(U-\mathcal{R}_h U, -\Delta s_{-j}).$$

Further, there hold the optimal estimates

$$\|u-\mathcal{R}_h^{-I}u\| + \sum_{i=1}^{I}|K_i-K_i^{-I}| \le c\|U-\mathcal{R}_h U\|. \qquad (4.27)$$

where $\|.\|$ *denotes an arbitrarily weak norm.*

The identities (i)-(iii) immediately follow from the definition in (4.26). The estimate (4.27) is then obtained from (i) and (iii) by means of Hölder's inequality.

As for the iterative procedure (4.22) the resulting error order is determined only by the smoothness of U. Some convergence properties for smooth solutions have been reported at the end of Section 3. For instance, from (3.13) we have the optimal pointwise estimate

$$\|U-\mathcal{R}_h U\|_{L_\infty} \le ch^{m-\epsilon}, \quad \text{for } U \in C^m(\bar{\Omega}). \qquad (4.28)$$

This means that by the use of sufficiently many singular functions the pollution effect can be completely avoided in the DSFM. Further, for the stress intensity factors we can employ the improved error estimates in negative norms for higher order elements. For $m \ge 3$, the best possible order is $\mathcal{O}(h^{2m-2})$.

Let us briefly compare the SFM and the DSFM. As mentioned above, both methods essentially need the same computational work while the error behavior is quite different.

In virtue of Theorem 4.3 and Theorem 4.5, one should prefer the DSFM for the pure computation of the stress intensity factors. Further, the pointwise error is optimal in the DSFM, as in the usual finite element method. On the other hand, at least theoretically, the SFM may sometimes produce a smaller error with respect to the L_2-norm because of the improved projection property (4.8). Inserting the SFM-solution into the representation formulas for K_i may then give somewhat improved results. In practice, this observation seems to be of minor importance. In all test calculations, the interior error for the DSFM was small compared to the SFM, see [17] for second order equations, and [18] for the plate bending problem.

Both methods seem to be applicable with a reasonable amount of work only in problems with *isolated* singularities. This is the case in two space dimensions and also for 3D-problems if the corners are formed by smooth surfaces. In polyhedral domains, however, where "every" point along a reentrant edge produces singular components of the solution, the necessary number of singular functions increases for smaller mesh size, and the implementation of the methods in the present form seems not sensible.

The idea of employing the representation formulas of Theorem 2.1 in a numerical procedure is of course not restricted to finite element approximations. Each discretization procedure which is convergent also in the presence of singularities can be used in the iterative scheme (4.22) as well as in the final version of the DSFM, such as finite difference approximations or solutions obtained by collocation methods. Since for the representation formulas there are only needed the values in the interior, one should look for a simple discretization scheme which has good approximation properties *away* from the corner point while the accuracy near z may be bad. Inserting the corresponding discrete solution into (4.21) would return good approximate K_i. However, it is an open question how to construct such a modified procedure in general. A simple modification of the usual five point difference scheme for problem (2.1) is found in [26].

4.4 Approximate representation formulas and a generalized iteration scheme

The evaluation of the representation formula (ii) of Theorem 2.1 certainly performs an optimal postprocessing and asymptotically gives the best possible information about the stress intensity factors which can be extracted from a discrete function u_h. On the other hand, the calculations may become costly if the formulas get more complicated as, e.g., in nonlinear equations. In the following, we describe a much simpler possibility to obtain information on the stress intensity factors from u_h which is based on the analysis of the interior pointwise error for the standard finite element method in Section 3.

We again consider the simplest model problem

$$-\Delta u = f \ , \ \text{in } \Omega \ , \quad u = 0 \ , \text{ on } \partial\Omega \ , \tag{4.29}$$

where Ω contains one significant corner z. For the smooth remainder U we assume C^2-regularity, i.e., we choose $I = 2$ in expansion (2.16) if $2/3 \leq \alpha < 1$, and $I = 3$ for $1/2 \leq \alpha < 2/3$, $\alpha = \pi/\omega$. Further, by suitable growth assumptions on f, we may require that $r^{-2}U$ is bounded. From (3.13), we already know that $\|U-\mathcal{R}_h U\|_\infty = \mathcal{O}(h^{2-\epsilon})$ for the Ritz projection onto piecewise linear elements, for any $\epsilon > 0$. Further, we have the following pointwise estimate for the singular functions,

$$|(s_i-\mathcal{R}_h s_i)(x)| \le c r^{\max\{-\alpha, i\alpha-2\}} h^{\min\{2,(i+1)\alpha\}-\epsilon} \quad , \quad r \ge h \, , \tag{4.30}$$

see Theorem 3.3 and the representation (3.15). Here, we have used that $\alpha_i = i\alpha$, for the model problem (4.29). The estimate (4.30) shows in which way the interior order of convergence decreases if x approaches the corner point.
In [13], the information (4.30), for $i = 1$, has been used for a very simple method to determine approximate stress intensity factors directly from the Ritz projection. The following results which make use of a generalized iteration scheme and on a stabilized variant of this approach have been presented in [5].

We fix two constants C_1, C_2, $0 < C_1 < C_2 < 1$, and, for $R > 0$, we set

$$(v,w)_R = \int_{A_R} vw \, dx \, . \tag{4.31}$$

Here, A_R is the annulus $C_1R < r < C_2R$. The corresponding norm is denoted by $\|.\|_R$. For any given function v , we now define approximate stress intensity factors by the formula

$$K_i(v,R) = (v,s_i)_R \,/\, (s_i,s_i)_R \, . \tag{4.32}$$

Since $(s_i,s_j)_R = 0$, for $i \ne j$, these values are characterized as the minimum of the functional

$$\|v - \sum_{i=1}^{I} l_i s_i\|_R \, , \quad l_i \in \mathbb{R} \, . \tag{4.33}$$

Later on, formula (4.32) will be applied to approximate solutions $u_h \in V_h^I$ which satisfy a pointwise estimate of the type (4.30), namely

$$|(u-u_h)(x)| \le c r^{-\beta} h^{\gamma} + O(h^2) \, , \quad r \ge h \, , \tag{4.34}$$

for certain constants $\beta, \gamma > 0$. By (4.30), for the usual Ritz projection $u_h = \mathcal{R}_h u$ there holds (4.34), with $\beta = \alpha$ and $\gamma = 2\alpha-\epsilon$, if u contains the leading singularity. By testing the solution u with s_i and observing (2.16), we obtain the estimate, for $R \ge h$,

$$\begin{aligned} |K_i - K_i(u_h,R)| &\le (\|u-u_h\|_R + \|U\|_R) \,/\, \|s_i\|_R \\ &\le c\,(R^{-\beta}h^{\gamma} + R^2) \,/\, R^{i\alpha} \, . \end{aligned} \tag{4.35}$$

We now choose the optimal value

$$R_{opt}(u,u_h) = h^{\gamma/(2+\beta)} , \qquad (4.36)$$

for which the right hand side in (4.35) is minimized. The resulting order of convergence is

$$|K_i - K_i(u_h, R_{opt}(u,u_h))| = O(h^{\gamma(2-i\alpha)/(2+\beta)}) . \qquad (4.37)$$

For the Ritz projection $\mathcal{R}_h u$, this only gives us the bad rate $O(h^{2\alpha(2-\alpha)/(2+\alpha)-\epsilon})$ for K_1. As a general principle, the accuracy of the stress intensity factors is determined by the difference of the exponents of two subsequent singular functions, as is the case in the SFM, since this determines the characteristic values β and γ. However, in contrast to the latter, we now have an explicit formula for K_i, namely (4.32), which can be efficiently used in the generalized iteration scheme described in [5]. The procedure is a modification of (4.22), where Step 3 is replaced by the following.

Step 3'. Define corrected stress intensity factors by (4.38)

$$\tilde{K}_i^j = F_h^j(f, u_h^j) .$$

As indicated by the notation, the functional F_h^j may now depend on the mesh size, and on the iteration count j, while in (4.22) we used a fixed functional F. Therefore, the iterative scheme (4.22),(4.38) usually will not converge to a limit as is the case for the DSFM. Nevertheless, the rate of convergence is improved in every step.

In our situation, taking as F_h^j the formula (4.32), with $R = R_{opt}$, the crucial identity (4.23) gives us, after one step of iteration, the following,

$$|(u-u_h^1)(x)| \le ch^{2\alpha(2-\alpha)/(2+\alpha)-\epsilon} h^{2\alpha-\epsilon} r^{-\alpha} + O(h^2) \qquad (4.39)$$

Thus, $\beta = \alpha$ remains fixed in (4.34) while the convergence rate γ is improved. Correspondingly, the values of R_{opt}, defined by (4.36), are decreasing in every step. We stop the procedure after J iterations if $R_{opt} \approx h$, or $\gamma_J/(2+\beta) \approx 1$.

<u>Theorem 2.6</u>: *For the iterate solutions of the generalized iteration scheme, based on formula* (4.32), *there holds*

$$\|u-u_h^J\|_{L_\infty} = O(h^{2-\epsilon}) , \quad |K_i - \tilde{K}_i^J| = O(h^{2-i\alpha-\epsilon}) , \quad i=1,\dots,I ,$$

where J *is as chosen above.*

For the solution u, this result is the best possible as in the DSFM.

The estimate for the coefficients is slightly worse compared to Theorem 4.5, but is much better than for the SFM.

4.5 Asymptotic error expansions and Richardson extrapolation

In this final subsection, we want to describe a quite different and very simple approach to remove the pollution effect, namely the application of Richardson extrapolation. As is well known, the theoretical basis for increasing the accuracy by extrapolation or defect correction techniques is the validity of an asymptotic expansion of the discretization error with respect to the mesh size parameter. For instance, for smooth solutions on convex polygons, the Ritz projection onto linear finite elements, for problem (2.1), admits an expansion

$$(u-\mathcal{R}_h u)(x) = e_1(x;u)h^2 + e_2(x;u)h^4 + o(h^4), \qquad (4.40)$$

in interior nodal points, if the triangulation is kept piecewise uniform [20]. Certainly, in view of Theorem 3.1, such a result cannot hold in the presence of corner singularities, in particular on nonconvex polygonal domains, since the best possible order is $O(h^{2\alpha})$, $\alpha < 1$.

In [21], this problem is analyzed, for the Dirichlet problem of the Laplacian on general polygonal domains, allowing several reentrant corners z_n, $n = 1,\dots N$. Based on the singular representation (1.2), we have the following error identity,

$$(u-\mathcal{R}_h u)(x) = \sum_{n=1}^{N}\sum_{i=1}^{I_n} K_{n;i}(s_{n;i}-\mathcal{R}_h s_{n;i})(x) + (U-\mathcal{R}_h U)(x). \qquad (4.41)$$

For U, we require the growth condition (3.12), at z_n, for some $\gamma_n < -2+2\alpha_{n;1}$. By application of Theorem 3.2 it can by shown that the last term on the right is of the order $O(h^2|\log(h)|)$, in the interior. To estimate the pointwise error for the singular functions in the sum, we again use the representation (3.15), obtaining

$$(s_{n;i}-\mathcal{R}_h s_{n;i})(x) = \sum_{m=1}^{N}\sum_{j=1}^{I_m} s'_{m;-j}(\nabla(s_{n;i}-\mathcal{R}_h s_{n;i}),\nabla(s_{m;j}-\mathcal{R}_h s_{m;j})) + \\ + (\nabla(s_{n;i}-\mathcal{R}_h s_{n;i}),\nabla(G_x-\mathcal{R}_h G_x)). \qquad (4.42)$$

The remainder term in (4.42) and the terms in the sum, for $n \neq m$, are of the order $O(h^2|\log(h)|)$, and $O(h^2)$, respectively. The same holds true, for $n = m$, if $\alpha_{n;i}+\alpha_{n;j} \geq 2$. Thus, (4.41) reduces to

$$(u-\mathcal{R}_h u)(x) = \sum_{n=1}^{N} \sum_{\alpha_{n;i}+\alpha_{n;j}<2} K_{n;i} s'_{n;-j}(x)(\nabla(s_{n;i}-\mathcal{R}_h s_{n;i}),\nabla(s_{n;j}-\mathcal{R}_h s_{n;j})) + O(h^2|\log(h)|)\,, \tag{4.43}$$

on interior subdomains. The inner products in (4.43) can be estimated by (3.16). In order to convert the estimate into an error *expansion*, we assume that the triangulation is kept *uniform* in some fixed neighborhood Ω_n, of each reentrant corner z_n, see [21] for a precise definition. We denote the "local" mesh size in Ω_n, by h_n, and h_0 is the size of the elements in the remaining part, $h = \max\{h_0, h_1, \ldots, h_N\}$. Then, following the stretching argument from [21], one arrives at the identity, for $\alpha_{n;i}+\alpha_{n;j} < 2$,

$$(\nabla(s_{n;i}-\mathcal{R}_h s_{n;i}),\nabla(s_{n;j}-\mathcal{R}_h s_{n;j})) = A^{(n)}_{i,j}\, h_n^{\alpha_{n;i}+\alpha_{n;j}} + O(h^2)\,, \tag{4.44}$$

with certain constants $A^{(n)}_{i,j}$. By combination of all results, we arrive at the following,

<u>Theorem 4.7</u>: *In interior subdomains of* Ω, *the error for the Ritz projection on locally uniform meshes admits the asymptotic expansion*

$$(u-\mathcal{R}_h u)(x) = \sum_{n=1}^{N} \sum_{\alpha_{n;i}+\alpha_{n;j}<2} A^{(n)}_{i,j} K_{n;i}\, s'_{n;-j}(x)\, h_n^{\alpha_{n;i}+\alpha_{n;j}} + O(h^2|\log(h)|)\,. \tag{4.45}$$

If, in addition, the mesh is symmetric in Ω_n, *then* $A^{(n)}_{1,2} = 0$.

The local symmetry of the mesh guarantees that $A^{(n)}_{1,2} = 0$ since s_1 is mirror-symmetric and s_2 is mirror-antisymmetric. All numerical test examples, however, indicate that this assumption is not needed in practice and that (4.45) goes over to the even simplified form

$$(u-\mathcal{R}_h u)(x) = \sum_{n=1}^{N} e^{(n)}(x;u)\, h_n^{2\alpha_{n;1}} + O(h^2|\log(h)|)\,, \tag{4.46}$$

with some error coefficients $e^{(n)}$ which are bounded in subdomains away from the corner points.

On the basis of (4.46) we now can remove the pollution effect by means of Richardson extrapolation. Notice that only the local mesh size parameters, h_n, enter the expansion, which may be chosen independent of each other as long as the size is comparable to some fixed h. Now we proceed as follows.

Step 1. Calculate the usual Ritz projection $\mathcal{R}_h u$, on a triangulation with local meshsize h_n, $n=1,\dots N$, at the reentrant corners.

Step 2. Calculate a second Ritz projection $\mathcal{R}_{h'} u$, using a refined mesh with the same uniform structure, in each Ω_n, but local mesh size $h'_n = \lambda_n h_n$, where $\lambda_n^{2\alpha_{n;1}} = q$, and $q < 1$ is fixed. (4.47)

Step 3. Form the linear combination
$$\tilde{\mathcal{R}}_h u = (q^{-1}\mathcal{R}_{h'} u - \mathcal{R}_h u) / (q^{-1}-1) .$$

Clearly, by (4.46), in interior subdomains there holds
$$(u-\tilde{\mathcal{R}}_h u)(x) = \mathcal{O}(h^2|\log(h)|) .$$

Step 2 of algorithm (4.47) is performed very easily if all reentrant corners have the same inner angle, leading to the same powers of all h_n in the expansion (4.46). Then, all values λ_n are the same, i.e., we have to perform a usual subdivision. Further we note that one also gets the stress intensity factors with the optimal order by inserting $\tilde{\mathcal{R}}_h u$ into one of the representation formulas, since only the values of the approximate functions in the interior are needed.

5. CORNER PROBLEMS IN THE BOUNDARY ELEMENT METHOD

5.1 Gårding's inequality for boundary integral operators on polygonal domains

In this final section we want to discuss some techniques for removing or reducing the pollution effect in the context of boundary element approximations. We concentrate on the boundary element Galerkin method in order to have available similar convergence results as presented for the Ritz projection in the usual finite element method.

The basis for the error analysis in Section 3 is the relation (3.4) which gives us convergence of the Ritz projection even under the weak assumption that u has only finite energy, $u \in V$. In the boundary element method the correponding result holds true as soon as we have available a suitable Gårding inequality for the boundary operators, see [24].

As an example we again consider the mixed boundary value problem for the Laplacian, in the form

$$\begin{aligned} -\Delta u &= 0 \,, \quad \text{in } \Omega \,, \\ u &= b_1 \,, \quad \text{on } \Gamma_u \,, \\ \frac{\partial u}{\partial n} &= b_2 \,, \quad \text{on } \Gamma_\sigma \,. \end{aligned} \tag{5.1}$$

The *direct* methods are based on the Green identity

$$\begin{aligned} u(x) &= -\int_{\partial\Omega} \frac{\partial}{\partial n_y} \gamma(x,y)u(y)ds_y + \int_{\partial\Omega} \gamma(x,y)\frac{\partial u}{\partial n}(y)ds_y = \\ &= -\mathcal{K}u(x) + \mathcal{V}(\frac{\partial u}{\partial n})(x) \,, \end{aligned} \tag{5.2}$$

valid for $x \in \Omega$, where

$$\gamma(x,y) = \frac{1}{2\pi} \log|x-y| \tag{5.3}$$

is a fundamental solution for problem (5.1), with source point at x .

There are two reasonable procedures to obtain boundary equations from the Green identity.

1. Let x approach the boundary on both sides of (5.2). Then, by the well known jump conditions of the double layer potential $\mathcal{K}$, we have that
$$\frac{1}{2}u = -\mathcal{K}u + \mathcal{V}(\frac{\partial u}{\partial n}) \tag{5.4}$$
on the boundary $\partial\Omega$.
2. Form the directional derivative $\partial/\partial n_x$ on both sides of (5.2) and then, let x tend to $\partial\Omega$. Again, using the jump conditions for the adjoint operator $\mathcal{K}'$ of $\mathcal{K}$, the normal derivative of the single layer potential, we arrive at
$$\frac{1}{2}\frac{\partial u}{\partial n} = \mathcal{D}u + \mathcal{K}'(\frac{\partial u}{\partial n}) \,. \tag{5.5}$$
Here, $\mathcal{D}$ denotes the *hypersingular* operator defined by
$$\mathcal{D}u(x) = -\frac{\partial}{\partial n_x} \int_{\partial\Omega} \frac{\partial}{\partial n_y} \gamma(x,y)u(y)ds_y \,. \tag{5.6}$$

Both relations can serve as the basis of an approximation procedure.

First, we take only (5.4) to obtain the following pair of boundary equations, for the mixed problem (5.1),

$$\begin{aligned} (\tfrac{1}{2}I+\mathcal{K})u &= \mathcal{V}b_2 \qquad , \text{ on } \Gamma_\sigma \,, \\ \mathcal{V}(\frac{\partial u}{\partial n}) &= (\tfrac{1}{2}I+\mathcal{K})b_1 \,, \text{ on } \Gamma_u \,. \end{aligned} \tag{5.7}$$

As mentioned above, we are interested in a Gårding inequality for the boundary operators on the left. On smoothly bounded domains, only containing points of changing boundary conditions with $\omega = \pi$, the operator $\mathcal{K}$ is compact such that the first equation in (5.7) is a Fredholm integral equation of the second kind. Abbreviating the operators on the left of (5.7) by $\mathcal{A}$, we get the desired result from [30],

$$(\mathcal{A}(u,\frac{\partial u}{\partial n}),(u,\frac{\partial u}{\partial n}))_{L_2} \geq \lambda\{\|u\|^2_{L_2(\Gamma_\sigma)}+\|\frac{\partial u}{\partial n}\|^2_{\tilde{H}^{-1/2}(\Gamma_u)}\} - \mathcal{C}(u,\frac{\partial u}{\partial n}) . \quad (5.8)$$

Here $\tilde{H}^{-1/2}(\Gamma_u)$ is the dual space of $H^{1/2}(\Gamma_u)$, the space of traces of functions in $H^{1,2}(\Omega)$ on the boundary, and $\mathcal{C}$ denotes some compact perturbation. Notice that the norm on the right is not the *natural* energy norm of the problem, which should involve $\|u\|_{1/2;\Gamma}$ instead of the L_2-norm. Unfortunately, the estimate (5.8) does not carry over to polygonal domains, although the single layer potential $\mathcal{V}$ is still coercive, since here the kernel of $\mathcal{K}$ is singular, and $\mathcal{K}$ is not compact any more; see [6], where also a modification of the scheme (5.7) is proposed.

A different approach to problem (5.1) is derived by application of the identity (5.5), for the unknown function u on the Neumann boundary,

$$\begin{aligned} \mathcal{D}u &= (\tfrac{1}{2}I-\mathcal{K}')b_2 \text{ , on } \Gamma_\sigma , \\ \mathcal{V}(\frac{\partial u}{\partial n}) &= (\tfrac{1}{2}I+\mathcal{K})b_1 \text{ , on } \Gamma_u . \end{aligned} \quad (5.9)$$

Now, both integral equations are of the first kind, governed by a hypersingular operator on Γ_σ. The ellipticity properties for this kind of operators are derived from the theory presented in [31], which holds on general polygonal domains,

$$(\mathcal{A}(u,\frac{\partial u}{\partial n}),(u,\frac{\partial u}{\partial n}))_{L_2} \geq \lambda\{\|u\|^2_{H^{1/2}(\Gamma_\sigma)}+\|\frac{\partial u}{\partial n}\|^2_{\tilde{H}^{-1/2}(\Gamma_u)}\} + \mathcal{C}(u,\frac{\partial u}{\partial n}) , \quad (5.10)$$

see also [32] for a survey on this topic. Again, by $\mathcal{A}$ we mean the operators on the left in (5.9). Since the coerciveness result (5.10) is generally available and also the "correct" energy norm is dominated, the formulation (5.9) can be considered as the natural boundary integral analogue of the usual displacement formulation (2.2), on the domain Ω.

Similar results hold true for other elliptic problems. Let us, as a second example, consider the system of equations arising from linear elasticity

$$\mu\Delta u + (\lambda+\mu)\text{grad div } u = 0 \text{ , on } \Omega . \quad (5.11)$$

Proceeding as above, one derives the two boundary identities,

$$u(x) = \frac{1}{2}u(x) - \int_{\partial\Omega} (T(y,x)u(y)-F(x,y)t(y))ds_y , \quad (5.12)$$

and

$$Tu(x) = -T_x \int_{\partial\Omega} T(y,x)u(y)ds_y + \frac{1}{2}t(x) + \int_{\partial\Omega} T(x,y)^T t(y)ds_y . \quad (5.13)$$

Here, $t = Tu_{|\partial\Omega}$, and $Tu = \lambda n\mathrm{div}(u) + 2\mu\partial_n u + \mu n\times\mathrm{curl}(u)$ is the traction operator. Further, by $F(x,y)$ we mean the fundamental solution of the equation (5.11) and $T(y,x) = (T_y F(y,x))^T$, where the index indicates the argument of differentiation.

Let us consider the Neumann problem with prescribed tractions on the boundary. Then similar to the preceding discussion for the Laplacian, only the second equation (5.13) which leads to a boundary integral formulation of the first kind satisfies a Gårding inequality (with respect to the natural energy norm) on domains with corners. The corresponding results for the equation of the second kind, arising from (5.12), can be justified only on smoothly bounded domains ([24],[31]).

The two preceding examples indicate that integral equation formulations of the first kind are less sensitive to corner singularities and discretizations based on these equations will show a similar convergence behavior as the Ritz projection in the usual finite element method. In particular, the discrete method will always be convergent without any further modification.

5.2 Improved convergence of boundary element methods

Let us, finally, report some convergence results for the boundary element Galerkin method and show how some of the modifications discussed in Section 4 carry over to this type of discretization. For ease of presentation, we restrict ourselves to the Dirichlet problem for (5.1), $\Gamma_\sigma = \emptyset$. The results can easily be shown also for more general problems as long as a Gårding inequality with respect to the energy norm holds true.

In virtue of (5.9), the unknown function $\sigma = \partial_n u$ on $\partial\Omega = \Gamma_u$ satisfies the equation

$$\mathcal{V}\sigma = (\tfrac{1}{2}I+\mathcal{K})b , \text{ on } \partial\Omega . \quad (5.14)$$

For simplicity, we require $\mathrm{diam}(\Omega) < 1$ in order to exclude possible eigensolutions of $\mathcal{V}$. Further, there hold the ellipticity and continuity estimates

$$(\mathcal{V}\sigma,\sigma) \geq \lambda\|\sigma\|_{-1/2} - \mathcal{C}(\sigma,\sigma) , \quad \forall\, \sigma \in H^{-1/2}(\partial\Omega) , \quad (5.15)$$

$$(\mathcal{V}\sigma,\tau) \leq c\|\sigma\|_{-1/2}\|\tau\|_{-1/2} , \quad \forall\, \sigma,\tau \in H^{-1/2}(\partial\Omega) . \quad (5.16)$$

We now approximate the equation (5.14) by conformal finite elements $V_h \subset H^{-1/2}(\partial\Omega)$ consisting, e.g., of piecewise constant functions. The Galerkin solution $\sigma_h \in V_h$ is defined by

$$(\mathcal{V}(\sigma-\sigma_h),\varphi_h) = 0 \quad , \quad \forall \varphi_h \in V_h \ . \qquad (5.17)$$

By the estimates (5.15) and (5.16) this gives us the optimal convergence result

$$\|\sigma-\sigma_h\|_{-1/2} \leq c \inf_{\varphi_h \in V_h} \|\sigma-\varphi_h\|_{-1/2} \ , \qquad (5.18)$$

if h is sufficiently small, see, e.g., [24].

Let us now again assume that there exists only one significant corner z and that the expansion (2.16) holds, for u. This carries over to a singular representation for the unknown function σ, on the boundary, of the form

$$\sigma = \sum_{i=1}^{I} K_i t_i + \Sigma \ , \qquad (5.19)$$

where we set $t_i = \partial_n \bar{s}_i$. For the coefficients K_i, there hold the representation formulas from Theorem 2.1. The mean square estimates for the Ritz projection in Theorem 3.1 now have the following analogue.

Theorem 5.1: *For $\alpha < 1$, the solution of the boundary element Galerkin method satisfies the estimates*

$$\|\sigma-\sigma_h\|_{-3/2} + h^{\alpha}\|\sigma-\sigma_h\|_{-1/2} = O(h^{2\alpha-\epsilon}) \ . \qquad (5.20)$$

for any $\epsilon > 0$.

Here, in virtue of (5.18), the approximation properties of finite elements in negative Sobolev norms are used for the energy estimate while the improved convergence in $H^{-3/2}$ again is obtained by a duality argument as (3.6), see also [24]. The estimates (5.20) can not be improved as long as σ contains the leading singularity $t_1 \sim r^{\alpha-1}$.

For removing the pollution effect, we now may use the techniques described in Section 4, although not for all of them the theoretical foundation is available in this context as, for example, for the application of Richardson extrapolation. Up to now, mostly mesh refinements have been used in connection with boundary elements for which the error analysis with respect to mean square norms yields similar results as given in Section 4.1. Moreover, the SFM has been analyzed and implemented for several corner and crack problems, see [6],[7], and [30]. Here, we shall restrict ourselves to the description of the DSFM and the SFM, which can be written in a form similar to that given in Section 4.

The basis for the three variants of the DSFM is the possibility to calculate approximate stress intensity factors by a representation formula. Here, we prefer formula (i) from Theorem 2.1 since it contains only boundary integrals if the right hand side f vanishes,

$$K_i = (Bu, S\bar{s}_{-i})_{\partial\Omega} - (S_* u, B_* \bar{s}_{-i})_{\partial\Omega} . \tag{5.21}$$

Bu stands for the *given* Cauchy data whereas $T_* u$ is determined, approximately, by the boundary element equations. For our specific problem (5.14), the formula (5.21) reads as

$$K_i = -(b, \partial_n \bar{s}_{-i})_{\partial\Omega} + (\sigma, \bar{s}_{-i})_{\partial\Omega} . \tag{5.22}$$

An application of the corresponding formula in plane elasticity is found in [33].

Now, as in Section 4.3, we may replace the unknown function σ by some approximation. If we insert the Galerkin solution $\sigma_h \in V_h$ of (5.17) we immediately arrive at the estimate

$$K_i - K_i^h = (\sigma - \sigma_h, \bar{s}_{-i})_{\partial\Omega} \le c\|\sigma - \sigma_h\|_{-3/2} . \tag{5.23}$$

Here we have use the fact that $B_* \bar{s}_{-i} = \bar{s}_{-i}$ vanishes on the edges near the corner point and is smooth away from it such that we may control the error for the stress intensity factors by an arbitrary weak norm of the approximation error.

As in the procedure (4.22) we may now employ formula (5.22) for an iterative correction of the solution σ_h. The iterates converge to a limit solution, $\bar{\sigma}_h^I$, which is characterized by the following scheme.

Step 0. Determine the boundary element solutions $t_{i,h}$ of t_i by (5.17) and calculate the matrix

$$\bar{a}_{ij}^I = -(\bar{s}_i, \partial_n \bar{s}_{-j})_{\partial\Omega} + (t_{i,h}, \bar{s}_{-j})_{\partial\Omega} , \quad i,j=1,\dots,I .$$

Step 1. For given b, calculate $\sigma_h \in V_h$.

Step 2. Define $\bar{K}_i^I$ by solving (5.24)

$$\sum_{i=1}^{I} \bar{a}_{ij}^I \bar{K}_i^I = -(b, \partial_n \bar{s}_{-j})_{\partial\Omega} + (\sigma_h, \bar{s}_{-j})_{\partial\Omega} .$$

Step 3. Set $\bar{\sigma}_h^I = \sigma_h + \sum_{i=1}^{I} \bar{K}_i^I (t_i - t_{i,h})$.

This procedure is formally equivalent to (4.26) and is analyzed in the same way.

<u>Theorem 5.2</u>: *For the solution of the DSFM in the boundary element Galerkin method, using I dual singular functions, there hold the estimates*

$$\|\sigma-\sigma_h^{-I}\| + \sum_{i=1}^{I} |K_i-K_i^{-I}| \leq c\|\Sigma-\Sigma_h\| ,$$

where $\|.\|$ *denotes an arbitrary weak norm.*

Thus, again the rate of convergence for the smooth remainder Σ completely determines the error for the solution, and for the stress intensity factors. For this we may apply the analogue of the estimate (3.9), if $\Sigma \in H^{1/2}(\partial\Omega)$, whereas an optimal order result like the weighted norm estimates in Theorem 3.2 seems not yet available for boundary element approximations.

As in the finite element method the implementation of the DSFM only requires the solution of linear equations for the usual system matrix. In particular, the singular functions in Step 0 are approximated by the equations

$$(\mathcal{V}t_{i,h},\varphi_h)_{\partial\Omega} = (\mathcal{V}t_i,\varphi_h)_{\partial\Omega} = ((\tfrac{1}{2}I+\mathcal{K})\bar{s}_i,\varphi_h)_{\partial\Omega} . \quad (5.25)$$

The last identity holds due to our definition of t_i in (5.19), not involving a cut-off function τ. The right hand side now is easily evaluated, since $\bar{s}_i$ is a smooth function on the boundary.

Let us conclude with some remarks on the SFM which makes use of the augmented spaces

$$V_h^I = V_h \oplus \{t_1,\dots,t_I\}$$

as trial and test spaces in the Galerkin equation (5.17). In virtue of the scheme (4.16) we may use the analogous form of (5.24) instead of explicitly forming the augmented Galerkin matrices. The I×I-matrix a_{ij}^{-I} in Step 0 and Step 2 must be replaced by

$$a_{ij}^I = (\mathcal{V}(t_i-t_{i,h}),t_j)_{\partial\Omega} = (t_i-t_{i,h},(\tfrac{1}{2}I+\mathcal{K})\bar{s}_j)_{\partial\Omega} . \quad (5.26)$$

Thus, as in formula (5.25) we avoid the evaluation of boundary integral operators applied to functions that become singular near the corner, by our definition of t_i .

We see that again the SFM and the DSFM require the same amount of computational work if, for the latter, we do not count the calculation of the integral $(b,\partial_n\bar{s}_{-j})_{\partial\Omega}$ on the right hand side of Step 2, which has to be handled with care, near z .

The error analysis for the SFM in boundary element methods was given, e.g., in [7],[30], yielding estimates for the solution and for the coefficients K_i simultaneously. The mean square estimates naturally show the same optimal results as stated in Theorem 4.2 for the Ritz projection. For the stress intensity factors, one may also follow the lines of the analysis in Section 4.2, starting from the error identity

$$\sigma-\sigma_h^I = \sum_{i=1}^{I}(K_i-K_i^I)(t_i-t_{i,h}) + (\Sigma-\Sigma_h) . \qquad (5.27)$$

For instance in the case $I = 1$ we obtain, as in (4.14),

$$K_1-K_1^1 = -(\mathcal{V}(\Sigma-\Sigma_h),t_1-t_{1,h})_{\partial\Omega} / (\mathcal{V}(t_1-t_{1,h}),t_1-t_{1,h})_{\partial\Omega} . \qquad (5.28)$$

The nominator may be estimated by (5.16) which gives us

$$(\mathcal{V}(\Sigma-\Sigma_h),t_1-t_{1,h})_{\partial\Omega} \le c\|\Sigma-\Sigma_h\|_{-1/2}\|t_1-t_{1,h}\|_{-1/2} = \mathcal{O}(h^{1+\alpha-\varepsilon}) , \qquad (5.29)$$

if piecewise constant elements are used. Since the denominator is of the order $\mathcal{O}(h^{2\alpha-\varepsilon})$ the combined estimate yields $\mathcal{O}(h^{1-\alpha-\varepsilon})$ which is slightly worse than the result in Theorem 4.3. The complete correspondence of results for boundary elements and finite elements would again require the validity of some analogue of the weighted estimate in Theorem 3.2.

REFERENCES

1. Kondrat'ev, V.A.: Boundary value problems for elliptic equations in domains with conical or angular points, Trans. Moscow. Math. Soc. 16 (1967), 227-313
2. Maz'ja, V.G. and B.A. Plamenevskij: Coefficients in the asymptotics of the solutions of elliptic boundary value problems, J. Sov. Math. 9 (1978), 750-764
3. Fix, G.J., Gulati, S. and G.I. Wakoff: On the use of singular functions with finite element approximations, J. Comput. Physics 13 (1973), 209-228
4. Blum, H. and M. Dobrowolski: On finite element methods for elliptic equations on domains with corners, Computing 28 (1982), 53-63
5. Dobrowolski, M.: On the numerical treatment of elliptic equations with singularities, Report 8505, Universität der Bundeswehr München, FB Informatik, 1985
6. Costabel, M. and E. Stephan: Boundary integral equations for mixed boundary value problems in polygonal domains and Galerkin approximation, Mathematical models and methods in mechanics, Banach Center Publ. 15, Warsaw 1985
7. Stephan, E.P. and W.L. Wendland: An augmented Galerkin procedure for the boundary integral method applied to two-dimensional screen and crack problems, Appl. Anal. 18 (1984), 183-220
8. Tolksdorf, P.: On the Dirichletproblem for quasilinear equations in domains with conical boundary points, Comm. PDE 8 (1983), 773-810

9. Dobrowolski, M.: On finite element methods for nonlinear elliptic problems on domains with corners, in: Singularities and Constructive Methods for Their Treatment, Oberwolfach 1983, Lecture Notes in Math. 1121, Springer 1985
10. Melzer, H. and R. Rannacher: Spannungskonzentrationen in Eckpunkten der Kirchhoffschen Platte, Bauingenieur 55 (1980), 181-184
11. Blum, H. and R. Rannacher: A note on Herrmann's second mixed plate element, Preprint, Universität Saarbrücken, in preparation
12. Babuska, I., Kellogg, R.B. and J. Pitkäranta: Direct and inverse error estimates for finite elements with mesh refinements, Numer. Math. 33 (1979), 447-471
13. Schatz, A.H. and L.B. Wahlbin: Maximum norm estimates in the finite element method on plane polygonal domains I,II, I: Math. Comp. 32 (1978), 73-109, II: Math. Comp. 33 (1979), 465-492
14. Babuska, I. and W.C. Rheinboldt: Error estimates for adaptive finite element computations, SIAM J. Numer. Anal. 15(4) (1978), 736-754
15. Bank, R.E. and A. Weiser: Some a posteriori error estimates for elliptic partial differential equations, Math. Comp. 44 (1985), 1-19
16. Rank, E.: A posteriori Fehlerabschätzungen und adaptive Netzverfeinerung für Finite-Element und Randintegralmethoden, Mitteilungen aus dem Institut für Bauingenieurwesen I, Heft 16, TU München 1985
17. Dobrowolski, M.: Numerical approximation of elliptic interface and corner problems, Habilitationsschrift, Universität Bonn 1981
18. Blum, H.: Der Einfluß von Eckensingularitäten bei der numerischen Behandlung der biharmonischen Gleichung, Bonner Math. Schr. 140, 1982
19. Moussaoui, M.A.: Sur l'approximation des solution du problème de Dirichlet dans un ouvert avec coins, in: Singularities and Constructive Methods for Their Treatment, Oberwolfach 1983, Lecture Notes in Math. 1121, Springer 1985
20. Blum, H., Lin, Q. and R. Rannacher: Asymptotic error expansion and Richardson extrapolation for linear finite elements, Numer. Math. 49 (1986), 11-37
21. Blum, H. and R. Rannacher: Extrapolation techniques for reducing the pollution effect of reentrant corners in the finite element method, submitted to Numer. Math.
22. Blum, H.: On the approximation of linear elliptic systems on polygonal domains, in: Singularities and Constructive Methods for Their Treatment, Oberwolfach 1983, Lecture Notes in Math. 1121, Springer 1985
23. Grisvard, P.: Elliptic problems in nonsmooth domains, Pitman 1985
24. Wendland, W.L.: Boundary element methods and their asymptotic convergence, in: Theoretical Acoustics and Numerical Techniques (Ed. P. Filippi), CISM Courses and Lectures, Springer 1983
25. Babuska, I.: Finite element method for domains with corners, Computing 6 (1970), 264-273

26. Zenger, C. and H. Gietl: Improved difference schemes for the Dirichlet problem of Poisson's equation in the neighbourhood of corners, Numer. Math. 30 (1978), 315-332
27. Schatz, A.H.: Lectures on the treatment of corner singularities, Universität Bonn 1980 (unpublished)
28. Schatz, A.H.: A weak discrete maximum principle and stability in the finite element method in L^{∞} on plane polygonal domains, Math. Comp. 34 (1980), 77-91
29. Frehse, J. and R. Rannacher: Eine L^{1}-Fehlerabschätzung für diskrete Grundlösungen in der Methode der finiten Elemente, Bonner Math. Schr. 89 (1976), 92-114
30. Wendland, W.L., Stephan, E. and G.C. Hsiao: On the integral equation method for the plane mixed boundary value problem of the Laplacian, Math. Meth. in the Appl. Sci. 1 (1979), 265-321
31. Costabel, M.: Starke Elliptizität von Randintegraloperatoren erster Art, Habilitationsschrift, TH Darmstadt 1984
32. Costabel, M.: Principles of boundary element methods, Preprint 998, TH Darmstadt 1986
33. Kuhn, G.: Boundary element technique in elastostatics and linear fracture mechanics, this volume
34. Blum, H.: A simple and accurate method for the determination of stress intensity factors and solutions for problems on domains with corners, in: The Mathematics of Finite Elements and Applications IV (Ed. J.R. Whiteman), Academic Press 1982
35. Dobrowolski, M.: On quasilinear elliptic equations in domains with conical boundary points, Report 8506, Universität der Bundeswehr München, FB Informatik, 1985
36. Nitsche, J.A.: Zur lokalen Konvergenz von Projektionen auf finite Elemente, Approximation Theory, Bonn 1976, Lecture Notes in Math. 556, Springer 1976

PLASTIC ANALYSIS BY BOUNDARY ELEMENTS

G. Maier, G. Novati, U. Perego
Politecnico di Milano, Milan, Italy.

1. INTRODUCTION

1.1 Problems in quasi-static, small deformation plasticity and purposes of this survey

In the traditional context of small deformation, quasi-static plasticity, we can distinguish the following kinds of problems (for details, see e.g. refs. [1] [2]).

(a) Incremental analysis, understood as the determination of geometric changes and stresses along a hystory of external actions. From the mathematical standpoint problems of this kind are centered on the time integration of the nonlinear relations governing the boundary value (b.v.) problem in terms of rates.

(b) Safety analysis, namely the determination of the load factor at which a critical situation occurs, which encompasses: (i) limit analysis when the "basic" loads affected by that factor are fixed and the structural crisis means plastic collapse; (ii) shakedown analysis when the basic loads are variable and repeated and the critical event is represented either by incremental collapse or alternating plasticity. In both cases one is led to mathematical programming, i.e. optimization of a function under inequality constraints.

(c) Bounding techniques, intended to provide upper bounds on hystory-dependent quantities without performing an incremental, evolutive analysis. To this purpose nonlinear mathematical programming plays a key role.

(d) Optimum limit design, i.e. the determination of design variables such that the total weight (or some other "cost function") attains its minimum under the condition that the system does not collapse. Once again linear or nonlinear programming represent the ideal mathematical tool.
(e) Structural identification in which experimental data on the overall response to given loads must be used to identify local properties such as constitutive constants or damages. This is an inverse problem amounting to the estimation of parameters.

Boundary Element (BE) methods have been fairly extensively developed for incremental elastic-plastic analysis since the pioneering work of Swedlow and Cruse, Riccardella, Mendelson, Rizzo et. al. in the seventies, e.g. [3] [4] [5]. For a conspectus of the developments of the subject the reader is referred to comprehensive books on BE methods, [6] [7] [8], where also linear elasticity problems are dealt with, as they are in more compact surveys such as [9]. Only recently BE approaches have been considered for limit and shakedown analysis and bounding techniques. To the authors' knowledge no experience has been gained so far in the last two problem classes in the above list. Therefore these will not be considered in the present survey, which will be primarily devoted to the first category of problems (incremental analysis).
Both large-deformation and dynamic plastic analysis by BEs are beyond the purpose and space limitations of this paper, although remarkable contributions have been published in the former area, e.g. [10].

Focus will be here on what is believed to represent main ideas and general procedures in the present (1986) intersection area between BE methods and small-strain quasistatic plasticity. Detailed developments of even important aspects, implementation in computer codes, computational experience, examples and engineering applications are not dealt with in the present survey but can be found disseminated in the literature, a comprehensive sample of which is represented by the reference list. Even within such severe limitations completeness is not our purpose and could be hardly achieved. Attention is preferably paid here to the most popular "direct" BE method and to some approaches which, though still under development, in the writers' opinion are promising or meaningful either from the computational standpoint or from the point of view of applied plasticity theory.

The resulting table of contents is as follows.

1. Introduction
2. Integral equations for elastic-plastic analysis (Somigliana equations allowing for inelastic strains; Integral equations for strains or stresses)
3. Discretizations and integrations in space (Discretizations; Algebrization of the boundary integral equation; Algebrization of the equation for stresses)
4. Essential features of elastic-plastic material models (General constitutive laws in terms of rates; Piecewiselinear approximations of plastic laws)
5. Incremental elastic-plastic BE analysis (The basic procedure; Integration in time)

6. On symmetrizations of the rate problem (Formulation of plastic laws for cells; Direct symmetrization of the BE equations; A Galerkin formulation and procedure)
7. Limit and shakedown analysis and bounding techniques (Plastic collapse analysis as a linear programming problem; Shakedown analysis and bounds on residual quantities).
8. Variants and alternative formulations (Variants resting on morphological peculiarities; Alternative formulations)
9. References

1.2 Notation

Some of the notions and symbols most frequently used in what follows are specified below.

Cartesian coordinates x_i $(i = 1,2,3)$, Cartesian tensors and index summation convention are adopted; δ_{ij} = Kronecker symbol. Ω = open domain, Γ = boundary of Ω (Γ_u with given displacements, Γ_p with given tractions, so that $\Gamma_u + \Gamma_p = \Gamma$); n_i = outward normal to Γ, possibly not uniquely defined (but no cusps: Lipschitz boundary).
Time derivatives are marked by a dot; space derivatives by a comma preceeding the index involved. Matrices (and column vectors) are indicated by underscored symbols; a tilde means transpose; vector inequalities apply componentwise; $\underline{I}$ = unit matrix and $\underline{0}$ = matrix with all zero components. Field point and load point are represented by $\underline{x} = \{x_i\}$, $\xi = \{\xi_i\}$, respectively. Indices concerning ξ are Greek and located at first place; indices concerning $\underline{x}$ are Latin and in second place. An outward normal at ξ will be denoted by $\underline{\nu}$. Equations which are definitions of new symbols make use of $\equiv$.

2. INTEGRAL EQUATIONS FOR ELASTIC-PLASTIC ANALYSIS

2.1 Somigliana equations allowing for inelastic strains

The traditional starting point of the BE direct methods in linear elasticity is the so-called Somigliana identity which relates the displacements in any interior point of the domain Ω, to the displacements and tractions of the boundary Γ and to body forces and imposed "initial" (such as thermal) strains in Ω (fig.2.1). In the present, inelastic context, the (unknown) plastic strains will preliminarily be regarded as imposed in order to formulate the basic integral equations ("elastic phase" of elastoplastic BE analysis); subsequently, they will be related to stresses in incremental terms in order to account for the inelastic (nonlinear, path-dependent) constitutive laws of materials ("plastic

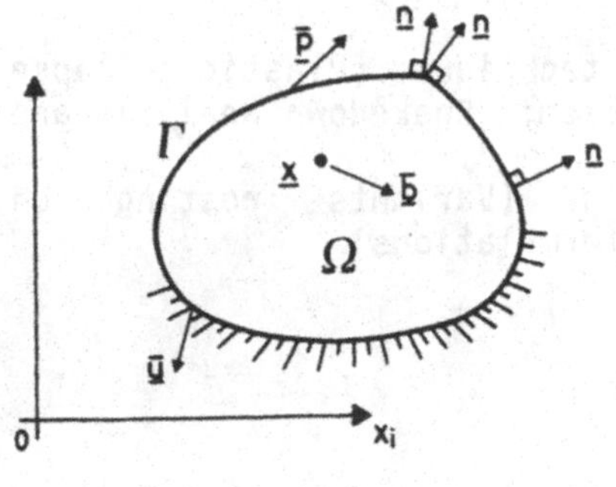

Fig.2.1

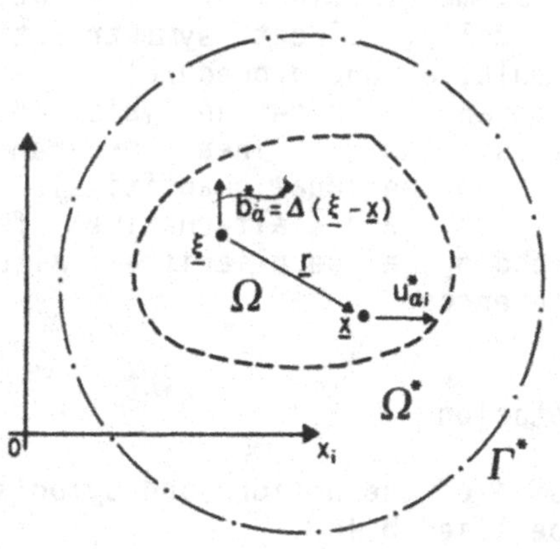

Fig.2.2

phase"). The Somigliana equation can be derived from Betti reciprocity theorem written in a generalized form which allows for possible inelastic strains.First, this generalized Betti theorem is obtained below.

Consider two different elastic solutions (relevant to different sets of external actions) defined over the same domain Ω filled with a given elastic material. Each one encompasses displacements u_i, body forces b_i, (total) strains ε_{ij}, plastic strains ε^p_{ij}, stresses σ_{ij} and boundary tractions p_i. We will denote with primed and unprimed symbols such sets of quantities for the two solutions. Due to the symmetry of the elastic moduli tensor $E_{ijhk} = E_{hkij}$, the relation

$$(\varepsilon_{ij} - \varepsilon^p_{ij})\, \sigma'_{ij} = (\varepsilon'_{ij} - \varepsilon^{p'}_{ij})\, \sigma_{ij} \qquad (2.1)$$

holds true at every point of Ω; hence, integrating it over the domain, one obtains:

$$\int_\Omega \varepsilon_{ij}\, \sigma'_{ij}\, d\Omega - \int_\Omega \varepsilon^p_{ij}\, \sigma'_{ij}\, d\Omega = \int_\Omega \varepsilon'_{ij}\, \sigma_{ij}\, d\Omega - \int_\Omega \varepsilon^{p'}_{ij}\, \sigma_{ij}\, d\Omega \qquad (2.2)$$

The first integrals on both l.h.s. and r.h.s. are then processed making use of the strain-displacement relations, integrations by parts and the equilibrium equations. Thus the sought Betti theorem is arrived at in the form:

$$\int_\Omega u_i\, b_i'\, d\Omega + \int_\Gamma u_i\, p_i'\, d\Gamma + \int_\Omega \varepsilon_{ij}^{p'}\, \sigma_{ij}\, d\Omega = \tag{2.3}$$

$$= \int_\Omega u_i'\, b_i\, d\Omega + \int_\Gamma u_i'\, p_i\, d\Gamma + \int_\Omega \varepsilon_{ij}^{p}\, \sigma_{ij}'\, d\Omega$$

Starting from eq.(2.3), the Somigliana identity is obtained as follows. The unprimed quantities are interpreted as the ones relevant to the elastoplastic boundary value (b.v.) problem to be solved; those denoted by a prime are chosen as the ones characterizing the elastic solution, defined over a domain $\Omega^* \supset \Omega$, due to a unit concentrated force acting at point $\underline{\xi}$ in direction α, in the absence of imposed inelastic strains. Namely $b_i'(\underline{x})$ becomes $\delta_{\alpha i}\ \Delta(\underline{x}-\underline{\xi})$, Δ denoting the Dirac distribution associated to point $\underline{\xi}$, $\delta_{\alpha i}$ the Kronecker symbol (= 1 for $i = \alpha$; = 0 for $i \neq \alpha$); a star will be used from now onward, to mark quantities associated to such fictitious elastic solution; see fig. 2.2. Using the sifting property of the delta "function", eq.(2.3) gives:

$$u_\alpha(\underline{\xi}) + \int_\Gamma p_{\alpha i}^*(\underline{\xi},\underline{x})\, u_i(\underline{x})\, d\Gamma = \int_\Gamma u_{\alpha i}^*(\underline{\xi},\underline{x})\, p_i(\underline{x})\, d\Gamma + \tag{2.4}$$

$$+ \int_\Omega u_{\alpha i}^*(\underline{\xi},\underline{x})\, b_i(\underline{x})\, d\Omega + \int_\Omega \sigma_{\alpha ij}^*(\underline{\xi},\underline{x})\, \varepsilon_{ij}^p(\underline{x})\, d\Omega$$

It is worth noting the peculiar "two-points" ($\underline{x}$ and $\underline{\xi}$) nature of the fictitious solution adopted ("fundamental solution").

For the fictitious domain Ω^* containing Ω, as in BE elastic analysis, the most convenient and popular choice in many cases is the unbounded space Ω_∞, with the elastic properties of Ω if this is homogeneous. When Ω exhibits elastic isotropy, the starred kernels in eq. (2.4) are the classical fundamental solution of Kelvin for three-dimensional situations ($\alpha, i = 1,2,3$), see for details [6] [7] [8]:

$$u_{\alpha i}^* \equiv \frac{1}{2kG}\,\frac{1}{r}\,[(3-4\nu)\,\delta_{\alpha i} + r_{,\alpha}\, r_{,i}] \tag{2.5}$$

$$p_{\alpha i}^* \equiv -\frac{1}{k}\,\frac{1}{r^2}\,[((1-2\nu)\,\delta_{\alpha i} + 3r_{,\alpha}\, r_{,i}) + \tag{2.6}$$

$$- (1-2\nu)\,(r_{,\alpha}\, n_i - r_{,i}\, n_\alpha)]$$

$$\overset{*}{\sigma}_{\alpha ij} \equiv -\frac{1}{k}\frac{1}{r^2}\left[(1-2\nu)\left(\delta_{\alpha i}\, r_{,j} + \delta_{\alpha j}\, r_{,i} - \delta_{ij}\, r_{,\alpha}\right) + \right.$$
$$\left. + 3r_{,i}\, r_{,j}\, r_{,\alpha}\right] \tag{2.7}$$

having set:

$$k \equiv 8\pi(1-\nu) \; ; \qquad r_i \equiv x_i - \xi_i \; ; \qquad r \equiv (r_i r_i)^{1/2} \tag{2.8}$$

These two-points functions are singular at $\underline{x} = \underline{\xi}$ (i.e. for $r = 0$) and, hence, the last two integrals in eq.(2.4) are so (the other integrals are not singular for $\underline{\xi}$ in Ω, as Ω is open, i.e. Ω and Γ are disjointed sets).

A function $f(r)$ singular at $r = 0$ will be said to behave like $O(r^{-n})$, briefly $f(r) \cong O(r^{-n})$, if $f(r)\, r^n$ is bounded; hence, the exponent n will be thought of as a measure of the "strength" of the singularity. The kernels $\overset{*}{u}_{\alpha i}$ and $\overset{*}{\sigma}_{\alpha ij}$ are singular of order 1 and 2 respectively. Passing to spherical coordinates centered on $\underline{\xi}$, one realizes that the singularities of the integrands in both domain integrals of (2.4) disappear. Therefore both these integrals can be understood in the usual sense.

If the load point $\underline{\xi}$ is taken to the boundary Γ, the Somigliana identity (2-4) gives rise to the boundary integral equation:

$$c_{\alpha\beta}(\underline{\xi})\, u_\beta(\underline{\xi}) + \int_\Gamma \overset{*}{p}_{\alpha i}(\underline{\xi},\underline{x})\, u_i(\underline{x})\, d\Gamma = \int_\Gamma \overset{*}{u}_{\alpha i}(\underline{\xi},\underline{x})\, p_i(\underline{x})\, d\Gamma +$$
$$+ \int_\Omega \overset{*}{u}_{\alpha i}(\underline{\xi},\underline{x})\, b_i(\underline{x})\, d\Gamma + \int_\Omega \overset{*}{\sigma}_{\alpha ij}(\underline{\xi},\underline{x})\, \varepsilon^p_{ij}(\underline{x})\, d\Gamma \tag{2.9}$$

where, since $\overset{*}{p}_{\alpha i} \cong O(r^{-2})$, the first boundary integral is to be understood in the sense of Cauchy principal value and $c_{\alpha\beta}$ denotes the 'free-term' coefficient. This depends on the geometry of Γ in the neighbourhood of $\underline{\xi}$ and is $\frac{1}{2}\,\delta_{\alpha\beta}$ if Γ is smooth in $\underline{\xi}$ (see e.g. [6] [7] [8]).

The last integral in eq.(2.9) can be re-written in the form:

$$\int_\Omega \overset{*}{\varepsilon}_{\alpha ij}(\underline{\xi},\underline{x})\, \sigma^p_{ij}(\underline{x})\, d\Omega \, , \qquad \text{with:} \quad \sigma^p_{ij} \equiv E_{ijhk}\, \varepsilon^p_{hk} \tag{2.10a,b}$$

$\overset{*}{\varepsilon}_{\alpha ij}$ being the fundamental solution in terms of strains, and σ^p_{ij} the so-called "initial" or "plastic" stresses (sometimes this denomination is used for the opposite of the above σ^p_{ij}). The purely formal variant (2.10) imply formal advantages in 2D situations, [8].

In elastic analysis, the domain integrals can be advantageously transformed into boundary integrals under practically weak conditions on the body force and imposed (thermal) strain distributions (the distinction being immaterial, as the latter can be taken care of as fictitious body forces and tractions), see e.g. [11] [6]. In inelastic analysis the distribution of the (unknown) inelastic strain response does not satisfy in general the above conditions. Therefore, the transformation into boundary integral, if any, must be confined to the first domain integral, but will not be pursued here.

In contrast to what happens in (thermo) elasticity, the collocation of eq.(2.9) at any point $\underline{\xi}$ on Γ, together with the enforcement of the boundary conditions, is not sufficient to define the (boundary) solution of the b.v. elastic-plastic problem. The plastic constitutive law is clearly required to complete the set of governing relations and so is an integral equation relating plastic strains to stresses (or strains) through the elastic and geometric properties of the body. This equation will be discussed in the next Section.

Several b.v. problems occurring in engineering (especially in geotechnical and offshore engineering) are defined over unbounded domains. The generalization of the above developments to such situations can easily be carried out by the very same path of reasoning followed in linear elasticity. This rests on an usual formulation within a fictitious spherical surface whose radius is taken to the infinite and on weak ("regularity") conditions on the asymptotic behaviour of the solution. These conditions, in turn, can be a priori rigorously shown to be satisfied if the body forces are confined to a bounded region, see e.g. [12]. The integrals over the fictitious spherical boundary turn out to vanish in the limit, so that the integral eqs.(2.4)(2.10) and their consequences in the next Section hold unaltered for unbounded domains [8].

2.2 Integral equations for strains or stresses

The Somigliana equation (2.4) concerning displacements in interior points $\underline{\xi}$ in Ω, provides an expression for $\varepsilon_{ij}(\underline{\xi})$ if differentiated with respect to the load point coordinates according to the (linear) compatibility operator. The kernel $\overset{*}{\sigma}_{\alpha ij}$, eq.(2.7) exhibits a singularity of order $O(r^{-2})$, which cannot be suppressed by recourse to spherical coordinates centered on $\underline{\xi}$, since ξ_i are the very variables in the differentiation. As pointed out by Bui [13], in order to differentiate with respect to ξ_α the last integral in eq.(2.9), one has to split its integration domain Ω into an infinitesimal spherical neighbourhood Ω' of $\underline{\xi}$, centered in $\underline{\xi}$, and the remainder $\Omega-\Omega'$. The differentiation of the integral over the former subdomain can be shown to give a contribution rigorously negligible under very weak regularity conditions on $\varepsilon^p_{ij}(\underline{x})$. The differentiation of the latter integral over $\Omega - \Omega'$ gives rise to two addends [13]:

$$\frac{\partial}{\partial \xi_\beta} \int_{\Omega-\Omega'} \sigma^*_{\alpha ij}(\underline{\xi},\underline{x})\varepsilon^p_{ij}(\underline{x})d\Omega = \int_{\Omega-\Omega'} \sigma^*_{\alpha ij,\beta}\, \varepsilon^p_{ij}\, d\Omega - \varepsilon^p_{ij}(\underline{\xi}) \int_{\Gamma'} \sigma^*_{\alpha ij}\, n_\beta\, d\Gamma \quad (2.11)$$

where Γ' is the surface of the sphere and n_β is its outward normal. The noteworthy latter addend ("convected" term) on the r.h.s. of eq.(2.11) is expected since the integration domain $\Omega-\Omega'$ depends on the differentiation variable ξ_β (cf. the classical Leibnitz formula for $\frac{d}{d\xi}\int_a^b f(x,\xi)dx$ when $b = b(\xi)$). The integral equation for stresses is generated from the one for strains through subtraction of the plastic strains and multiplication by the elastic moduli tensor. It reads:

$$\sigma_{\alpha\beta}(\underline{\xi}) = \int_\Gamma u^*_{\alpha\beta i}(\underline{\xi},\underline{x})p_i(\underline{x})d\Gamma - \int_\Gamma p^*_{\alpha\beta i}(\underline{\xi},\underline{x})u_i(\underline{x})d\Gamma +$$

$$+ \int_\Omega u^*_{\alpha\beta i}(\underline{\xi},\underline{x})b_i(\underline{x})d\Omega + \int_\Omega \sigma^*_{\alpha\beta ij}(\underline{\xi},\underline{x})\varepsilon^p_{ij}(\underline{x})d\Omega + f_{\alpha\beta\phi\psi}\, \varepsilon^p_{\phi\psi}(\underline{\xi}) \quad (2.12)$$

The last, nonintegral "free" term arises through manipulations here omitted for brevity (see e.g. [7] [8]) from the convective latter addend in eq.(2.12) and reads:

$$f_{\alpha\beta\phi\psi}\, \varepsilon^p_{\phi\psi} = \frac{-2G}{15(1-\nu)} [(7-5\nu)\varepsilon^p_{\alpha\beta} + (1+5\nu)\, \delta_{\alpha\beta}\, \varepsilon^p_{\chi\chi}] =$$

$$= \frac{-2G}{15(1-\nu)} [(7-5\nu)\, \delta_{\alpha\phi}\, \delta_{\beta\phi} + (1+5\nu)\, \delta_{\alpha\beta}\, \delta_{\chi\phi}\, \delta_{\chi\psi}]\, \varepsilon^p_{\phi\psi} \quad (2.13)$$

The kernels in eq.(2.12), obtained from the corresponding ones in eq.(2.4), are given below:

$$u^*_{\alpha\beta i}(\underline{\xi},\underline{x}) = \frac{1}{k}\, \frac{1}{r^2} \{(1-2\nu)(r_{,\alpha}\, \delta_{\beta i} + r_{,i}\, \delta_{\alpha\beta} - r_{,\beta}\, \delta_{\alpha i}) +$$

$$+ 3r_{,\alpha}\, r_{,\beta}\, r_{,i}\} \quad (2.14)$$

$$p^{*}_{\alpha\beta i}(\underline{\xi},\underline{x}) = \frac{-2G}{k}\frac{1}{r^3}\{3\frac{\partial r}{\partial n}[(1-2\nu)\delta_{\alpha i}\, r_{,\beta} + \nu(\delta_{\beta i}\, r_{,\alpha} + \delta_{\alpha\beta}\, r_{,i}) - 5\, r_{,\alpha}\, r_{,\beta}\, r_{,i}] + 3\nu(n_i\, r_{,\alpha}\, r_{,\beta} + n_\alpha\, r_{,\beta}\, r_{,i}) + (1-2\nu)(3n_\beta\, r_{,\alpha}\, r_{,i} + n_\alpha\, \delta_{\beta i} + n_i\, \delta_{\alpha\beta}) - (1-4\nu)n_\beta\, \delta_{\alpha i}\} \tag{2.15}$$

$$\sigma^{*}_{\alpha\beta ij}(\underline{\xi},\underline{x}) = \frac{2G}{k}\frac{1}{r^3}\{3(1-2\nu)(\delta_{\alpha\beta}\, r_{,i}\, r_{,j} + \delta_{ij}\, r_{,\alpha}\, r_{,\beta}) + 3\nu(\delta_{\alpha i}\, r_{,\beta}\, r_{,i} + \delta_{\beta i}\, r_{,\alpha}\, r_{,j} + \delta_{\alpha i}\, r_{,\beta}\, r_{,j} + \delta_{\beta j}\, r_{,\alpha}\, r_{,i}) - 15\, r_{,\alpha}\, r_{,\beta}\, r_{,i}\, r_{,j} + (1-2\nu)(\delta_{\alpha i}\, \delta_{\beta i} + \delta_{\beta i}\, \delta_{\alpha i}) - (1-4\nu)\delta_{\alpha\beta}\, \delta_{ij}\} \tag{2.16}$$

The last kernel (2.16) exhibits a singularity $O(r^{-3})$. The last integral in (2.12) is understood in the sense of Cauchy principal value, i.e. as the following limit:

$$\lim_{\rho\to 0} \int_{\Omega-\Omega_\rho} \sigma^{*}_{\alpha\beta ij}(\underline{\xi},\underline{x})\, \varepsilon^{p}_{ij}(\underline{x})\, d\Omega \tag{2.17}$$

where Ω_ρ is a small sphere of radius ρ centered at load point $\underline{\xi}$ in Ω. The existence of such limit can be formally proved through the path of reasoning outlined below.

In view of the peculiar structure of the kernel, by using spherical coordinates r, ϕ, θ centered at $\underline{\xi}$ (ϕ being the "longitude" and θ the "latitude" angle), the above integral can be given the form:

$$\int_{\Omega-\Omega_\rho} \frac{1}{r^3}\psi^{*}_{\alpha\beta ij}(\phi,\theta)\, [\varepsilon^{p}_{ij}(\underline{x}) - \varepsilon^{p}_{ij}(\underline{\xi})] \cos\theta\, r^2\, dr\, d\phi\, d\theta + \varepsilon^{p}_{ij}(\underline{\xi}) \int_{\Omega-\Omega_\rho} \frac{1}{r^3}\psi^{*}_{\alpha\beta ij}(\phi,\theta) \cos\theta\, r^2\, dr\, d\phi\, d\theta \tag{2.18}$$

Let us assume that $\varepsilon^{p}_{ij}(\underline{x})$ satisfies a Hölder condition at $\underline{\xi}$, namely that there exist three positive constants c, A and α such that

$$|\varepsilon_{ij}^{p}(\underline{x}) - \varepsilon_{ij}^{p}(\underline{\xi})| \leq A r^{\alpha} \qquad \forall \underline{x} \text{ for which } |\underline{x} - \underline{\xi}| \leq c \qquad (2.19)$$

This suffices for the limit of the former integral in eq.(2.18) to be finite [14]. In order to show that the second integral also has a finite limit, we rewrite it in the form:

$$\int_{\Omega-\Omega_\rho} \frac{1}{r} \psi^{*}_{\alpha\beta ij}(\phi,\theta) \cos\theta \, dr \, d\phi \, d\theta =$$

$$= \int_{0}^{2\pi} \int_{-\pi/2}^{\pi/2} \psi^{*}_{\alpha\beta ij}(\phi,\theta) \cos\theta \left(\int_{\rho}^{R(\phi,\theta)} \frac{1}{r} dr\right) d\phi \, d\theta = \qquad (2.20)$$

$$= \ln\rho \int_{0}^{2\pi} \int_{-\pi/2}^{\pi/2} \psi^{*}_{\alpha\beta ij}(\phi,\theta) \cos\theta \, d\phi \, d\theta - \int_{0}^{2\pi} \int_{-\pi/2}^{\pi/2} \psi^{*}_{\alpha\beta ij}(\phi,\theta) \ln R \cos\theta \, d\phi \, d\theta$$

The first, singular term in the last expression drops due to the vanishing of the integral of $\psi^{*}_{\alpha\beta ij}$ over a spherical surface, for any combination of the four indices. Such intrinsic property of $\psi^{*}_{\alpha\beta ij}$, which is crucial for the existence of the integral (2.17) in the principal value sense, is not difficult to be verified. In fact, note from eq.(2.16) that $\psi^{*}_{\alpha\beta ij}$ depends on direction cosines only which can be easily expressed in terms of the spherical coordinate ϕ and θ.

If the source point belongs to the boundary Γ, eq.(2.12) is not applicable. Note that the second integral becomes singular as $O(r^{-3})$ and, hence, cannot be integrated over a surface.

Numerical difficulties should be expected when stresses are determined by (2.12) at interior points $\underline{\xi}$ which are close to the boundary, because the strong singularities of the kernels in the integrals over Γ amplify any cause of inaccuracy (truncation, modelling and/or integration errors, see Sec.3). Clearly, such "boundary layer effect" arises also in elastic analysis for possible stress evaluations subsequent to the boundary solution, see e.g. [8] [15]. However, this effect is more important in the present context where stress evaluations represent part of the main solution process.

An alternative way of calculating stresses at $\underline{\xi}$ on Γ is to relate through (Cauchy) equilibrium and Hooke's law the six stress components to the three tractions and the three strains in the plane tangent to Γ. The latter have to be generated by differentiating the boundary displacements modelled by suitable interpolations. Details can be found in [8] with reference to two-dimensional situations.

3. DISCRETIZATIONS AND INTEGRATIONS

3.1 Discretizations

The first move toward the implementation of the BE method is the subdivision of the boundary into elements $\Gamma_e (e=1...n_e)$ and of the potentially plastic subdomain Ω_p into cells $\Omega_c (c=1...n_c)$ and the approximation of the variables by suitable interpolations over each element and cell. The recourse to mappings of simple regions in a nondimensional coordinate space onto elements and cells permits more flexibility and greatly simplifies the integrations, see fig. 3.1. Therefore the notion of isoparametric element is taken over from the FE method. The matrix operators arising from the discretization process are basic ingredients in the elastic-plastic analysis procedure expounded later.

We discuss below the fundamentals of such process focusing on plasticity related aspects. With reference to three-dimensional situations, the geometry of elements and cells will be given the parametric representations, respectively:

$$x_i(\eta_1,\eta_2) = \bar{N}^n(\eta_1,\eta_2)\,(x_i)_e^n\,, \qquad i = 1,2,3\ ; \qquad e = 1...n_e \quad (3.1)$$

$$x_i(\zeta_1,\zeta_2,\zeta_3) = \bar{M}^m(\zeta_1,\zeta_2,\zeta_3)\,(x_i)_c^h\,, \quad i = 1,2,3\ ; \quad c = 1...n_c \quad (3.2)$$

where indices n and m run over the node sets of elements Γ_e and cell Ω_c, respectively; $(x_i)_e^n$ and $(x_i)_c^m$ are the coordinates of the nodes which govern the relevant meshes ("mesh points"); η_h and ζ_k are the nondimensional local coordinates. The approximations through polynomial interpolating nodal values concern boundary displacements and tractions and plastic strains, namely:

$$u_i(\underline{\eta}) = N^n(\underline{\eta})\,(u_i)_e^n\,, \quad p_i(\underline{\eta}) = [\tilde{N}^n(\underline{\eta})]\,(p_i)_e^n \qquad (3.3a,b)$$

$$i = 1,2,3\ ; \quad e = 1...n_e$$

$$\varepsilon_{ij}^p(\underline{\zeta}) = M^m(\underline{\zeta})\,(\varepsilon_{ij}^p)_c^m, \quad i,j = 1,2,3; \quad c = 1...n_c \qquad (3.3c)$$

Here indices n and m run over the sets of nodes where we define the values $(u_i)_e^n,(p_i)_e^n,(\varepsilon_{ij})_c^m$ which govern the relevant fields. Usually the interpolation functions $\tilde{N}^n$ and N^n are identical. Whether it would be more consistent to use one-degree higher interpolation functions for

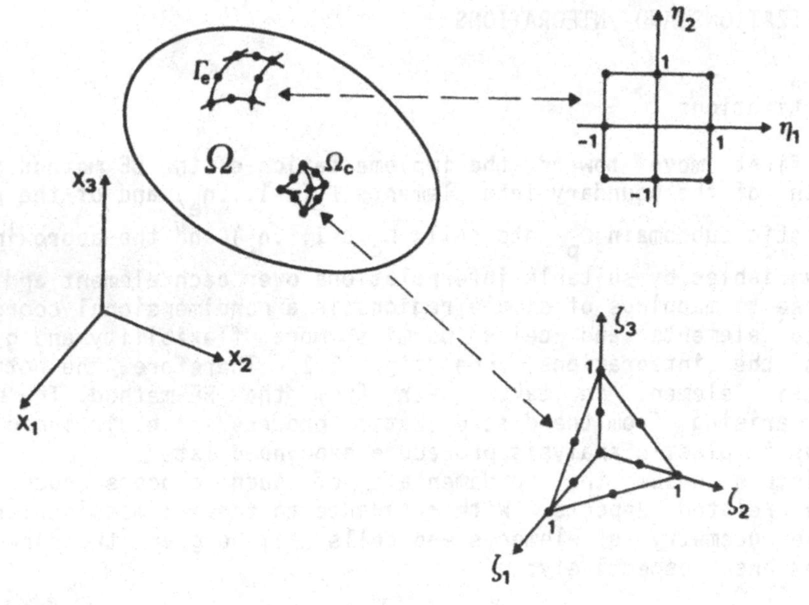

Fig. 3.1

displacements than for traction, is still a research issue, see e.g. [16].

In most cases $\bar{N}^n = N^n$ and in some cases $\bar{M}^m = M^m$, i.e. geometries and fields are modelled by the same interpolating polynomials (isoparametric representation) and mesh points and field nodes coincide, see e.g. [6] [8].

3.2 Algebrization of the boundary integral equation

Eq.(2.9) generates a system of linear algebraic equations as soon as it is collocated by taking as load point $\underline{\xi}$ each of the above defined field nodes (say $\underline{\xi}^r$) on Γ and the integrations over elements and cells are performed after the above described discretizations. These algebraic equations read, in general terms:

$$c_{\alpha\beta}(\underline{\xi}^r)u_\beta(\underline{\xi}^r) + \sum_e \sum_n [\iint p^*_{\alpha i}(\underline{\xi}^r, \underline{x}(\underline{\eta}))N^n(\underline{\eta})G(\underline{\eta})d\eta_1 d\eta_2]_e (u_i)^n_e =$$

$$= \sum_e \sum_n [\iint u^*_{\alpha i}(\underline{\xi}^r, \underline{x}(\underline{\eta}))\bar{N}^n(\underline{\eta})G(\underline{\eta})d\eta_1 d\eta_2]_e (p_i)^n_e + \tag{3.4a}$$

$$+ \sum_c \sum_m [\iiint \sigma^*_{\alpha ij}(\underline{\xi}^r, \underline{x}(\underline{\zeta}))M^m(\underline{\zeta})J(\underline{\zeta})d\zeta_1 d\zeta_2 d\zeta_3]_c (\varepsilon^p_{ij})^m_c + B_\alpha(\underline{\xi}^r)$$

where:

$$G(\underline{\eta}) \equiv d\Gamma / d\eta_1 d\eta_2 \ , \quad J(\underline{\eta}) \equiv d\Omega / d\zeta_1 d\zeta_2 d\zeta_3 \tag{3.4b}$$

are by-products of the adopted coordinate transformations (3.1)(3.2). $B_\alpha(\underline{\xi}^r)$ represents the result of the domain integration involving a given distribution of body forces. Usually nodes and mesh points on Γ and sometimes nodes points and mesh points in Ω^p coincide, so that continuity of the approximating fields is enforced at the interfaces of elements and cells, respectively. Clearly, it is desirable to permit possible traction discontinuities along BE interfaces and this can be achieved by double nodes [6] [8]. Nodal variable identification at interface nodes is expressed when assembling the 3N algebraic equations (3.4) for all N boundary nodes $\underline{\xi}$ involving the 6N boundary nodal variables.

These equations, together with the boundary conditions, would define all boundary unknowns if the plastic strains were known. In a compact matrix form they read:

$$\underline{H}\ \underline{u} = \underline{G}\ \underline{p} + \underline{D}\ \underline{\varepsilon}^p + \underline{B} \tag{3.5}$$

where vectors $\underline{u}$ and $\underline{p}$ gather all independent nodal variables on Γ, vector $\underline{\varepsilon}^p$ all the independent nodal plastic strains; vector $\underline{B}$ all domain data. Matrices $\underline{H}$, $\underline{G}$ and $\underline{D}$ are generated by assembling the integrals of products of appropriate fundamental solution, shape function and Jacobian.

An equation system formally analogous to (3.5) can be generated by choosing field nodes on Γ different from the mesh points ("non-conforming" BEs) [8][17]. The following remarks on non-conforming elements compared to conforming ones, hold true for both elastic and inelastic analysis.

(a) The evaluation of the free terms in eq.(3.4) is unnecessary, because $c_{\alpha\beta} = \frac{1}{2}\,\delta_{\alpha\beta}$, since Γ is smooth at all load points $\underline{\xi}$.

(b) No assembling by identification of nodal variables is needed.

(c) Interface field discontinuities inherent to non-conforming BEs confer more flexibility to the loading description but introduce fictitious displacement jumps (which, however, tend to vanish as the boundary mesh grows finer).

(d) The treatment of corners is straightforward in all circumstances (without double nodes and, without the need for auxiliary equations [18] even at corners interior to the constrained boundary.

(e) The computer implementation of the BE method with nonconforming elements is much simpler as a consequence of the remarks (b) and (d).

(f) At equal number of boundary unknowns, the accuracy of results is in general lower [17].

As for volume cells, it is convenient to govern the distribution of plastic strains by their values in points ("strain points") where the plastic constitutive law will be enforced (see Sec.5). When these nodes are interior to the cell, computational advantages arise similar to the above (b) and (e) for BE, besides the avoidance of calculating stresses on the boundary as in Sec.5. All these facts justify the frequent

adoption of strain points coincident with the nodes and distinct from the mesh points in Ω^p ("nonconforming cells"), e.g. as in a constant-strain cell with a strain point in its centroid.
When the integrations indicated in eq.(3.4) are performed numerically according to a Gauss scheme as in the FE method, the coefficients in eqs.(3.5) are calculated, e.g., in the form:

$$[\iint p^*_{\alpha i}(\underline{\xi}^r,\underline{x}(\underline{\eta}))N^n(\underline{\eta})G(\underline{\eta})d\eta_1 d\eta_2]_e = \qquad (3.6)$$
$$= \Sigma_h\, w_h(\underline{\eta}_h)[p^*_{\alpha i}(\underline{\xi},\underline{x}(\underline{\eta}_h))N^n(\underline{\eta}_h)G(\underline{\eta}_h)]_e$$

Since the kernels are nonpolynomial in the integration variables, errors are always implied by Gauss integrations like (3.6) and increase as the collocation point $\underline{\xi}^r$ approaches the element Γ_e on which the integration is performed. These errors can be reduced by either increasing the number of Gauss points or using subelements [18] [19]; in both cases such refinements are governed by criteria sensitive to a suitably defined "distance" between $\underline{\xi}^r$ and Γ_e. When $\underline{\xi}^r$ belongs to Γ_e special provisions are needed. Precisely, for $\underline{\xi}^r$ coinciding with the n-th node of Γ_e, the sum of the free-term coefficient and the first integral (in Cauchy principal value sense) contained in (3.4) (i.e. the principal diagonal entry in matrix $\underline{H}$) can be evaluated, to within the sign, by summing up the other integrals for the same α-th displacement component in $\underline{\xi}^r$(i.e. the other integrand entries in the same row of $\underline{H}$), according to the familiar rigid body motion scheme [7] [8]. In the second term of eq.(3.4a) the integrand, when singular, is of order $O(r^{-1})$ and hence, its singularity is removed by transition to polar coordinates centered in the image of $\underline{\xi}^r$ on the master element in the plane η_1, η_2, [6] [8]. Similar provision holds for the volume integral whose kernel becomes singular of order $O(r^{-2})$, when $\underline{\xi}^r$ on Γ coincides with the n-th node on the c-th cell.

The above provisions permit to generate the matrices $\underline{H}$, $\underline{G}$ and $\underline{D}$ by numerical integrations. However, analytical integrations are possible and have been advantageously used in special cases (triangular BEs with linear interpolations; triangular, constant strain cells, see e.g. [18] [19] [20]).

3.3 Numerical treatment of the integral expression for stresses

Eq.(2.12) gives rise to linear algebraic equations when it is collocated in interior strain points $\underline{\xi}^s$ and the integrations are carried out account taken of the geometry and field modelling of Subsec.3.1 (stresses at $\underline{\xi}$ on Γ are to be calculated "locally"):

$$\sigma_{\alpha\beta}(\underline{\xi}^s) = \sum_e \sum_n [\iint u^*_{\alpha\beta i}(\underline{\xi}^s, \underline{x}(\underline{\eta}))N^n(\underline{\eta})G(\underline{\eta})d\eta_1 d\eta_2]_e (p_i)^n_e -$$

$$- \sum_e \sum_n [\iint p^*_{\alpha\beta i}(\underline{\xi}^s, \underline{x}(\underline{\eta}))\bar{N}^n(\underline{\eta})G(\underline{\eta})d\eta_1 d\eta_2]_e (u_i)^n_e +$$

$$+ \sum_c \sum_m [\iiint \sigma^*_{\alpha\beta ij}(\underline{\xi}^s, \underline{x}(\underline{\zeta}))M^m(\underline{\zeta})J(\underline{\zeta})d\zeta_1 d\zeta_2 d\zeta_3]_c (\varepsilon^p_{ij})^m_c +$$

$$+ f_{\alpha\beta\phi\psi}\ \varepsilon^p_{\phi\psi}(\underline{\xi}^s) + B_{\alpha\beta}(\underline{\xi}^s) \qquad (3.7)$$

where $B_{\alpha\beta}(\underline{\xi}^s)$ is the addend arising from the body force integral. If there are interface nodes the identification of nodal variables is implied in assembling the equations (3.7) for all strain points $\underline{\xi}^s$ not lying on the boundary. All these equations can be represented in the following compact form:

$$\underline{\sigma}' = \underline{H}'\underline{u} + \underline{G}'\underline{p} + \underline{D}'\ \underline{\varepsilon}^p + \underline{B}' \qquad (3.8)$$

with the symbols equal to or counterparts of those in eq.(3.5). Clearly, the comments of Subsec.3.2 on "non-conforming" elements and cells apply unaltered to (3.8), except remark (a).

As for the integrations over elements Γ_e , in eq.(3.7) no singularity shows up (since $\underline{\xi}^s$ does not lie on Γ). Therefore Gauss schemes like eq.(3.6) is generally applicable, and will be applied unless analytical integration is possible.

Also the integrations over cells not containing the strain point $\underline{\xi}^s$ can be carried out numerically, preferably with Gauss point numbers depending on a "distance" of Ω_c from $\underline{\xi}^s$.

When $\underline{\xi}^s$ belongs to the integration cell Ω_c, the kernel $\sigma^*_{\alpha\beta ij}$ is singular as $O(r^{-3})$ and the improper integral cannot be understood in the usual sense. The existence of such integral in the Cauchy principal value sense (with the esclusion of an infinitesimal sphere Ω_ε centered in ξ^s) was discussed in Subsec.2.2. As for its numerical evaluation a recently proposed procedure is outlined below (ref.[21]). It is hinged on the use of so-called "special polar" coordinates, which permit not only to reduce the order of singularity (a goal achievable by usual spherical

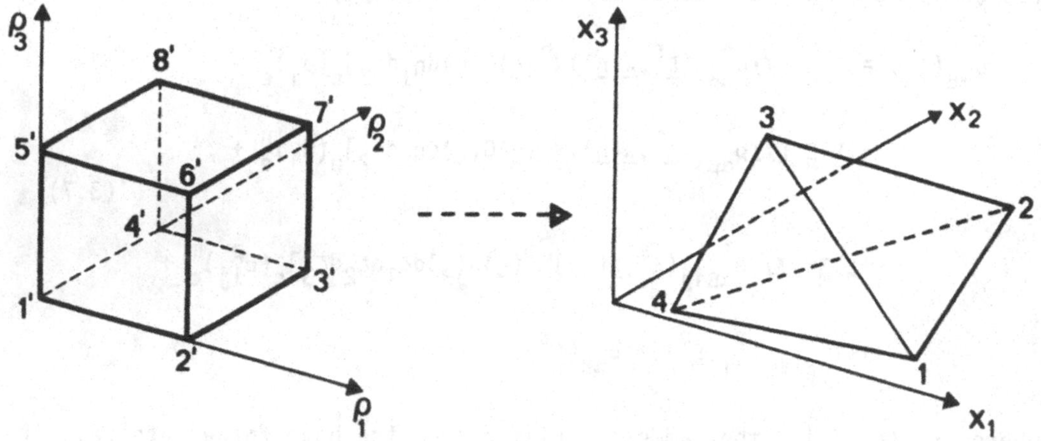

Fig. 3.2

coordinates as well) but also to perform the integrations over simple master elements. Consider for simplicity a tetrahedrical cell, with linear interpolations for both geometry and plastic strain field. With reference to fig.3.2 and assuming vertex 1 to be the singular point, the following nonlinear transformation maps a cube of unit side length onto the cell (at discrepancy from the one-to-one mapping of the tetrahedron onto tetrahedron hinted at in fig.3.1):

$$x_i = (1 - \rho_1)\, x_i^{(1)} + \rho_1 (1 - \rho_2)\, x_i^{(2)} + \rho_1 \rho_2 (1 - \rho_3)\, x_i^{(3)} + \rho_1 \rho_2 \rho_3\, x_i^{(4)}; \qquad i = 1, 2, 3 \tag{3.9}$$

where ρ_1, ρ_2 and ρ_3 are the nondimensional tetrahedron polar coordinates. The following features of transformation (3.9) are worth noting (see fig.3.2):

(a) The cube face (1'4'8'5') in the plane $\rho_1 = 0$ collapses into vertex 1, the face in the plane $\rho_2 = 0$ into edge 12; the other four cube faces are mapped into the tetrahedron faces.

(b) The distance $r = |\underline{x} - \underline{x}^{(1)}|$, through (3.9), becomes the product of ρ_1 times a function $f(\rho_2, \rho_3)$ and, consequently, the derivatives $r_{,i}$ (i = 1, 2, 3) do not depend on ρ_1.

(c) The Jacobian of (3.9) turns out to be $J = 6\rho_1^2\, \rho_2 \Omega_c$ denoting by Ω_c the cell volume.

Let us consider first the integral over cell Ω_c in eq.(3.7), with collocation in $\underline{\xi} = \underline{x}^{(1)}$, let us adopt as nondimensional coordinates ζ_i the above special polar coordinates $\rho_i (i = 1, 2, 3)$. By virtue of mapping (3.9) and of the above observed properties, we notice first that the kernel $\overset{*}{\sigma}_{\alpha\beta ij}$, in view of (b) and of the nature of its dependence on r, acquires the form, for $\underline{\xi}^r = \underline{\xi}^{(1)}$ at vertex 1:

$$\overset{*}{\sigma}_{\alpha\beta ij}(\rho_1\rho_2\rho_3) = [\rho_1 f(\rho_2,\rho_3)]^{-3}\ \bar{\overset{*}{\psi}}_{\alpha\beta ij}(\rho_2,\rho_3) \tag{3.10}$$

where $[f(\rho_2,\rho_3)]^{-3}$ is not singular in $\underline{\xi}^{(1)}$. The linearity of the plastic strain modelling over the cell combined with property (b) of the coordinates ρ_i, permits to confer to the interpolation functions the form:

$$M^m[\underline{x}(\underline{\rho})] = \delta_{1m} + \rho_1\mu^m(\rho_2,\rho_3) \tag{3.11}$$

As a consequence of (3.11)(3.10), through remark (c) the integral over Ω_c becomes:

$$\Sigma_m \int_{\Omega_c} \overset{*}{\sigma}_{\alpha\beta ij}(\underline{\xi}^1,\underline{x})M^m(\underline{x})d\Omega\ (\varepsilon^p_{ij})_c = \int_{\Omega_c} \overset{*}{\sigma}_{\alpha\beta ij}(\underline{\xi}^1,\underline{x})d\Omega\ (\varepsilon^p_{ij})^1_c +$$

$$+ \int_0^1\int_0^1\int_0^1 [\rho_1 f(\rho_2,\rho_3)]^{-3}\bar{\overset{*}{\psi}}_{\alpha\beta ij}(\rho_2,\rho_3)[\rho_1\Sigma_m\ \mu^m(\rho_2,\rho_3)(\varepsilon^p_{ij})^m_c]6\rho_1^2\ \rho_2\ \Omega_c d\rho_1 d\rho_2 d\rho_3 =$$

$$= \int_{\Omega_c} \overset{*}{\sigma}_{\alpha\beta ij}(\underline{\xi}^1,\underline{x})d\Omega\ (\varepsilon^p_{ij})^1_c + \tag{3.12}$$

$$+ 6\Omega_c \int_0^1\int_0^1\int_0^1 g_{\alpha\beta ij}(\rho_2,\rho_3)\Sigma_m\ \mu^m(\rho_2,\rho_3)(\varepsilon^p_{ij})^m_c\ d\rho_1 d\rho_2 d\rho_3$$

having set:

$$g_{\alpha\beta ij}(\rho_2,\rho_3) \equiv [f(\rho_2,\rho_3)]^{-3}\ \bar{\overset{*}{\psi}}_{\alpha\beta ij}(\rho_2,\rho_3)\rho_2 \tag{3.13}$$

Since the latter integral in eq.(3.12) is regular, focus is on the former, to be interpreted in the sense of Cauchy principal value. For calculating it (over all adjacent cells with the exclusion of the

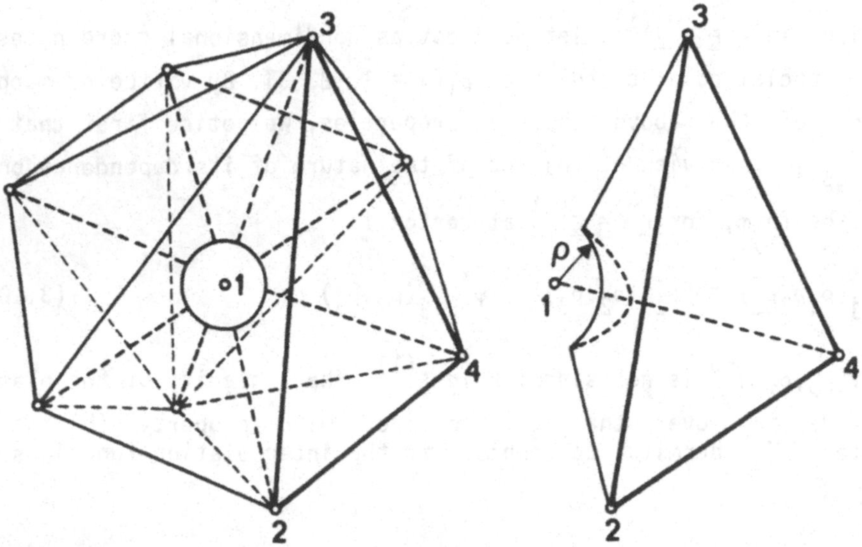

Fig. 3.3

infinitesimal sphere centered in $\underline{\xi}^1$) a formal way has been outlined in Subsec.2.2, see e.g. [8] for two-dimensional cases. An alternative, more efficient way is suggested in [21]: after the recognition of the integrand as a sum of space derivatives, the integrals over cells adjacent to $\underline{\xi}^1$ (see fig.3.2) are transformed into surface integrals. Among such integrals those over cell interfaces containing $\underline{\xi}^1$ cancel out in pairs; those over all facets of the exclusion sphere give a zero sum, see Subsec.2.2; those over the remaining facets are not singular and can be evaluated numerically over the corresponding faces of the unit cube.

Turning now to the domain integral in the boundary integral eq.(3.4), since the integrand is singular as $O(r^{-2})$, the above recourse to "special polar coordinates" removes the singularity by means of its jacobian and, hence, the Gauss integration over the master cube can be carried out.

Note that in [21] the special polar coordinates are also shown to be useful for carrying out the integrations (in common with elastic analysis) over the elements on Γ. Besides they can advantageously be used for quadratic BEs and cells though at the price of formal complications [21]. For example, if quadratic BEs and cells are dealt with, first they are mapped onto a cube as usual; second tetrahedrical portions of the cube (which contain the singular point $\underline{\xi}^1$ as a vertex) are mapped on to the unit cube through the nonlinear transformation, eq.(3.9).

Advanced procedures for the numerical treatments of the integrations resting on nonlinear mappings have also been developed by Mustoe and Banerjee et al., [18] [19] [20].

4. ESSENTIAL FEATURES OF ELASTIC-PLASTIC MATERIAL MODELS

4.1 General constitutive laws of incremental elastoplasticity

A broad class of plastic material models is described by the traditional relationships which follow, see e.g. [1] [2] and fig.4.1:

$$\phi_h(\sigma_{ij}, \chi_\alpha) \le 0 \qquad (h = 1...y) \tag{4.1}$$

$$\dot{\varepsilon}^p_{ij} = \sum_1^y{}_h \frac{\partial\psi_h}{\partial\sigma_{ij}}(\sigma_{ij}, \chi_\alpha)\dot{\lambda}_h \quad , \qquad \dot{\lambda}_h \ge 0 \tag{4.2a,b}$$

$$\phi_h \dot{\lambda}_h = 0 \quad , \quad \dot{\phi}_h \dot{\lambda}_h = 0 \qquad \text{(no summation on h)} \tag{4.3a,b}$$

$$\sigma_{ij} = E_{ijrs}(\varepsilon_{rs} - \varepsilon^p_{rs} - \Theta_{rs}) \tag{4.4}$$

The y inequalities (4.1) define the current elastic region (and its boundary called yield locus or yield surface) in the stress space, ϕ_h being the h-th (differentiable) yield function corresponding to the "yield mode" $\phi_h = 0$. The parameters χ_α are "internal variables" which depend on the whole past history of yielding processes: in a sense, they keep track of such history and are chosen so that they represent some measure of the yielding processes occurred (e.g. plastic work $\int\sigma_{ij}\dot{\varepsilon}_{ij}dt$), see e.g. [22].

Eqs.(4.2) expresses the plastic strain rates as linear combination through nonnegative scalars (or "plastic multipliers" $\dot{\lambda}_h$) of gradients of "plastic potentials" ψ_h, one for each yield mode h. The derivatives marked by dots are with respect to an arbitrary increasing function t of the physical time: this function merely acts as ordering variable and its arbitrariness reflects the time independent ("inviscid") nature of the plastic constitution. Note that, in contrast to $\dot{\lambda}_h$, its integral λ_h has not necessarily a physical meaning.

Eq.(4.3a) requires $\dot{\lambda}_h = 0$ if $\phi_h < 0$ and $\phi_h = 0$ if $\dot{\lambda}_h > 0$; i.e. it rules out contributions to plastic strain rates by a yield mode not in contact with the current stress point, and lack of contact for a mode currently yielding (or "active"). Eq.(4.3a) and (4.2) define the "flow rule" at the current instant t. Eq.(4.3b) is a consequence of the preceding relations, evidenced for its convenience when only infinitesimal processes are referred to.

Eq.(4.4) is Hooke's law, if the additivity of elastic and plastic strains is postulated, Θ_{rs} denoting possible imposed (such as thermal) strains to be ignored henceforth for brevity.

Let us focus now on the infinitesimal processes (described in terms of rates) starting from a given state of the material at time t, with stresses represented by a point $\underline{\sigma}$ at a "corner" of the current yield surface where y'(< y) yield modes intersect. We make recourse below to matrix notation, being understood that all matrices and vectors concern here merely a single material point (or a homogeneous specimen) and the above y' yield modes which can be "active" in the infinitesimal process.

The problem of finding $\dot{\underline{\varepsilon}}$ for given $\dot{\underline{\sigma}}$ can be formulated as follows on the basis of eqs.(4.1)-(4.4).

$$\dot{\underline{\varepsilon}} = \underline{E}^{-1}\dot{\underline{\sigma}} + \underline{\Psi}\,\dot{\underline{\lambda}} \tag{4.5}$$

$$\dot{\underline{\phi}} = \tilde{\underline{\Phi}}\,\dot{\underline{\sigma}} - \underline{H}\,\dot{\underline{\lambda}} \le \underline{0} \;, \quad \dot{\underline{\lambda}} \ge \underline{0} \;, \quad \tilde{\dot{\underline{\phi}}}\,\dot{\underline{\lambda}} = 0 \tag{4.6a,b,c}$$

by setting: $\tilde{\dot{\underline{\psi}}} \equiv \{\psi_1 \ldots \psi_y\}$, similarly for $\underline{\phi}$ and $\dot{\underline{\lambda}}$, and

$$\underline{\Psi} \equiv [\cdot\cdot\frac{\partial\psi_h}{\partial\underline{\sigma}}\cdot\cdot] \;, \quad \underline{\Phi} \equiv [\cdot\cdot\frac{\partial\phi_h}{\partial\underline{\sigma}}\cdot\cdot] \quad (h = 1 \ldots y') \tag{4.7a}$$

$$H_{hk} \equiv \frac{\partial\phi_h}{\partial\chi_\alpha}\,\frac{\partial\chi_\alpha}{\partial\varepsilon^p_{ij}}\,\frac{\partial\psi_k}{\partial\sigma_{ij}} \;, \quad \text{or} \quad \underline{H} \equiv [H_{hk}] = \frac{\partial\underline{\phi}}{\partial\tilde{\underline{\chi}}}\,\frac{\partial\underline{\chi}}{\partial\tilde{\underline{\varepsilon}}^p}\,\frac{\partial\tilde{\underline{\psi}}}{\partial\underline{\sigma}} \tag{4.7b}$$

In fact, (4.6a) is obtained by differentiating the yield functions taking into account the dependence of the internal variables on plastic strains and the flow rule (4.2a). The sign constraint on $\dot{\underline{\phi}}$ arises from inequalities (4.1) and from the fact that $\phi_h = 0$ for h = 1...y'.

The converse problem of finding $\dot{\underline{\sigma}}$ for given $\dot{\underline{\varepsilon}}$ is readily given the formulation:

$$\dot{\underline{\sigma}} = \underline{E}\,\dot{\underline{\varepsilon}} - \underline{E}\,\underline{\Psi}\,\dot{\underline{\lambda}} \tag{4.8}$$

$$\dot{\underline{\phi}} = \tilde{\underline{\Phi}}\,\underline{E}\,\dot{\underline{\varepsilon}} - (\tilde{\underline{\Phi}}\,\underline{E}\,\underline{\Psi} + \underline{H})\,\dot{\underline{\lambda}} \le \underline{0} \;, \quad \dot{\underline{\lambda}} \ge \underline{0} \;, \quad \tilde{\dot{\underline{\phi}}}\,\dot{\underline{\lambda}} = 0 \tag{4.9a,b,c,d}$$

where eq.(4.9a) flows from (4.6a) substituting (4.8) in it.

The remarks which follow concisely elucidate what precedes.

(a) The coincidence of yield functions and plastic potentials ($\phi_h = \psi_h$) makes (4.2a) a normality requirement (in generalized sense at a corner) of plastic strain rates with respect to the yield surface ("associative" flow rule).

(b) The hardening matrix $\underline{H}$ governs the influence of yielding on the yield modes in infinitesimal processes. It will be assumed here as symmetric ("reciprocal interaction"). The positive definiteness of $\underline{H}$ characterizes strictly stable hardening behaviour and existence and uniqueness of $\dot{\underline{\varepsilon}}$ for any $\dot{\underline{\sigma}}$ via (4.5)(4.6), [23]; existence and uniqueness of $\dot{\underline{\sigma}}$ for any $\dot{\underline{\varepsilon}}$ is ensured by the much weaker condition of positive definiteness of the matrix in brackets in eq.(4.9). In view of its broader range of uniqueness the converse $\dot{\underline{\varepsilon}} \rightarrow \dot{\underline{\sigma}}$ relationship will be central to the subsequent use of plastic constitution. Unstable behaviour or softening is characterized by the fact that the quadratic form associated to $\underline{H}$ (i.e. the plastic work of second order) becomes negative for some $\dot{\underline{\lambda}}$ [23]; $\underline{H} = \underline{0}$ means (at least instantaneous) perfectly plastic behaviour. All the above statements become almost self-evident for y = 1 (smooth point of the yield surface with unique outward normal). Drucker-Prager yield criterion provides a classical example of yield function usually associated with perfect plasticity [1]:

$$\phi = (3-\sin\omega)^{-1}[2\sin\omega\, I_1 + \sqrt{3}(3-\sin\omega)\sqrt{J_2} - 6C\cos\omega] \leq 0 \qquad (4.10)$$

I_1 and J_2 are the first invariant and the second deviatoric invariant of stresses, respectively; ω denotes the internal friction angle and C the cohesion. For nonfrictional materials such as metals, $\omega = 0$, C is one-half the yield stress in uniaxial test and (4.10) reduces to von Mises criterion.

(c) Mathematically, eqs.(4.8)(4.9) represent a set of nonlinear differential equations and inequalities to be integrated in time for a given strain path $\underline{\varepsilon}(t)$. This path-dependent or "nonholonomic" character of material behaviour is a central feature of plasticity. It carries over to b.v. problems formulated by associating eqs.(4.8)(4.9) or (4.5)(4.6) to equilibrium and compatibility equations, the given path being in the space of the external actions. In the present context of "small deformations", the nonlinearity in rates reduces to the "linear complementarity problem" (4.9), or alternatively (4.6). This can be described as an orthogonality requirement of two sign-constrained and linearly related vectors; clearly, it plays a central role, implicitly or explicitly, in elastoplastic analysis conceived as a time integration process (Subsec.5.3) whatever the approach adopted may be for the discretization in space, including BE methods.

4.2 Piecewise-linear approximations of constitutive laws

In classical (rigid-perfectly plastic) limit analysis, cf. Subsec. 7.1, an effective simplification is often achieved by linearizing the yield functions in the stresses, i.e. by approximating the yield locus by a polyhedrical (piecewise-linear) surface. We adopt here such linearization for finite stress increments $\Delta\underline{\sigma}$ (starting from a state $\underline{\sigma}_o$) by preserving hardening but assuming as linear the dependence of the yield functions and plastic potential on variables $\Delta\lambda_\alpha$ (one for each yield mode plane) which measure the contributions to plastic strain increments $\Delta\underline{\varepsilon}^p$ of the modes in the finite step $\Delta\underline{\sigma}$ (see fig.4.2). Then we may write:

$$\underline{\phi} = \underline{\phi}_o + \underline{\tilde{\Phi}}\,\Delta\underline{\sigma} - \underline{H}\,\Delta\underline{\lambda} \leq \underline{0} \quad , \quad \Delta\underline{\lambda} \geq \underline{0} \tag{4.11a,b}$$

$$\Delta\underline{\varepsilon} = \underline{E}^{-1}\Delta\underline{\sigma} + \Delta\underline{\varepsilon}^p = \underline{E}^{-1}\Delta\underline{\sigma} + \underline{\Psi}\,\Delta\underline{\lambda} \tag{4.12}$$

when matrices $\underline{\Phi},\underline{H},\underline{\Psi}$ may depend on the plastic history prior to state $\underline{\sigma}_o$, $\underline{\varepsilon}^p_o$, but not on increments.

Now supplement eqs.(4.11)(4.12) first with (4.3a, b), second with:

$$\underline{\tilde{\phi}}\,\Delta\underline{\lambda} = 0 \tag{4.13}$$

In the former case one merely arrives at a special case of the general nonholonomic constitution (4.1)(4.4). In the latter case we obtain the corresponding holonomic material model over the stress step $\Delta\underline{\sigma}$, i.e. a path-independent nonlinear relationship between $\Delta\underline{\sigma}$ and $\Delta\underline{\varepsilon}$. In fact, since $\underline{\tilde{\phi}}\,\dot{\underline{\lambda}} + \dot{\underline{\tilde{\phi}}}\,\Delta\underline{\lambda} = \frac{d}{dt}(\underline{\tilde{\phi}}\,\Delta\underline{\lambda})$, when $\dot{\underline{\tilde{\phi}}}\,\Delta\underline{\lambda} = 0$ at any $t > t_o$ (i.e. there is no "unloading" of any yield mode), then for the stress path from $\underline{\sigma}_o$ to $\underline{\sigma}_o+\Delta\underline{\sigma}$, the holonomic version of the model provides the same $\Delta\underline{\varepsilon}$ as the nonholonomic version.

A similar conclusion can be drawn for the formulation for strain increments. Namely, its "stepwise holonomic" version which follows provides $\Delta\underline{\sigma}$ (and $\Delta\underline{\varepsilon}^p$) on the basis of the "step" $\Delta\underline{\varepsilon}$ (and not also of the strain path along which it grows from $\underline{\varepsilon}_o$):

$$\underline{\phi} = \underline{\phi}_o + \underline{\tilde{\Phi}}\,\underline{E}\,\Delta\underline{\varepsilon} - (\underline{\tilde{\Phi}}\,\underline{E}\,\underline{\Psi} + \underline{H})\,\Delta\underline{\lambda} \leq \underline{0} \;,\quad \Delta\underline{\lambda} \geq \underline{0} \;,\quad \underline{\tilde{\phi}}\,\Delta\underline{\lambda} = 0 \tag{4.14a,b,c}$$

$$\Delta\underline{\varepsilon}^p = \underline{\Psi}\,\Delta\underline{\lambda} \;,\qquad \Delta\underline{\sigma} = \underline{E}(\Delta\underline{\varepsilon} - \Delta\underline{\varepsilon}^p) \tag{4.15}$$

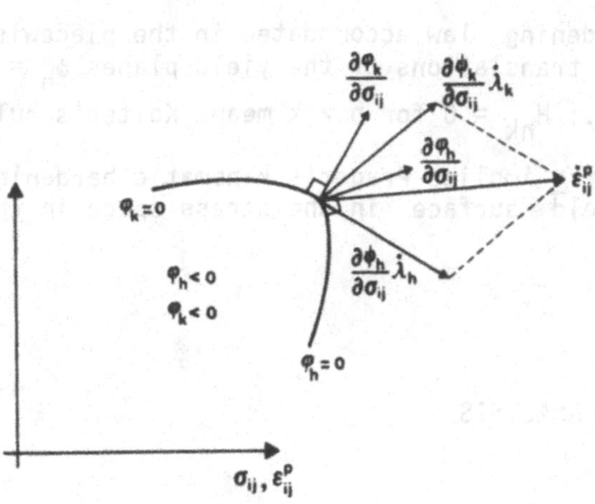

Fig. 4.1

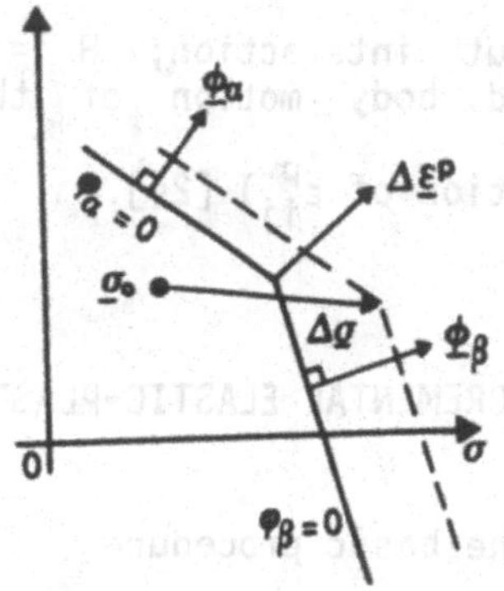

Fig. 4.2

The following remarks may help to clarify the meaning and coverage of the above simplified formulations.

(a) If the increments are regarded as infinitesimal, eqs.(4.9) are formally recovered from (4.14). In fact, modes potentially active in infinitesimal processes are only those for which $\phi_h^o = 0$: hence, $\phi_h = \Delta\phi_h = \dot{\phi}_h\delta t$. In other, more general terms, a piecewise linearized, stepwise holonomic material model is analogous to the general elastic-plastic rate relationship at a corner of the yield surface. Both are mathematically described by a linear complementarity problem [23] [24].

(b) If the strain (or stress) step starts from an "initial" stressless state and the piecewise linearization of yield functions and plastic potentials is carried out "a priori" once for all and for all modes $y' = y$, then eqs.(4.14)(4.15) or (4.11)(4.12) become descriptions of holonomic (fully reversible, nonlinear elastic) material models. Then $\underline{\phi}_o = \underline{R}$, if $\underline{R}$ denotes a vector gathering the "yield limits", i.e. the initial distances of the yield planes from the origin of the stress axes.

(c) The fully irreversible, dissipative nature of plastic flow is restored in the piecewise linear context simply replacing (4.14b,c) by (4.2b)(4.3a), i.e. by:

$$\dot{\underline{\lambda}} \geq \underline{0}\ , \quad \tilde{\underline{\phi}}\,\dot{\underline{\lambda}} = 0 \tag{4.16}$$

Thus for $\underline{H} = \underline{0}$ the perfectly plastic model with piecewise linearized yield criteria traditional in limit analysis is arrived at.

(d) Matrix $\underline{H}$ defines the hardening law accomodated in the piecewise linear model, i.e. it governs the translations of the yield planes $\phi_h = 0$ due to plastic processes. E.g.: $H_{hk} = 0$ for $h \neq k$ means Koiter's rule without interaction; $\underline{H} = h\, \tilde{\underline{\phi}}\, \underline{\phi}$ implies Prager's kinematic hardening (rigid body motion of the yield surface in the stress space in the direction of $\dot{\varepsilon}^p_{ij}$) [24].

5. INCREMENTAL ELASTIC-PLASTIC BE ANALYSIS

5.1 The basic procedure

We outline below a typical solution procedure by the direct BE method in incremental elastoplasticity, see e.g. [7] [8] [25]. The purpose is to gather and employ in a sequence of conceptual steps the ideas separately considered in what precedes and to set a scenario for subsequent developments and supplementary information and variants.

Matrix (instead of index) notation will be adopted henceforth. Namely, vectors $\tilde{\underline{u}}(\underline{x}) \equiv \{u_i(\underline{x})\}$, $\tilde{\underline{p}}(\underline{x}) \equiv \{p_i(\underline{x})\}$, $\tilde{\underline{\varepsilon}}^p(\underline{x}) \equiv \{\varepsilon^p_{11}(\underline{x}),\ \varepsilon^p_{22}(\underline{x}),\ \varepsilon^p_{33}(\underline{x}),\ 2\varepsilon^p_{12}(\underline{x}),\ 2\varepsilon^p_{23}(\underline{x}),\ 2\varepsilon^p_{31}(\underline{x})\}$, etc.; free term matrices $\underline{C} \equiv [c_{\alpha\beta}]$ of order 3 and $\underline{F} \equiv [f_{\alpha\beta\phi\psi}]$ of order 6; kernel matrices of order 3: $\underline{V} \equiv [u^*_{\alpha i}]$, $\underline{T} \equiv [p^*_{\alpha i}]$, $\alpha, i = 1, 2, 3$ whose entries are e.g. Kelvin's two-point functions (2.5)(2.6); the rectangular kernel matrix $\underline{W}(\underline{\xi},\underline{x})$ of size 3 X 6, whose entries in the α-th row are the six independent components of the fundamental stress tensor $\sigma^*_{\alpha ij}(\underline{\xi},\underline{x})$, see e.g. eq.(2.7). The kernels in the integral equation (2.12) for stresses will be ordered in the new matrices: $\underline{Q} \equiv [u^*_{\alpha\beta i}]$ of size 6 X 3; $\underline{R} \equiv [p^*_{\alpha\beta i}]$ of size 6 X 3 from eqs.(2.14)(2.15); $\underline{S} \equiv [\sigma^*_{\alpha\beta ij}]$ of order 6 from eq.(2.16).

1. With reference to an elastically homogeneous domain Ω where plastic strains $\underline{\varepsilon}^p$ may develop, choose an appropriate fundamental solution for unit concentrated force (Subsec. 2.1.) and write the boundary integral equation (2.9), which in the above defined matrix notation reads:

$$\underline{c}(\underline{\xi})\, \underline{u}(\underline{\xi}) + \int_\Gamma \underline{T}(\underline{\xi},\underline{x})\, \underline{u}(\underline{x})\, d\Gamma = \int_\Gamma \underline{V}(\underline{\xi},\underline{x})\, \underline{p}(\underline{x})\, d\Gamma + \int_\Omega \underline{W}(\underline{\xi},\underline{x})\, \underline{\varepsilon}^p(\underline{x})\, d\Omega + \int_\Omega \underline{V}(\underline{\xi},\underline{x})\, \underline{b}(\underline{x})\, d\Omega \tag{5.1}$$

Imposed strains (such as thermal) possibly present among the external actions are thought of as accomodated in the body forces $\underline{b}$ and boundary tractions $\underline{p}$ (see Subsec.2.1).

2. Write the integral equation (2.12) for stresses at interior points. With the above symbology:

$$\underline{\sigma}(\underline{\xi}) + \int_\Gamma \underline{R}(\underline{\xi},\underline{x})\, \underline{u}(\underline{x})\, d\Gamma = \int_\Gamma \underline{Q}(\underline{\xi},\underline{x})\, \underline{p}(\underline{x})\, d\Gamma + + \int_\Omega \underline{S}(\underline{\xi},\underline{x})\, \underline{\varepsilon}^p(\underline{x})\, d\Omega + \underline{F}\, \underline{\varepsilon}^p(\underline{\xi}) + \int_\Omega \underline{Q}(\underline{\xi},\underline{x})\, \underline{b}(\underline{x})\, d\Omega \tag{5.2}$$

3. Subdivide the boundary Γ into elements and the potentially plastic region Ω into cells; after choosing suitable nodes, model displacements and tractions over Γ, plastic strains over $\hat{\Omega}$, eqs.(3.3a,b) and (3.3c). The geometry of boundary elements and/or cells can be generated by suitable mapping of a master element in the space of adimensional coordinates (often using the isoparametric concept), eqs.(3.1)(3.2).

4. Collocate the discretized integral eq.(5.1) in each boundary node. Perform over elements and cells the integrations of integrands represented by products of kernels, interpolation functions and jacobian, according to Sec.3. These integrations will be analytical or numerical with suitable choice of integration points. Assembling the discretized equations for all nodes on Γ (i.e. imposing matching conditions at interfaces of BE, if conforming), obtain the algebraic equations (3.5):

$$\underline{H}\, \underline{u} = \underline{G}\, \underline{p} + \underline{D}\, \underline{\varepsilon}^p + \underline{B} \tag{5.3}$$

where: vectors $\underline{u}$ and $\underline{p}$ (symbols without argument) gather all boundary nodal variables; vector $\underline{\varepsilon}^p$ (without argument) all nodal plastic strains; vector $\underline{B}$ collects domain data. The main diagonal entries of $\underline{H}$ can be generated by the usual rigid body procedure, see e.g. [6] [8].

5. Choose the "strain points", i.e. points where the plastic constitutive law will be enforced in the potentially plastic region $\hat{\Omega}$ and, possibly, on the portion $\hat{\Gamma}$ of its boundary in common with the boundary Γ. Strain points will always coincide in practice with the nodes adopted at step 3 to model the plastic strain field. E.g. constant strain tetrahedrical cells would imply a strain point and modelling node in the centroid of the cell, whereas the four vertices would be the nodes used to define the geometry.

6. Collocate the discretized integral eq.(5.2) in each strain point in the potentially plastic region $\hat{\Omega}$ to express the stresses there (gathered in vector $\underline{\sigma}'$) as functions of boundary nodal variables ($\underline{u}$,$\underline{p}$) and nodal

plastic strains ($\underline{\varepsilon}^p$). Perform the (analytical and numerical) integrations over elements and cells as at point 4. In strain points located on the boundary Γ, if any, the stresses (collected in vector $\underline{\sigma}''$) will be evaluated on a local basis through Hooke's law and equilibrium, as mentioned at the end of Subsec.2.2. By assembling all the equations of the two kinds, obtain the algebraic equations:

$$\underline{\sigma} = \begin{Bmatrix} \underline{\sigma}' \\ \underline{\sigma}'' \end{Bmatrix} = \begin{bmatrix} \underline{H}' \\ \underline{H}'' \end{bmatrix} \underline{u} + \begin{bmatrix} \underline{G}' \\ \underline{G}'' \end{bmatrix} \underline{p} + \begin{bmatrix} \underline{D}' \\ \underline{D}'' \end{bmatrix} \underline{\varepsilon}^p + \begin{Bmatrix} \underline{B}' \\ \underline{B}'' \end{Bmatrix} \tag{5.4}$$

7. Suppose to have reached a completely known statical situation along a history of external actions involving plastic flow. Consider a finite loading step starting from this situation. Introduce the relevant increments of boundary nodal data $\Delta\bar{\underline{u}}$ and $\Delta\bar{\underline{p}}$, on Γ_u and Γ_p, respectively, into eqs.(5.3) and (5.4). Rearranging them after having collected in vector $\Delta\underline{X}$ the increments of boundary unknowns, the two equations can be written as:

$$\underline{A}\, \Delta\underline{X} = \Delta\underline{F} + \underline{D}\, \Delta\underline{\varepsilon}^p \tag{5.5}$$

$$\Delta\underline{\sigma} = \bar{\underline{A}}\, \Delta\underline{X} + \Delta\bar{\underline{F}} + \bar{\underline{D}}\, \Delta\underline{\varepsilon}^p \tag{5.6}$$

where matrices $\underline{A}$ and $\bar{\underline{A}}$ consist of the columns pertaining to unknowns, vectors $\Delta\underline{F}$ and $\Delta\bar{\underline{F}}$ contain both the boundary data and the domain data, such as body forces and thermal strains.

8. Solve (5.5) with respect to the boundary unknowns $\Delta\underline{X}$ (which would thus be determined if there were no plasticity) and substitute into (5.6) to obtain:

$$\Delta\underline{\sigma} = \Delta\underline{\sigma}^E + \underline{Z}\, \Delta\underline{\varepsilon}^p \tag{5.7}$$

having set:

$$\Delta\underline{\sigma}^E \equiv \bar{\underline{A}}\, \underline{A}^{-1}\, \Delta\underline{F} + \Delta\bar{\underline{F}} \quad ; \quad \underline{Z} = \bar{\underline{A}}\, \underline{A}^{-1}\underline{D} + \bar{\underline{D}} \tag{5.8a,b}$$

where the new vector (of data by now) represents the stress response to the loading step in all strain points, if the system were purely elastic; matrix $\underline{Z}$ transforms plastic strain increments at strain points into consequent self-equilibrated elastic stresses there.

9. In order to compensate for the unknown nature of $\Delta\underline{\varepsilon}^p$ in eq.(5.7), enforce (or "collocate") in all strain points the plastic constitutive laws. Namely, derive from the plastic material model a relationship (or

an algorithm) leading either (a) from the (finite) stress increments $\Delta\underline{\sigma}$ to plastic strain increments $\Delta\underline{\varepsilon}^p$ or, conversely (b) from total strain increments $\Delta\underline{\varepsilon}$ to $\Delta\underline{\varepsilon}^p$. Symbolically (and expressing for convenience these local, constitutive relations in terms of the vectors concerning all strain points):

$$\Delta\underline{\varepsilon}^p = \Delta\underline{\varepsilon}^p(\Delta\underline{\sigma}) \quad , \quad \Delta\underline{\varepsilon}^p = \Delta\underline{\varepsilon}^p(\Delta\underline{\varepsilon}) \qquad (5.9a,b)$$

For hardening materials uniqueness (besides existence) of $\Delta\underline{\varepsilon}^p$ is ensured in both alternatives (a) and (b), the former being more straightforward for the present form of the integral equation. In a more general plasticity context (encompassing also perfect nonhardening behaviours and softening) the latter (b) option is needed as it is the only one for which uniqueness is guaranteed (except in subcritical softening case), see Sec.4. Clearly, in this case eqs.(5.7) and (5.8b) must be supplemented by Hooke law (for all strain points simultaneously):

$$\Delta\underline{\varepsilon} = \underline{E}^{-1}\Delta\underline{\sigma} + \Delta\underline{\varepsilon}^p \qquad (5.10)$$

where $\underline{E}$ is the block-diagonal matrix containing the elastic moduli matrix of order 6 as repeated diagonal block. Equivalently but more directly, without (5.10), eq.(5.9b) could be associated to the discretized version of the integral equation for total strains (instead of that for stresses used above). Note that this is the one directly emanating from Somigliana equation through suitable differentiations (Sec.2).

10. Solve the discretized b.v. problem for the current loading step, dealing either (a) with relation set (5.7)(5.9a) or (b) with the relation set (5.7)(5.9b)(5.10). This crucial solution process represents a step of an approximate time integration of the governing nonlinear differential equations centered on the plastic flow rules (in terms of rates); it will be the subject of Subsec.5.3. Lack of solution (usually lack of convergence in an iterative procedure) is normally assumed as an indication of plastic collapse within the loading step.

11. By means of the increments resulting from phase 9, update the variables in order to fully define the statical situation, starting from which the response to new increments of external actions has to be evaluated. To this purpose go back to phase 7 as the preceding phases can be carried out once for all loading steps.

The procedure outlined in what precedes may give rise to the following comments (among others).

(a) Some authors use the expressions "initial strain" and "initial stress" to distinguish the ways in which inelastic strains are accomodated in the integral equations, i.e. directly as here in eqs.(2.4) and (2.12) or through "plastic stresses" defined by eq.(2.10). It is worth noting that the same expressions have been introduced in the early

FE literature with reference to a different and more substancial circumstance, namely to the use of the constitution from strain to stress increments ("initial stress" approach) or viceversa ("initial strain" approach), i.e. in parallel to the alternatives (a) and (b) mentioned in Subsec.5.1., phase 9.

(b) The vector of the elastic incremental stress response $\Delta\underline{\sigma}^E$ in all strain points to all external actions, when these increase proportionally, can be calculated once for all. Then, certain values of the load factor α will define loading steps from the onset of plasticity α_E up to the overall collapse limit α_L.

Matrix $\underline{Z}$ is also a result of once for all preliminary calculations based only on the elastic properties α, on the geometry and discretization of the system and of its potentially plastic region Ω.

(c) In the above outlined BE plastic analysis and in its further developments in limit and shakedown analysis (Sec.7), an important role is played by the matrix $\underline{Z}$, eq.(5.7)(5.8b). Its entries can be conceived as influence coefficients of plastic strains over consequent self-stresses in strain points. However neither symmetry nor negative semidefiniteness are in general exhibited by matrix $\underline{Z}$ arising as described in the preceding subsection. These properties, which are desirable and even necessary to some purposes, can be restored by the alternative approaches outlined in the next Section.

It is worth noting here the generation of $\underline{Z}$ according to eq.(5.8b), which requires the inversion of matrix $\underline{A}$ of order equal to the number of nodal unknowns. The counterpart to $\underline{Z}$ in the FE method with displacement modelling would require more matrix multiplications and the inversion of the overall elastic stiffness matrix.

5.3 Integration in time

5.3.1 As shown in the previous section, the discretization of the integral equations yields the set of eqs. (5.7) for stress increments in the adopted "strain points". In (5.7) only the elastic response of the structure is involved for given increments of external forces, displacements and imposed strains. The actual constitution of the material, which can be formally represented by eqs.(4.8,9), must be associated to eqs.(5.7), written in terms of rates, in order to express a further relationship between stress and plastic strain rates. The two sets of eqs.(4.8,9) and (5.7), completely define the incremental response of the structure for given external action rates. In order to obtain results in finite terms, a time integration must be performed over the whole load history.

The load history is given as a function of a parameter t monotonically increasing with time. Often this load path is not completely known: only some sample points are assigned and linear interpolation is assumed.

There are very few cases for which analytical integrations of the constitutive laws are available. In general it is necessary to make use of numerical techniques which require time steps of finite size.

Let a load path be given as in fig. 5.1. It is sampled at a suitable number n of points t_i defining a set of finite load increments. The true loading history within a step is immaterial since generally no information are available on it. The size of each load step is governed by accuracy and convergence requirements and must be carefully considered in relation to the stability of the adopted integration algorithm and to the degree of nonlinearity in the structural response. If the initial solution is very distant from the final one, convergence may occur (if ever) only after
a time consuming number of iterations.

At this stage the problem is formulated as a stepwise problem consisting of finding for a given increment of the external actions (say $\Delta\underline{P}$), the response of the system starting from a completely known state at a given time t^n, ($\underline{\varepsilon}_o$, $\underline{\varepsilon}_o^p$, and, hence, $\underline{\sigma}_o$ through Hooke's law). Over each step the solution can be sought by means of some numerical procedure which implies to transit iteratively, with improved solutions, from (5.7) to (4.8,9) and vice-versa, untill convergence is reached.

For the (n+1)th step it is possible to write:

$$\Delta\underline{\varepsilon} = \Delta\underline{\varepsilon}^E + \underline{M}\,\Delta\underline{\varepsilon}^p \,, \qquad \Delta\underline{\sigma} = \Delta\underline{\sigma}^E + \underline{Z}\,\Delta\underline{\varepsilon}^p \qquad (5.11a,b)$$

Equation (5.11b) provides a "prediction" $\Delta\underline{\sigma}$ of the stress increments. These are equilibrated with the load increments (in the approximate sense consistent with the adopted interpolations), since they are the sum of the linear elastic stress responses to the increments of external actions and to the increments of plastic strains conceived as imposed dislocations. In general, the stress $\underline{\sigma}_o + \Delta\underline{\sigma}$ will violate the plastic consistency condition (4.1). Hence a correction is obtained by imposing the suitably integrated constitutive law.

Two distinct phases are recognized in the procedure: the first one will be referred to as the "predictor" phase and will be considered in detail later, while the second one will be referred to as the "corrector" phase.

It is worth stressing that accuracy is affected only by the "corrector". In fact, the choice of a particular integration scheme implies assumptions on the constitutive law (e.g. on the flow rule) which can heavily affect the results. In other words, using a numerically integrated constitutive law is equivalent to assume a particular material which behaves, in general, differently from the actual one. The prediction of the stress increments affects only the speed of convergence, i.e. the number of iterations needed to attain a pre-assigned accuracy. If the algorithm will converge, it will converge to

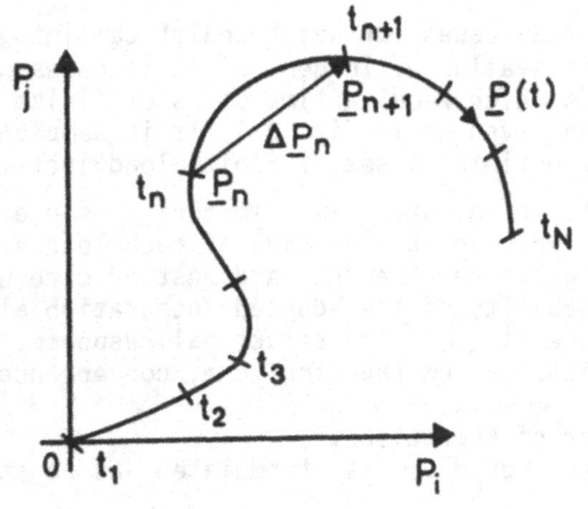

Fig. 5.1

the solution consistent with the adopted "corrector" scheme, whatever the "predictor" is.

Recently Ortiz and Popov [26] presented a comprehensive analysis of the "generalized trapezoidal" and "generalized midpoint" rules. The most popular integration methods can be obtained from these general formulations wich can be summarized as follows (the total strain increment $\Delta\underline{\varepsilon}$ is regarded as an assigned data; quantities marked by a zero subscript refer to the state at the beginning of the step; the initial stress point $\underline{\sigma}_0$ is assumed to be on the yield surface; β is a scalar to be suitably chosen in the interval (0,1)).

(a) Generalized trapezoidal rule:

$$\Delta\underline{\varepsilon}^p \simeq [\underline{\Psi}(\underline{\sigma}_0,\underline{\chi}_0)(1-\beta) + \underline{\Psi}(\underline{\sigma}_0+\Delta\underline{\sigma},\ \underline{\chi}_0+\Delta\underline{\chi})\ \beta]\ \Delta\underline{\lambda} \qquad (5.12a)$$

$$\Delta\underline{\chi} \simeq [\underline{\omega}(\underline{\sigma}_0,\underline{\chi}_0)(1-\beta) + \underline{\omega}(\underline{\sigma}_0+\Delta\underline{\sigma},\ \underline{\chi}_0+\Delta\underline{\chi})\ \beta]\ \Delta\underline{\lambda} \qquad (5.12b)$$

$$\underline{\Phi}(\underline{\sigma}_0+\Delta\underline{\sigma},\ \underline{\chi}_0+\Delta\underline{\chi}) \le \underline{0} \quad ; \quad \Delta\underline{\lambda} \ge \underline{0} \quad ; \quad \tilde{\underline{\Phi}}\ \Delta\underline{\lambda} = 0 \qquad (5.12c,d,e)$$

$$\Delta\underline{\sigma} = \underline{E}\ (\Delta\underline{\varepsilon} - \Delta\underline{\varepsilon}^p) \qquad (5.12f)$$

where it has been set $\dot{\underline{\chi}} = \underline{\omega}(\underline{\sigma},\underline{\chi})\ \dot{\underline{\lambda}}$.

(b) Generalized midpoint rule (fig. 5.3, with $\underline{\Psi} \equiv \underline{\Phi}$):

$$\Delta\underline{\varepsilon}^p \simeq \underline{\Psi}(\underline{\sigma}_0+\beta\Delta\underline{\sigma},\ \underline{\chi}_0+\Delta\underline{\chi})\ \Delta\underline{\lambda} \qquad (5.13a)$$

$$\Delta\underline{\chi} \simeq \underline{\omega}(\underline{\sigma}_0+\beta\Delta\underline{\sigma},\ \underline{\chi}_0+\beta\Delta\underline{\chi})\ \Delta\underline{\lambda} \qquad (5.13b)$$

$$\phi(\underline{\sigma}_o + \Delta\underline{\sigma},\ \underline{\chi}_o + \Delta\underline{\chi}) \leq 0 \quad ; \quad \Delta\underline{\lambda} \geq \underline{0} \quad ; \quad \tilde{\underline{\phi}}\ \Delta\underline{\lambda} = 0 \qquad (5.13c,d,e)$$

$$\Delta\underline{\sigma} = \underline{E}\ (\Delta\underline{\varepsilon} - \Delta\underline{\varepsilon}^p) \qquad (5.13f)$$

The following remarks appear appropriate:

(a) According to ref.[26], both methods are conditionally stable (depending on the step size) for $\beta < 1/2$. The "best" accuracy (2° order) is exhibited for $\beta = \frac{1}{2}$. For $\beta \geq \frac{1}{2}$ the stability of the first method is conditioned by the curvature of the yield surface: high curvatures or, even worse, corners in the yield surface, can deeply affect the stability. On the contrary, the midpoint rule exhibits convergence characteristics not affected by the curvature.
(b) If the starting stress point $\underline{\sigma}_o$ is in the elastic field, both methods require finding the 'contact point' $\hat{\underline{\sigma}}$ where the yield locus is firstly arrived at, and assuming it as the starting stress point. The determination of $\hat{\underline{\sigma}}$ may be cumbersome for multisurface constitutive laws. Moreover, in order to consistently obtain symmetric tangent stiffness matrices of the material, the 'midpoint rule' requires plastic consistency to be imposed at the point $\underline{\sigma}_o + \beta\Delta\underline{\sigma}$ and not at the final point [27]. This implies that the final stress state will not satisfy plastic consistency, in general.
(c) For $\beta = 0$ and $\beta = 1$ the two methods coincide and the two above mentioned drawbacks disappear.
(d) For $\beta = 0$ both formulations reduce to the well known Euler forward method, which has been extensively used so far, both in the FE and in the BE methods. If the plastic consistency condition in the final stress point is substituted by the linearized condition of continuous yielding (loading) over the step, $\Delta\underline{\phi} = \underline{0}$, where:

$$\Delta\underline{\phi} = \underline{\Phi}(\underline{\sigma}_o)\Delta\underline{\sigma} - \underline{H}(\underline{\sigma}_o,\underline{\chi}_o)\ \Delta\underline{\lambda} = \underline{0} \qquad (5.14)$$

then explicit equations are obtained. This allows to compute the stress and plastic strain increments at a minimum cost. On the other hand the final stress point will not lie on the yield surface.
(e) The assumption $\beta = 1$ leads to the backward difference method. This choice is gaining increasing popularity because of its characteristics of stability. For the von Mises case, it is coincident with the 'radial return' method. Several authors made comparisons with results obtained by exact integrations and always the backward difference method turned out to be the most accurate. Recently it has been successfully applied to more complex constitutive laws [28], and analysed in order to give it a theoretical basis [29].

(f) If piecewise linear yield surfaces and linear hardening are assumed, the coefficient β becomes immaterial and the distinction between forward and backward difference vanishes since $\underline{\Psi}$ and $\underline{H}$ have constant entries. In this case the integration is exact for the stepwise holonomic model described in sec. 4.2.

5.3.2 We will focus now on the backward difference integration scheme for the above fairly general class of nonassociative constitutive models with nonlinear yield functions and nonlinear hardening. Eqs. (5.12) or (5.13), written for $\beta = 1$, read:

$$\Delta\underline{\varepsilon}^p = \underline{\Psi}(\underline{\sigma}_0 + \Delta\underline{\sigma}, \underline{\chi}_0 + \Delta\underline{\chi})\,\Delta\underline{\lambda} \quad ; \quad \Delta\underline{\chi} = \underline{\omega}(\underline{\sigma}_0 + \Delta\underline{\sigma}, \underline{\chi}_0 + \Delta\underline{\chi})\,\Delta\underline{\lambda} \qquad (5.15a,b)$$

$$\underline{\phi}(\underline{\sigma}_0 + \Delta\underline{\sigma}, \underline{\chi}_0 + \Delta\underline{\chi}) = \underline{0} \quad ; \quad \Delta\underline{\lambda} \geq \underline{0} \quad ; \quad \tilde{\underline{\phi}}\,\Delta\underline{\lambda} = 0 \qquad (5.15c,d,e)$$

$$\Delta\underline{\sigma} = \underline{E}\,(\Delta\underline{\varepsilon} - \Delta\underline{\varepsilon}^p) \qquad (5.15f)$$

If, in the step, the yield functions depend linearly on $\Delta\underline{\sigma}$ and $\Delta\underline{\lambda}$, then this relation set turns out to be formally equivalent to (4.14), (4.15). Therefore, the solution of eqs. (5.15) coincides with the exact solution for the stepwise holonomic material which has a piecewise linear surface with a plane (or a cone of planes if the real yield surface has a corner in correspondence of $\underline{\sigma}_0 + \Delta\underline{\sigma}$) tangent to the actual yield surface in the point $\underline{\sigma}_0 + \Delta\underline{\sigma}$, with a linear hardening rule (where $\underline{H} = \underline{H}(\underline{\sigma}_0 + \Delta\underline{\sigma}, \underline{\chi}_0 + \Delta\underline{\chi})$) and a flow rule given by $\Delta\underline{\varepsilon}^p = \underline{\Psi}\,\Delta\underline{\lambda}$ with $\underline{\Psi} = \underline{\Psi}(\underline{\sigma}_0 + \Delta\underline{\sigma}, \underline{\chi}_0 + \Delta\underline{\chi})$.

This point of view provides a better understanding of the assumptions which are implied by the backward difference scheme. Using this method is recognized to be equivalent to neglect the path dependence of the material within the step, and to impose the non-reversibility of the plastic strains only at a finite number of situations along the loading history, namely at the end of each step. Clearly, large time steps can cause significant loss of accuracy; conversely, by decreasing the step size the actual material behaviour is recovered.

Rearranging eqs.(5.15), one obtains:

$$\Delta\underline{\sigma} = \underline{E}[\Delta\underline{\varepsilon} - \underline{\Psi}(\underline{\sigma}_0 + \Delta\underline{\sigma}, \underline{\chi}_0 + \Delta\underline{\chi})\Delta\underline{\lambda}] \; ; \; \Delta\underline{\chi} = \underline{\omega}(\underline{\sigma}_0 + \Delta\underline{\sigma}, \underline{\chi} + \Delta\underline{\chi})\Delta\underline{\lambda} \qquad (5.16a,b)$$

$$\underline{\phi}(\underline{\sigma}_0 + \Delta\underline{\sigma}, \underline{\chi}_0 + \Delta\underline{\chi}) = \underline{0} \quad ; \quad \tilde{\underline{\phi}}\,\Delta\underline{\lambda} = 0 \quad ; \quad \Delta\underline{\lambda} \geq \underline{0} \qquad (5.16c,d,e)$$

Eqs.(5.16) represent a system of non linear equations in the unknowns $\Delta\underline{\sigma}, \Delta\underline{\chi}$ and $\Delta\underline{\lambda}$, which can be solved iteratively according to various numerical schemes. This must be done for each 'strain point' where the trial stress $\underline{\sigma}_o + \underline{E}\ \Delta\underline{\varepsilon}$ does not comply with the plastic consistency requirements. In the three dimensional case there are six independent stress components, a certain number of internal variables and as many plastic multipliers as the yield surfaces which describe the yield loci. Solving system (5.16) in the general case can be very laborious and this is one of the main reasons for which the forward integration has been so extensively used thus far. Nevertheless, in many practical cases, the number of unknowns can be substantially reduced, and for some simple models (with linear hardening rules) system (5.16) can even be solved analytically. It can be shown [28] that, if isotropic elasticity is assumed and the plastic potentials $\underline{\Psi}$ are dependent only on the first invariant of the stress tensor and on the second invariant of the deviator (which is the case e.g. of von Mises and Drucker-Prager models), eqs. (15.6) can be expressed in terms of these two invariants only and of $\Delta\underline{\lambda}$. This drastically reduces the cost of the solution.

The application of the backward difference method (or of any "trapezoidal" or "midpoint" scheme) leads to the following two functions (explicit if analytical solution of (5.16) is available):

$$\Delta\underline{\sigma} = \Delta\underline{\sigma}(\Delta\underline{\varepsilon}) \ ; \quad \Delta\underline{\varepsilon}^p = \Delta\underline{\varepsilon}^p(\Delta\underline{\varepsilon}) \qquad (5.17a,b)$$

defined for all the 'strain points'.

5.3.3 Turning now to the specific BE context, the total strain increment $\Delta\underline{\varepsilon}$ (assumed as a given quantity so far) can be computed ("predicted") through the BE equations as linear function of the plastic strain increments $\Delta\underline{\varepsilon}^p$ when the elastic strain response to external action increments has been determined.

If the backward difference method is used, taking account of eqs. (5.11) and (5.17), the stepwise holonomic problem can be formulated as follows:

$$\Delta\underline{\varepsilon} = \Delta\underline{\varepsilon}^E + \underline{M}\ \Delta\underline{\varepsilon}^p(\Delta\underline{\varepsilon}) \qquad (5.18)$$

where the function $\Delta\underline{\varepsilon}^p(\Delta\underline{\varepsilon})$ is not usually known in explicit form. This is a system of nonlinear equations in the unknown vector $\Delta\underline{\varepsilon}$, which can be solved by making use of some numerical method like successive substitutions, steepest descent, Newton Raphson, etc.

The simplest one (and the only one which has been used so far in the literature on BEM, see e.g. [8] [30]) is the "successive substitution" method which can be summarized as follows.

(1) Given the linear elastic response $\Delta\underline{\varepsilon}^E$ of the considered load step, initialize the iteration sequence (for i = 0) by setting $\Delta\underline{\varepsilon}_p^p = \underline{0}$ (if a better guess is not available).
(2) Compute the total strain increment from (5.11) ("predictor"):

$$\Delta\underline{\varepsilon}_{i+1} = \Delta\underline{\varepsilon}^E + \underline{M}\,\Delta\underline{\varepsilon}_i^p \tag{5.19}$$

(3) Compute $\Delta\underline{\varepsilon}_{i+1}^p$ from the material constitution, through one of the proposed integration methods ("corrector");

(4) Check convergence: if $|\Delta\underline{\varepsilon}_{i+1} - \Delta\underline{\varepsilon}_i|\,(|\Delta\underline{\varepsilon}_i|)^{-1} < \gamma$, where γ is a given tolerance, then compute $\Delta\underline{\sigma}_{i+1}$ corresponding to $\Delta\underline{\varepsilon}_{i+1} - \Delta\underline{\varepsilon}_{i+1}^p$ through the elastic constitution and stop the iterative procedure;
(5) If convergence is not reached, set i = i+1 and go back to step (1);
(6) When convergence is reached, assign a new load increment, compute the linear elastic response $\Delta\underline{\sigma}^E$ and start iterating again from step (1).

This procedure is convenient inasmuch since no matrix inversions are required; however, the convergence is usually very slow. The application of the other, more efficient methods, must still be investigated, taking into account the non-symmetry and non-definitness of matrix $\underline{Z}$.

6. ON SYMMETRIZATION OF THE RATE PROBLEM

The discretized elastic-plastic b.v. problem formulated in Subsec.5.1 has been seen to be centered on a linear complementarity problem in terms of rates. It will be realized later that the symmetry and positive semi-definitness of its matrix are desirable features, necessary for the validity of important further developments. In this section two possible approaches are outlined apt to confer these features: in subsection 6.2. a "direct symmetrization" proposed in [31]; in subsection 6.3. a double integration, weighted residual formulation proposed in [32]. Both formulations imply the enforcement of the plastic flow rules in a suitable average sense over each cell. This issue will be considered first in the next subsection.

6.1 Formulation of plastic laws for cells

In what precedes the plastic strains and the stresses in specific points ("strain points") have been considered over the potentially plastic subdomain, related to each other by the material law and gathered in vectors $\underline{\varepsilon}_c^p$ and $\underline{\sigma}_c$ for each cell Ω_c. The plastic strain field over Ω_c is governed by vector $\underline{\varepsilon}_c^p$ through the interpolation function matrix $\underline{M}(\underline{x})$, eq.(3.3c).

For the sake of formal simplicity, Cartesian instead of nondimensional local coordinates will be used henceforth.

As a first step towards a symmetric formulation, let us model now both strains and stresses over Ω_c

$$\underline{\varepsilon}(\underline{x}) = \underline{\bar{M}}_\varepsilon(\underline{x})\,\underline{\bar{\varepsilon}}_c \quad , \quad \underline{\sigma}(\underline{x}) = \underline{\bar{M}}_\sigma(\underline{x})\,\underline{\bar{\sigma}}_c \tag{6.1a,b}$$

in such a way that the internal virtual work be represented by the dot product of the governing parameter vectors:

$$\tilde{\underline{\bar{\sigma}}}_c\,\underline{\bar{\varepsilon}}_c = \int_{\Omega_c} \tilde{\underline{\sigma}}(\underline{x})\,\underline{\varepsilon}(\underline{x})\,d\Omega \quad , \quad \text{for any} \quad \underline{\bar{\varepsilon}}_c\,,\,\underline{\bar{\sigma}}_c \tag{6.2}$$

When (6.2) is satisfied, the discretization will be called "consistent" and the components of $\underline{\bar{\varepsilon}}_c$ and $\underline{\bar{\sigma}}_c$ "generalized strains and stresses" of cell c (according to a terminology introduced by W. Prager). Assuming that stresses were originally interpolated by means of the same matrix $\underline{M}(\underline{x})$ adopted for plastic strains, square nonsingular transformation matrices $\underline{T}_\varepsilon$ and $\underline{T}_\sigma$ are to be determined which relate consistent cell variables to corresponding quantities in the strain points:

$$\underline{\varepsilon}_c = \underline{T}_\varepsilon\,\underline{\bar{\varepsilon}} \quad , \quad \underline{\sigma}_c = \underline{T}_\sigma\,\underline{\bar{\sigma}}_c \tag{6.3a,b}$$

Account taken of (6.1) and (6.3), eq. (6.2) can be continued as follows:

$$\begin{aligned}\tilde{\underline{\bar{\sigma}}}_c\,\underline{\bar{\varepsilon}}_c &= \int_{\Omega_c} \tilde{\underline{\sigma}}(\underline{x})\,\underline{M}(\underline{x})\,d\Omega\,\underline{T}_\varepsilon\,\underline{\bar{\varepsilon}}_c = \\ &= \tilde{\underline{\bar{\sigma}}}_c\,\tilde{\underline{T}}_\sigma \int_{\Omega_c} \tilde{\underline{M}}(\underline{x})\,\underline{\varepsilon}(\underline{x})\,d\Omega = \tilde{\underline{\bar{\sigma}}}_c\tilde{\underline{T}}_\sigma\int_{\Omega_c}\tilde{\underline{M}}(\underline{x})\,\underline{M}(\underline{x})\,d\Omega\,\underline{T}_\varepsilon\underline{\bar{\varepsilon}}_c\end{aligned} \tag{6.4}$$

Since they must hold for any arbitrarily chosen $\underline{\bar{\varepsilon}}_c$ and $\underline{\bar{\sigma}}_c$, these three identities imply, respectively:

$$\bar{\underline{\sigma}}_c = \tilde{\underline{T}}_\varepsilon \int_{\Omega_c} \tilde{\underline{M}}\, \underline{\sigma}\, d\Omega \;, \quad \bar{\underline{\varepsilon}}_c = \tilde{\underline{T}}_\sigma \int_{\Omega_c} \tilde{\underline{M}}\, \underline{\varepsilon}\, d\Omega \;, \quad \tilde{\underline{T}}_\sigma \int_{\Omega_c} \tilde{\underline{M}}\, \underline{M}\, d\Omega\, \underline{T}_\varepsilon = \underline{I} \tag{6.5a,b,c}$$

Eq. (6.5c) represents the only requirement on the choice of the transformation matrices. Two rather natural options are as follows:

$$\underline{T}_\varepsilon = \underline{I} \;, \text{ whence: } \underline{T}_\sigma = \left(\int_{\Omega_c} \tilde{\underline{M}}\, \underline{M}\, d\Omega\right)^{-1} \tag{6.6}$$

$$\underline{T}_\sigma = \underline{T}_\varepsilon \equiv \underline{T} = \tilde{\underline{T}} = \left(\int_{\Omega_c} \tilde{\underline{M}}\, \underline{M}\, d\Omega\right)^{-\frac{1}{2}} \tag{6.7}$$

Turning now to the fields which intervene in the plastic constitution, we follow below the same path of reasoning which led above to the generation of consistent models for strains and stresses. Let us discretize the fields of plastic multiplier rates and yield functions over cell Ω_c by means of the same interpolation matrix $\underline{L}(\underline{x})$. This will interpolate strain point values collected in vectors $\dot{\underline{\lambda}}_c$ and $\underline{\phi}_c$, respectively, to be transformed into "consistent" parameters (marked by bars):

$$\dot{\underline{\lambda}}(\underline{x}) = \underline{L}(\underline{x})\, \dot{\underline{\lambda}}_c = \underline{L}(\underline{x})\, \underline{T}_\lambda\, \bar{\dot{\underline{\lambda}}}_c \equiv \bar{\underline{L}}_\lambda(\underline{x})\, \bar{\dot{\underline{\lambda}}}_c \tag{6.8}$$

$$\underline{\phi}(\underline{x}) = \underline{L}(\underline{x})\, \underline{\phi}_c = \underline{L}(\underline{x})\, \underline{T}_\phi\, \bar{\underline{\phi}}_c \equiv \bar{\underline{L}}_\phi(\underline{x})\, \bar{\underline{\phi}}_c \tag{6.9}$$

The ("consistency") condition we impose, is analogous to (6.2):

$$\tilde{\bar{\underline{\phi}}}_c\, \bar{\dot{\underline{\lambda}}}_c = \int_{\Omega_c} \tilde{\underline{\phi}}(\underline{x})\, \dot{\underline{\lambda}}(\underline{x})\, d\Omega \;, \quad \text{for all } \bar{\underline{\phi}}_c \,,\, \bar{\dot{\underline{\lambda}}}_c \tag{6.10}$$

Through (6.8) (6.10) and consequent identities similar to (6.4), the implications of (6.10) are:

$$\bar{\underline{\phi}}_c = \tilde{\underline{T}}_\lambda \int_{\Omega_c} \tilde{\underline{L}}\, \underline{\phi}\, d\Omega \quad , \quad \bar{\dot{\underline{\lambda}}}_c = \tilde{\underline{T}}_\phi \int_{\Omega_c} \tilde{\underline{L}}\, \dot{\underline{\lambda}}\, d\Omega \tag{6.11a,b}$$

$$\tilde{\underline{T}}_\phi \int_{\Omega_c} \tilde{\underline{L}}\, \underline{L}\, d\Omega\, \underline{T}_\lambda = \underline{I} \tag{6.11c}$$

Two ways of satisfying this condition are anologous to, but not necessarily concomitant with, eqs. (6.6) and (6.7). As a final step, the above defined "generalized variables" for cell Ω_c will now be related to each other in order to generate a description of the overall behaviour of the cell. Such description stems from the material constitutive law enforced in some average sense, instead of locally everywhere in Ω_c.

Making use of the material normality rule and of the discrete model (6.8), eq.(6.5b), written for plastic strain rates, becomes:

$$\dot{\bar{\underline{\varepsilon}}}^p_c = \bar{\underline{\Phi}}_c \dot{\bar{\underline{\lambda}}}_c \ , \qquad \text{having set: } \bar{\underline{\Phi}}_c \equiv \tilde{\underline{T}}_\sigma \int_{\Omega_c} \tilde{\underline{M}}\, \underline{\Phi}\, \underline{L}\, d\Omega\, \underline{T}_\lambda \qquad (6.12a,b)$$

Substitute now the expression of the material yield function rates (4.6a) into (6.1a) and use the consistent model for stress rates, (6.3b) and plastic multipliers rates (6.11b), to obtain:

$$\dot{\bar{\underline{\phi}}}_c = \tilde{\bar{\underline{\Phi}}}_c \dot{\underline{\sigma}}_c - \bar{\underline{H}}_c \dot{\bar{\underline{\lambda}}}_c \ , \qquad \text{where: } \bar{\underline{H}}_c \equiv \tilde{\underline{T}}_\lambda \int_{\Omega_c} \tilde{\underline{L}}\, \underline{H}\, \underline{L}\, d\Omega\, \underline{T}_\lambda \qquad (6.13a,b)$$

The complementarity condition on the generalized variables complete the cell flow rule:

$$\dot{\bar{\underline{\lambda}}}_c \geq \underline{0} \ , \qquad \dot{\bar{\underline{\phi}}}_c \leq \underline{0} \ , \qquad \tilde{\dot{\bar{\underline{\phi}}}}_c \dot{\bar{\underline{\lambda}}}_c = 0 \qquad (6.14)$$

Note that, in the "cell plastic law", eq.(6.12) expresses the normality rule, eq.(6.13b) defines a symmetric hardening matrix (positive semidefinite if the material matrix $\underline{H}$ is so), eqs.(6.14) materialize the enforcement of constitutive complementarity in average only.

6.2 "Direct" symmetrization of the BE equations

The discretized integral equations for stresses (Secs. 3.2 and 3.3) provided a relation "through the system" between plastic strains and stresses in all the strain points:

$$\dot{\underline{\sigma}} = \dot{\underline{\sigma}}^E + \underline{Z}\, \dot{\underline{\varepsilon}}^p \qquad (6.15)$$

It will be assumed henceforth that all strain points are internal to the cells, so that the stresses contained in vector $\underline{\sigma}$ are not evaluated on the boundary Γ using boundary information ("nonconforming cells"); besides, all variables defined in cells remain independent in (6.15).

Pre-multiplying (6.15) by $\mathrm{diag}[\underline{T}_\sigma^{-1}]$ (block diagonal matrix of matrices $\underline{T}_\sigma$ for all cells) and using the transformations (6.3), we obtain the following relation in terms of generalized stress and plastic strain rates for cells as defined in Subsec. 6.1:

$$\dot{\bar{\underline{\sigma}}} = \dot{\bar{\underline{\sigma}}}^E + \bar{\underline{Z}}\,\dot{\bar{\underline{\varepsilon}}}^p = \dot{\bar{\underline{\sigma}}}^E + \dot{\bar{\underline{\sigma}}}^s \text{ , where: } \bar{\underline{Z}} \equiv \mathrm{diag}[\underline{T}_\sigma^{-1}]\,\underline{Z}\,\mathrm{diag}[\underline{T}_\varepsilon] \quad (6.16a,b)$$

Now consider two plastic strain fields (marked by ' and ") as given external actions and the relevant selfstresses σ_{ij}^s which would solve exactly the continuum b.v. problem. Write the virtual work equation:

$$\int_\Omega \sigma_{ij}^{s'}(\varepsilon_{ij}^{p''} + C_{ijrs}\,\sigma_{rs}^{s''})\,d\Omega = 0 \qquad (6.17)$$

If the two situations ' and " coincide, eq.(6.17) shows that:

$$-\frac{1}{2}\int_\Omega \sigma_{ij}^s\,\varepsilon_{ij}^p\,d\Omega = \frac{1}{2}\int_\Omega \sigma_{ij}^s\,C_{ijrs}\,\sigma_{rs}^s\,d\Omega \geq 0 \qquad (6.18)$$

If the two situations are distinct, eq. (6.17) together with its analogue obtained by swapping ' and ", in view of the symmetry of the elastic tensor, leads to Volterra reciprocity theorem:

$$\int_{\Omega^{p''}} \sigma_{ij}^{s'}\,\varepsilon_{ij}^{p''}\,d\Omega = \int_{\Omega^{p'}} \sigma_{ij}^{s''}\,\varepsilon_{ij}^{p'}\,d\Omega \qquad (6.19)$$

Here $\Omega^{p'}$ and $\Omega^{p''}$ denote the zones where there are plastic strains in the two situations. If "consistent models" in the sense of Subsec.6.1 (eq. 6.2) are adopted, eq.(6.18) implies:

$$-\frac{1}{2}\,\tilde{\bar{\underline{\varepsilon}}}^p\,\bar{\underline{Z}}\,\bar{\underline{\varepsilon}}^p \simeq \frac{1}{2}\,\tilde{\bar{\underline{\varepsilon}}}^p\,\tilde{\bar{\underline{Z}}}\,\mathrm{diag}[\int_{\Omega_c} \tilde{\bar{\underline{M}}}_\sigma\,\underline{C}\,\bar{\underline{M}}_\sigma\,d\Omega]\,\bar{\underline{Z}}\,\bar{\underline{\varepsilon}}^p \geq 0 \text{ , for all } \bar{\underline{\varepsilon}}^p \qquad (6.20)$$

$\underline{C}$ being the material elastic compliance matrix. Eqs.(6.19) and (6.16b) permit to write:

$$\tilde{\bar{\underline{\varepsilon}}}^{p'}\,\tilde{\bar{\underline{Z}}}\,\bar{\underline{\varepsilon}}^{p''} \simeq \tilde{\bar{\underline{\varepsilon}}}^{p''}\,\tilde{\bar{\underline{Z}}}\,\bar{\underline{\varepsilon}}^{p'}, \quad \text{for all } \bar{\underline{\varepsilon}}^{p'},\ \bar{\underline{\varepsilon}}^{p''} \qquad (6.21)$$

The approximation sign used in (6.20) and (6.21) reflects the circumstance that (6.17) and, hence, the equalities in (6.18) and (6.19) are invalidated by the local violations of equilibrium and compatibility implicit in the discretizations adopted for the response to imposed strains. However, it appears reasonable to conjecture that these equalities tend to be valid asymptotically as such "modelling errors" reduce by refining the discretization, provided convergence to the exact continuum solution be guaranteed in a local sense. We will refer to this conjecture by the term "almost" in the following conclusion: if generalized stress and strain variables (in the sense of Subsec. 6.1) are employed for all cells, matrix $\bar{\underline{Z}}$ (6.16b) relating selfstress rates to plastic strain rates is "almost" negative semidefinite as a consequence of (6.20), and "almost" symmetric in view of (6.21).

It is worth stressing that such conclusion would not hold if one were to deal with variables defined in the strain points, as in (6.15) according to the customary procedure. In fact, without (6.2), eqs. (6.18) and (6.19) would not entail (6.20) and (6.21), respectively, even in the absence of modelling errors. The above conclusion suggests and, in a still vague sense, legitimates to replace matrix $\bar{\underline{Z}}$ (6.16b) by its symmetric part:

$$\bar{\underline{Z}}_s \equiv \frac{1}{2}(\tilde{\bar{\underline{Z}}} + \bar{\underline{Z}}) \tag{6.22}$$

This reduction to the symmetric part of $\bar{\underline{Z}}$ is similar to what has been proposed in elasticity for BE stiffness matrices, see e.g. [8], and critically assessed as for accuracy deterioration [33].

Let us combine now the remarks expounded so far in this and in the preceding Subsec. 6.1. Substituting (6.12) into (6.16a) and this into (6.13a) and using (6.22) instead of $\bar{\underline{Z}}$, we obtain:

$$\dot{\bar{\underline{\phi}}} = \tilde{\bar{\underline{\Phi}}}\,\dot{\bar{\underline{\sigma}}}^E - (\bar{\underline{H}} - \tilde{\bar{\underline{\Phi}}}\,\bar{\underline{Z}}_s\,\bar{\underline{\Phi}})\,\dot{\bar{\underline{\lambda}}} \tag{6.23}$$

Associated with the complementarity condition (6.14), this is a reformulation of the BE analysis problem in terms of rates and in form of a linear complementarity problem (LCP). At difference from the corresponding matrix arising in the formulation of Sec.5, the matrix in parentheses turns out to be "almost" positive semidefinite in the above sense and symmetric by construction at the price of additional errors implied by replacing $\bar{\underline{Z}}$ by its symmetric part.

However, the computational merits of the above "direct" symmetrization is still doubious and the control of the errors due to the replacement of $\bar{\underline{Z}}$ by $\bar{\underline{Z}}_s$ is presently an open problem.

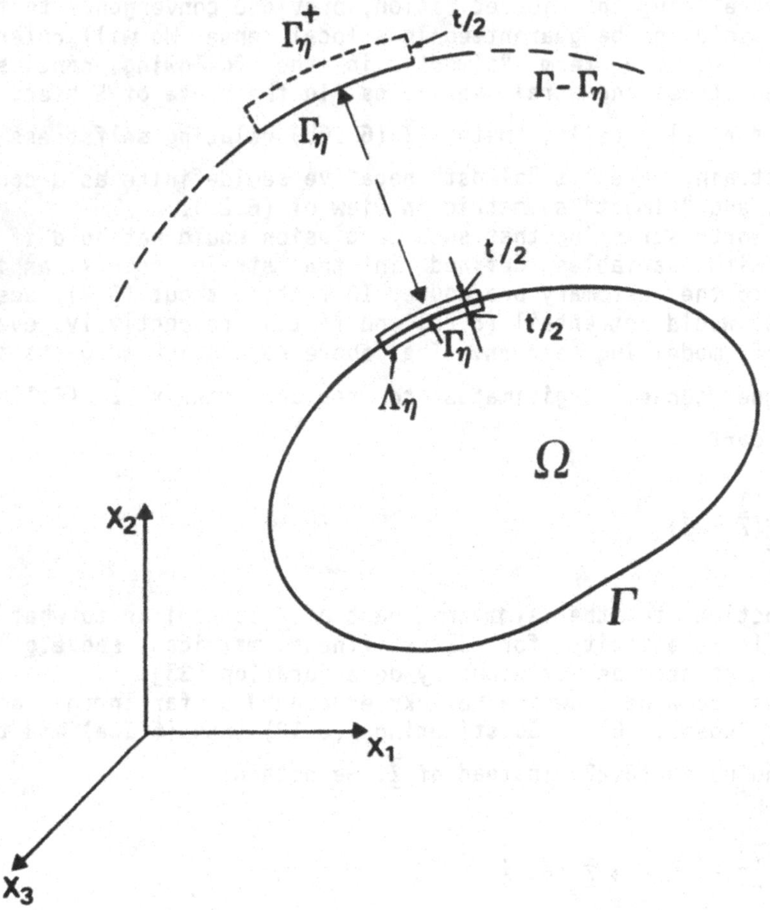

Fig. 6.1

6.3 A Galerkin formulation

A symmetric matrix operator relating plastic strain variables to consequent selfstress variables like in eq.(5.7) can be generated by the alternative approach [32], outlined below.

Let us apply Betti's theorem in the form (2.3) to the real quantities (unprimed) on one side and on the other side to distributions of fictitious external actions and to their relevant consequences (primed) in the elastic space Ω^*.

The fictitious external actions on Ω^* will be, in turn: (a) body forces b_i' over a region Λ_η ; (b) imposed strains $\varepsilon_{ij}^{o'}$ over a cell Ω_γ ; (c) imposed strains $\varepsilon_{ij}^{o'}$ over Λ_η. The region Λ_η will be the layer of constant thickness t shown in fig. 6.1 containing element Γ_η as its intersection with Γ; the body referred to will be augmented to include Λ_η so that its boundary becomes $\Gamma' \equiv \Gamma - \Gamma_\eta + \Gamma_\eta^+$, denoting by Γ_η^+ the boundary of Λ_η outside the original domain Ω, see fig.6.1. Thus Betti's equation (2.3) reads:

$$\int_{\Lambda_\eta} b_i' \, u_i \, d\Lambda + \int_{\Omega_\gamma} \sigma_{ij} \, \varepsilon_{ij}^{o\prime} \, d\Omega + \int_{\Lambda_\eta} \sigma_{ij} \, \varepsilon_{ij}^{o\prime} \, d\Lambda =$$
$$= \int_{\Gamma'} p_i \, u_i' \, d\Gamma - \int_{\Gamma'} p_i' \, u_i \, d\Gamma + \int_\Omega \sigma_{ij}' \, \varepsilon_{ij}^p \, d\Omega \tag{6.24}$$

(a) We will assume first the following special body forces alone:

$$b_\alpha'(\underline{\xi},\tau) = \hat{\underline{N}}(\underline{\xi}) \, \Delta(\tau-0) \, \underline{F}_\alpha^\eta \tag{6.25}$$

where $\underline{\xi}$ is meant to belong to Γ_η and τ is a coordinate varying from $-t/2$ to $t/2$ along the normal $\underline{\nu}$ to Γ_η with origin at $\underline{\xi}$ on Γ_η; Δ the Dirac function whose dimension is the inverse of length; vector $\underline{F}_\alpha^\eta$ collects the nodal values of the α-th component of force per unit surface for all nodes over element Γ_η; $\hat{\underline{N}}$ is a row vector of interpolation functions (no longer a shape matrix) and so are $\underline{N}$, $\underline{M}$ and $\hat{\underline{M}}$ used later in this Section.

After the usual discretization over the (plastic) domain Ω and boundary Γ, taking the limit $t \to 0$, eq.(6.24) becomes:

$$\tilde{\underline{F}}_\alpha^\eta \int_{\Gamma_\eta} \tilde{\hat{\underline{N}}}(\underline{\xi})\underline{N}(\underline{\xi})d\Gamma \underline{u}_\alpha^\eta = \tilde{\underline{F}}_\alpha^\eta \Sigma_e \int_{\Gamma_e}\int_{\Gamma_\eta} \tilde{\hat{\underline{N}}}(\underline{\xi})u_{\alpha i}^*(\underline{\xi},\underline{x})\hat{\underline{N}}(\underline{x})d\Gamma d\Gamma \underline{p}_i^e +$$
$$-\tilde{\underline{F}}_\alpha^\eta \sum_{e\neq\eta} \int_{\Gamma_e}\int_{\Gamma_\eta} \tilde{\hat{\underline{N}}}(\underline{\xi})p_{\alpha i}^*(\underline{\xi},\underline{x})\underline{N}(\underline{x})d\Gamma d\Gamma \underline{u}_i^e - \tilde{\underline{F}}_\alpha^\eta \int_{\Gamma_e^+}\int_{\Gamma_\eta} \tilde{\hat{\underline{N}}}(\underline{\xi})p_\alpha^*(\underline{\xi},\underline{x})\underline{N}(\underline{x})d\Gamma d\Gamma \underline{u}_i^\eta +$$
$$+ \tilde{\underline{F}}_\alpha^\eta \Sigma_c \int_{\Omega_c}\int_{\Gamma_\eta} \tilde{\hat{\underline{N}}}(\underline{\xi})\sigma_{\alpha ij}^*(\underline{\xi},\underline{x})\underline{M}(\underline{x})d\Gamma d\Omega \underline{\varepsilon}_{ij}^{pc} \tag{6.26}$$

On the r.h.s. of eq.(6.26) integrations with respect to $\underline{\xi}$ define functions of $\underline{x}$ which represent cumulative effects of forces distributed over element Γ_η through Kelvin's kernels playing the role of influence functions.

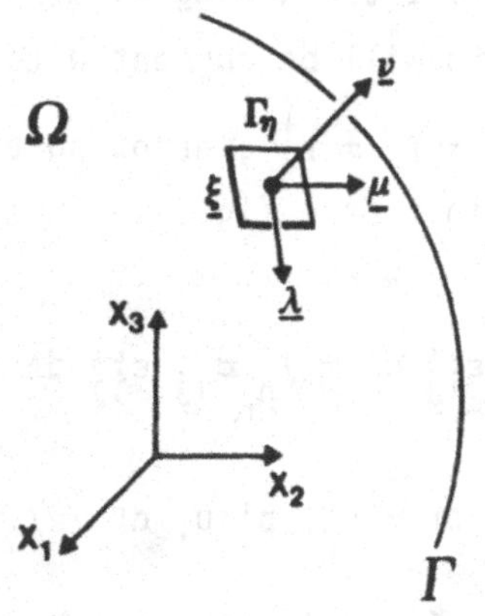

Fig. 6.2

(b) Second we will consider imposed strains over the γ-th cell Ω_γ modelled by:

$$\varepsilon^{\circ\prime}_{\alpha\beta}(\underline{\xi}) = \underline{M}(\underline{\xi})\underline{\varepsilon}^{\circ\gamma}_{\alpha\beta} \qquad (6.27)$$

where $\underline{\varepsilon}^{\circ\gamma}_{\alpha\beta}$ is the vector of nodal values of the inelastic strain component αβ. Completing the discretization over Ω and Γ and going to the limit for $t \to 0$, eq.(6.24) reads:

$$\tilde{\underline{\varepsilon}}^{\circ\gamma}_{\alpha\beta} \int_{\Omega_\gamma} \hat{\underline{M}}(\underline{\xi})\sigma_{\alpha\beta}(\underline{\xi})d\Omega = \tilde{\underline{\varepsilon}}^{\circ\gamma}_{\alpha\beta} \Sigma_e \int_{\Gamma_e}\int_{\Omega_\gamma} \hat{\underline{M}}(\underline{\xi})u^*_{\alpha\beta i}(\underline{\xi},\underline{x})\hat{\underline{N}}(\underline{x})d\Omega d\Gamma \underline{p}^e_i +$$

$$- \tilde{\underline{\varepsilon}}^{\circ\gamma}_{\alpha\beta} \Sigma_e \int_{\Gamma_e}\int_{\Omega_\gamma} \hat{\underline{M}}(\underline{\xi})p^*_{\alpha\beta i}(\underline{\xi},\underline{x})\underline{N}(\underline{x})d\Omega d\Gamma \underline{u}^e_i +$$

$$+ \tilde{\underline{\varepsilon}}^{\circ\gamma}_{\alpha\beta} \Sigma_c \int_{\Omega_c}\int_{\Omega_\gamma} \hat{\underline{M}}(\underline{\xi})\sigma^*_{\alpha\beta ij}(\underline{\xi},\underline{x})\underline{M}(\underline{x})d\Omega d\Omega \underline{\varepsilon}^{p^c}_{ij} \qquad (6.28)$$

The integrals with respect to $\underline{\xi}$ on the r.h.s. of (6.28) arise as cumulative effects due to imposed strains over Ω_γ. The influence functions indicated as kernels represent effects at $\underline{x}$ due to a concentrated imposed strain component in $\underline{\xi}$.

(c) We will now consider only a special distribution of imposed inelastic strains $\varepsilon^{\circ\prime}_{ij}$ over the layer Λ_η, such that it is described as follows in a local frame of axes $\underline{\nu}$ (outward normal in $\underline{\xi}$), $\underline{\lambda}$ and $\underline{\mu}$ (see fig.6.2):

$$\begin{Bmatrix} \varepsilon^{\circ\prime}_{\nu\nu}(\underline{\xi},\tau) \\ \varepsilon^{\circ\prime}_{\nu\lambda}(\underline{\xi},\tau) \\ \varepsilon^{\circ\prime}_{\nu\mu}(\underline{\xi},\tau) \end{Bmatrix} = \Delta(\tau-0)\mathrm{diag}[\underline{N}(\underline{\xi})] \begin{Bmatrix} \underline{D}^{\eta}_{\nu} \\ \underline{D}^{\eta}_{\lambda} \\ \underline{D}^{\eta}_{\mu} \end{Bmatrix} \tag{6.29a}$$

$$\varepsilon^{\circ\prime}_{\lambda\lambda} = \varepsilon^{\circ\prime}_{\mu\mu} = \varepsilon^{\circ\prime}_{\lambda\mu} = 0 \tag{6.29b}$$

The vector on the r.h.s. of (6.29a) contains nodal values of displacement dicontinuities across Γ_η, e.g. $\underline{D}^{\eta}_{\nu} \equiv \underline{u}^{+}_{\nu} - \underline{u}^{-}_{\nu}$, $\underline{u}^{+}_{\nu}$ and $\underline{u}^{-}_{\nu}$ being the nodal displacements in the ν direction at the outer (+) and inner (-) face of Γ_η respectively. After discretization and the limiting process $t \rightarrow 0$, in this case eq.(6.24) acquires the form:

$$\tilde{\underline{D}}^{\eta}_{\alpha} \int_{\Gamma_\eta} \hat{\underline{N}}(\xi)\hat{\underline{N}}(\xi)d\Gamma \underline{p}^{\eta}_{\alpha} = \tilde{\underline{D}}^{\eta}_{\alpha} \sum_{e\neq\eta} \int_{\Gamma_e}\int_{\Gamma_\eta} \hat{\underline{N}}(\underline{\xi})\overset{*}{u}{}^{\nu}_{\alpha i}(\underline{\xi},\underline{x})\hat{\underline{N}}(\underline{x})d\Gamma d\Gamma \underline{p}^{e}_{i} +$$

$$+ \tilde{\underline{D}}^{\eta}_{\alpha} \int_{\Gamma_e^+}\int_{\Gamma_\eta} \hat{\underline{N}}(\underline{\xi})\overset{*}{u}{}^{\nu}_{\alpha i}(\underline{\xi},\underline{x})\hat{\underline{N}}(\underline{x})d\Gamma d\Gamma \underline{p}^{\eta}_{i} - \tilde{\underline{D}}^{\eta}_{\alpha} \sum_{e} \int_{\Gamma_e}\int_{\Gamma_\eta} \hat{\underline{N}}(\underline{\xi})\overset{*}{p}{}^{\nu}_{\alpha i}(\underline{\xi},\underline{x})\underline{N}(\underline{x})d\Gamma d\Gamma \underline{u}^{e}_{i} +$$

$$+ \tilde{\underline{D}}^{\eta}_{\alpha} \sum_{c} \int_{\Omega_c}\int_{\Gamma_\eta} \hat{\underline{N}}(\underline{\xi})\overset{*}{\sigma}{}^{\nu}_{\alpha ij}(\underline{\xi},\underline{x})\underline{M}(\underline{x})d\Gamma d\Omega \underline{\varepsilon}^{p^c}_{ij} \tag{6.30}$$

Here the integrals with respect to $\underline{\xi}$ on the r.h.s. represent effects of a displacement discontinuity along the boundary element Γ_η depicted in fig.6.2. If Γ_η is flat and $\underline{N}(\underline{\xi})$ specializes to constant interpolations, the functions defined by these integrals reduce to those central in the displacement discontinuity method developed by Crouch et al., [34]. The kernels acting as influence functions now represent effects in $\underline{x}$ due to a concentrated displacement discontinuity at $\underline{\xi}$ across an elementary area of normal $\underline{\nu}$. They have arisen as consequences of assumption (6.29a) from the kernels describing effects of concentrated imposed strains used at (b). The l.h.s. in (6.30) is arrived at from the l.h.s. of (6.24) using local reference, eqs.(6.29) and Cauchy equilibrium equations; it can be also interpreted as the work done by the real tractions because of a crack stretching along the boundary element Γ_η but conceived as interior to the real domain Ω.

Betti's reciprocity theorem, applied this time to the unbounded elastic space Ω_∞ (under the usual regularity conditions on asymptotic behaviour) leads to the following typical symmetry relation in eq.(6.26):

$$\int_{\Gamma_e}\int_{\Gamma_\eta} \hat{N}^m(\underline{\xi})u^*_{\alpha i}(\underline{\xi},\underline{x})\hat{N}^n(\underline{x})d\Gamma d\Gamma = \int_{\Gamma_\eta}\int_{\Gamma_e} \hat{N}^n(\underline{x})u^*_{i\alpha}(\underline{x},\underline{\xi})\hat{N}^m(\xi)d\Gamma d\Gamma \qquad (6.31)$$

and the following typical link between eq.(6.26) and (6.30):

$$\int_{\Gamma_e}\int_{\Gamma_\eta} \hat{N}^m(\underline{\xi})p^*_{\alpha i}(\underline{\xi},\underline{x})N^n(\underline{x})d\Gamma d\Gamma = \int_{\Gamma_\eta}\int_{\Gamma_e} N^n(\underline{x})u^*_{i\alpha}(\underline{x},\underline{\xi})\hat{N}^m(\underline{\xi})d\Gamma d\Gamma \qquad (6.32)$$

In fact, eq.(6.31) can be interpreted as containing on the l.h.s. the work done by forces along axis i distributed over element Γ_e according to the interpolation function $\hat{N}^n$ of its node n, because of displacements due to forces along axis α distributed over element Γ_η according to the interpolation function $\hat{N}^m$ of its node m; on the r.h.s. the work done by forces along axis α distributed over element Γ_η according to interpolation $\hat{N}^m$, because of displacements due to forces along axis i distributed over Γ_e according to interpolation $\hat{N}^n$. An analogous path of reasoning would justify eq.(6.32) and other relations similar to (6.31) and (6.32).

Merely in order to simplify the notation, we refer here to nonconforming elements and cells, so that no variable identification is needed to assemble the equations.

Eqs.(6.26), (6.28) and (6.30) can be imposed for any arbitrary $\underline{F}^\eta_\alpha$, $\underline{\varepsilon}^{\circ\gamma}_{\alpha\beta}$ and $\underline{D}^\eta_\alpha$, respectively, which act as weights in the spirit of Galerkin weighted residual statements. Taking into account reciprocity relations like (6.31) and (6.32), equations (6.26)(6.28) and (6.30) can be written as follows in a more compact symbology which is clearly defined by term-to-term correspondences between their old and new versions:

$$\underline{\hat{T}}_{\eta\eta}\,\underline{u}_\eta = \Sigma_e\,\underline{V}_{\eta e}\,\underline{p}_e - \Sigma_e\,\underline{T}_{\eta e}\,\underline{u}_e + \Sigma_c\,\underline{W}_{\eta c}\,\underline{\varepsilon}^p_c \qquad (6.33)$$

$$\underline{\bar{\sigma}}_\gamma = \Sigma_e\,\underline{\tilde{W}}_{\gamma e}\,\underline{p}_e - \Sigma_e\,\underline{R}_{\gamma e}\,\underline{u}_e + \Sigma_c\,\underline{S}_{\gamma c}\,\underline{\varepsilon}^p_c \qquad (6.34)$$

$$\underline{\tilde{\hat{T}}}_{\eta\eta}\,\underline{p}_\eta = \Sigma_e\,\underline{\tilde{T}}_{\eta e}\,\underline{p}_e - \Sigma_e\,\underline{L}_{\eta e}\,\underline{u}_e + \Sigma_c\,\underline{\tilde{R}}_{\eta c}\,\underline{\varepsilon}^p_c \qquad (6.35)$$

The l.h.s. of eq.(6.34) defines generalized stresses in the sense of Subsec.6.1; they are consistent through eq.(6.5a) for $\underline{T}_\varepsilon = \underline{I}$ with the chosen interpolation functions for strains over cell Ω_γ used as Galerkin weight functions. Therefore $\underline{\varepsilon}_c^p$ represents the vector of generalized plastic strains for cell Ω_c in the sense of (6.2) and hence, will be denoted below by $\bar{\underline{\varepsilon}}_c^p$.

It is worth noting that:

$$\underline{V}_{\eta e} = \tilde{\underline{V}}_{e\eta}\ ,\quad \underline{L}_{\eta e} = \tilde{\underline{L}}_{e\eta}\ ,\quad \underline{S}_{\gamma c} = \tilde{\underline{S}}_{c\gamma} \tag{6.36a,b,c}$$

as consequences of reciprocity relations like (6.31) and (6.32).

Turning now to the boundary value problem to solve, we will write eq.(6.33) for all E_u boundary elements belonging to the constrained boundary $\Gamma_u(\eta = 1...E_u)$, eq.(6.35) for all E_p elements on the free portion $\Gamma_p(\eta = 1...E_p)$. The set of all these equations expressed in rates, separating (incremental) data from unknowns $\dot{\underline{X}}$, becomes in a compact form like (5.5) (primes mark the different origin):

$$\underline{A}'\ \dot{\underline{X}} = \dot{\underline{F}}' + \underline{D}'\dot{\underline{\varepsilon}}^p \tag{6.37}$$

In contrast to its counterpart (5.5), eq.(6.37) exhibits a symmetric coefficient matrix $\underline{A}'$, as it could be shown in view of eqs.(6.36). After similar rearranging in compact form, eq.(6.34) reads with self-explaining meaning of symbols:

$$\dot{\bar{\underline{\sigma}}} = \bar{\underline{A}}'\ \dot{\underline{X}} + \dot{\bar{\underline{F}}}' + \bar{\underline{D}}'\dot{\underline{\varepsilon}}^p \tag{6.38}$$

Substituting the boundary unknown vector $\dot{\underline{X}}$ into eq.(6.38) using (6.37), we obtain:

$$\dot{\bar{\underline{\sigma}}} = (\dot{\bar{\underline{\sigma}}}')^E + \bar{\underline{Z}}'\dot{\underline{\varepsilon}}^p \tag{6.39}$$

having set:

$$(\dot{\bar{\underline{\sigma}}}')^E \equiv \dot{\bar{\underline{F}}}' + \bar{\underline{A}}'(\underline{A}')^{-1}\dot{\underline{F}}'; \qquad \bar{\underline{Z}}' \equiv \bar{\underline{D}}' + \bar{\underline{A}}'(\underline{A}')^{-1}\ \underline{D}' \tag{6.40a,b}$$

It can be proved that $\bar{\underline{A}}' = -\tilde{\underline{D}}'$ once again as a consequence of reciprocity relations. On the other hand matrix $\bar{\underline{D}}'$ is symmetric in view of eq.(6.36c). Therefore matrix $\bar{\underline{Z}}$ is symmetric. It might be proved to be also negative semidefinite [35]. At last, combining eq.(6.39) obtained from the preceding Galerkin BE approach with the overall approximate plastic relations generated in Subsec.6.1 for cells, we arrive at a rate b.v. problem formulation in terms of a linear complementarity problem. In contrast to its counterpart (6.23), it exhibits a rigorously symmetric positive semidefinite matrix. Its equivalence to a convex quadratic

minimization (programming) problem in vector $\underline{\lambda}$ alone has been pointed out in [36], where the dual of such extremum characterization and the uniqueness of solution have also been discussed. If the plastic laws for cells are piecewise linearized (a priori or locally) as mentioned in Subsec.4.2, then the holonomic finite step problem acquires an identical linear complementarity problem formulation [36].

7. LIMIT AND SHAKEDOWN ANALYSIS AND BOUNDING TECHNIQUES

7.1 Plastic collapse analysis as a linear programming problem

Perhaps the most classical problem in structural plasticity is the limit analysis, namely the determination of the "carrying capacity" or plastic collapse treshold for a perfectly plastic solid (with associative flow rules) subjected to external forces affected by a common factor, say α. Together with the limit value s (or "safety factor") of α, the/a collapse mechanism and the/a stress state at incipient collapse are sought.

The solution of this problem is based on the pair of limit theorems, the former of which is the "safe" (or static, or lower bound) theorem: loads do not exceed the carrying capacity (i.e. with the above symbols, $\alpha \le s$), if and only if there exists a stress field which complies with equilibrium and yield conditions, see e.g. [1].

This statement permits to cast limit analysis into a linear programming (LP) problem, if the material yield functions are linear or linearized as an approximation (see Subsec.4.2). Computational gains are generally expected with respect to nonlinear programming formulations generated in the absence of such a priori piecewise linearization of the yield locus, see e.g. [2] [37].

In the BE context outlined in Secs.2-5, it appears quite natural to formulate the limit problem on the basis of the static theorem as follows:

$$\bar{s} = \max_{\alpha, \underline{\varepsilon}^p} \alpha \quad , \text{subject to: } \underline{\tilde{\Phi}}\, \underline{\sigma} \le \underline{R} \quad ; \quad \underline{\sigma} = \alpha\, \underline{\sigma}^E + \underline{Z}\, \underline{\varepsilon}^p \qquad (7.1a,b,c)$$

In fact, eq.(7.1b) enforces the material yield inequalities in the strain points (not necessarily everywhere); eq.(7.1c) ensures the fulfillment of equilibrium in the approximate sense of the discretization adopted over the boundary and generates stresses to substitute into (7.1b) due to plastic strains acting as optimization variables[38].

The LP formulation (7.1) gives rise to the following remarks.

(a) The optimal value $\bar{s}$ of (7.1) is only an approximation of the safety factor s for the continuum (not necessarily a lower bound) because of the above mentioned approximations and because one cannot generate all possible selfstresses in the continuum by varying the vector $\underline{\varepsilon}^p$ of plastic strains in the strain points.

(b) The dual of problem (7.1) reads, $\underline{\lambda}'$ being a vector of dual variables:

$$\bar{s} = \min_{\underline{\lambda}'} \underline{\tilde{R}}\,\underline{\lambda}', \text{ subject to: } \underline{\tilde{\sigma}}^E \underline{\Phi}\,\underline{\lambda}' = 1; \quad \underline{\tilde{Z}}\,\underline{\Phi}\,\underline{\lambda}' = \underline{0}\,; \quad \underline{\lambda}' \geq \underline{0} \quad (7.2a,b,c,d)$$

The constraints (7.2b,c) are clearly incompatible if det $\underline{Z} \neq 0$. In this case duality theory of LP ensures that $\bar{s} = \infty$ (the infeasibility of the dual implies unbounded feasible domain and unbounded solution for the primal). Matrix $\underline{Z}$ arising from the usual BE method generally exhibits a defect which is zero or anyway inadequate to provide a sufficiently large set of feasible vectors $\underline{\lambda}'$. This jeopardizes the practical usefulness of both LP formulations. However, the procedure applied below might be fruitful.

Consider the following problem:

$$\max_{\alpha,\underline{\phi},\underline{\lambda}} \alpha\,, \quad \text{subject to:} \quad (7.3)$$

$$\underline{\phi} = \underline{\tilde{\Phi}}\,\underline{\sigma}^E \alpha - (\underline{H} - \underline{\tilde{\Phi}}\,\underline{Z}\,\underline{\Phi})\,\underline{\lambda} - \underline{R} \leq \underline{0}, \quad \underline{\lambda} \geq \underline{0}, \quad \underline{\tilde{\phi}}\,\underline{\lambda} = 0 \quad (7.4a,b,c,d)$$

The contraints (7.4) are easily recognized to govern the response in terms of plastic strains $\underline{\lambda}(\alpha)$ of the continuum discretized in the sense of the BE method, subjected to proportional loading and endowed with the fully holonomic (instead of piecewise holonomic), a priori piecewise linearized material model defined in Subsec.4.2, remark (b); clearly, attention should be paid to the present more comprehensive meaning of the same symbols.

Mathematically, eqs.(7.4) is a parametric linear complementarity problem and its complete solution $\underline{\lambda}(\alpha)$ is a piecewise linear vector function of the load factor α, is achieved when the maximization (7.3) is carried out by the following procedure: start from $\alpha = 0$, $\underline{\lambda} = \underline{0}$; process by any LP algorithm resting on the Simplex method the LP problem obtained from (7.3)(7.4) by ignoring (7.4d), but enforce (7.4d) as an additional rule ("restricted basis entry") at each pivoting transformation. Thus the optimization problem (7.3)(7.4), which rigorously is a nonconvex nonlinear programming problem, computationally is amenable to slightly modified ("restricted basis") linear programming, as proposed in [38]. In contrast to the LP problem (7.1), problem (7.3)(7.4) defines at each step of the optimization process mechanically meaningful quantities (the inelastic strains at the current load factor α); moreover, for perfect plasticity ($\underline{H} = \underline{0}$) it can be terminated at a manifestation of plastic collapse such as large increases of some components of $\underline{\lambda}(\alpha)$ corresponding to small increments of α, and not necessarily at an unbounded set of feasible vectors $\underline{\lambda}$ representing a collapse mechanism. Thus the difficulty typical of BE limit analysis by (7.1) pointed out in remark (b) may be circumvented.

In the above procedure the irreversible nature of plastic reponse can be allowed for simply if one replaces the given vector $\underline{R}$ by the current $\underline{\phi} \equiv \underline{\phi}_o$ as soon as some λ tends to decrease, and proceeds by assuming as variables further increments of α and $\underline{\lambda}$.

When applied to nonhardening systems, the restricted basis LP method actually represents a combination of limit and evolutive elastoplastic analysis and, hence, provides much more information than conventional limit analysis. However, so far, it is corroborated by available theories (parametric convex quadratic programming) only when the matrix in parentheses in (7.4) is symmetric positive semidefinite and under such restriction it has been successfully tested numerically only for FE models of frames. This circumstance clearly advocates the symmetric BE formulations dealt with in Sec.6. Further motivation for them will arise in the next Subsection.

7.2 Shakedown analysis and bounds on residual quantities.

The conventional shakedown analysis of elastic-perfectly plastic solids and structural systems, see e.g. [1] [2], can be regarded as a generalization of plastic collapse analysis. For a given "loading domain" of variable-repeated (in particular cyclic) external actions in the space of parameters governing them, a classical problem is to determine the safety factor s_v with respect to lack of adaptation by incremental collapse or alternating plasticity, the/a collapse mechanism, the/a distribution of residual deformations at incipient inadaptation.

To solve this problem a traditional basis is provided by Melan's theorem which can be stated as follows: the safety factor s_v is the maximum of all load factors α such that the yield inequalities are fulfilled everywhere and at any time t by the superposition of the linear elastic stress response and some selfequilibrated stress distribution constant in time.

If Melan's concept has to be applied in a BE context where stresses are made explicitly dependent on plastic strains $\underline{\varepsilon}^p$, it seems quite natural to take $\underline{\varepsilon}^p$ (instead of stresses or redundant internal forces) as independent variable vector. With the piecewise linearized yield conditions used in Subsec.6.1, $\underline{\varepsilon}^p$ depends on $\underline{\lambda}$ through the normality rule with $\underline{\Psi} = \underline{\Phi}$, and it is possible and convenient to define, for each strain point $\bar{i}$ and each yield mode j a component of an "elastic stress envelope" vector $\underline{M}$ ($\underline{\Phi}_j^i$ being the outward normal to the corresponding yield plane):

$$M_j^i \equiv \max_t \{\underline{\tilde{\Phi}}_j^i \, \underline{\sigma}_i^E(t)\} \;, \quad \underline{M}^T \equiv \{\ldots M_j^i \ldots\} \qquad (7.5)$$

On this basis, the search for the safety factor with respect to inadaptation, could be made through Melan's theorem by solving the following linear programming problem [39]:

$$s_V = \max_{\alpha,\underline{\lambda}} \alpha, \text{ subject to:} \qquad (7.6)$$

$$\alpha\underline{M} + \underline{\tilde{\Phi}}\,\underline{Z}\,\underline{\Phi}\,\underline{\lambda} \leq -\underline{R} \;, \quad \underline{\lambda} \geq \underline{0} \qquad (7.7a,b)$$

However, it has been shown in [40] that Melan's theorem and its consequences like (7.6)(7.7) can be proved in the BE context only if the plastic b.v. problem is preliminarily given symmetric formulation in the sense of Sec.6.

Problem (7.6)(7.7) formulated in [39], straightforwardly reduces to limit analysis (7.1) when the loading domain shrinks to a point in the space of external actions so that $\underline{M} = \underline{\tilde{\Phi}}\,\underline{\sigma}^E$.

Clearly, remarks similar to (a) and (b) referred to problem (7.1) apply to problem (7.6)(7.7) as well. Therefore, as noted in [39], one should expect to obtain $s_V = \infty$ if the latter problem were to be processed by a linear programming algorithm.

However, generally speaking, structural safety is not guaranteed merely by adaptation in the sense of dissipated work bounded in time, as assumed in classical shakedown theory. One has also to ensure that residual deformations be admissible with respect to serviceability requirements and, often, with the limited ductility of materials or structural components. This circumstance makes methods for bounding from above residual deformations (and possible other post-shakedown quantities) an essential and conceptually and practically important supplement to traditional shakedown analysis.

Upper bounds to the present purposes have to be determined by avoiding the elastoplastic incremental analysis, not only because the computational burden required by it, but also because the real time history of external actions within the loading domain is often unknown.

Various bounding inequalities have been established with reference to continua and FE models under the restriction of associated plastic constitutive laws, e.g. [1] [2]. It is quite natural to transfer such category of classical and recent results of plasticity theory to the BE context. To bounding purposes once again a symmetrization in the sense of Sec.6 turns out to be a necessary premise, see e.g. [40] [41]. These promising possible developments in elastoplasticity by BE approaches are still in their infancy and will not be pursued here for space limitations.

8. VARIANTS AND ALTERNATIVE FORMULATIONS

8.1 Variants resting on morphological peculiarities

Boundary value problems occurring in engineering often exhibit special features which can be exploited to simplify the BE solution of inelastic as well as linear elastic b.v. problems. The simplifying provisions, being basically the same as in elasticity, are only sketchily outlined below. It is worth noting that they concern not only the modelling phase, but also the basic formulation.

(a) The arbitrariness of the choice of the fictitious domain Ω^*, under the condition that it contains the given one Ω, can sometimes be taken profit of by choosing Ω^* so that its boundary Γ^* coincides at least partly with the real boundary Γ and the boundary conditions on Γ^* over this intersection are homogeneous and of the same type of the actual ones. The portion of the intersection $\Gamma \cap \Gamma^*$ where this circumstance occurs contributes to the integral equations (2.4)(2.9)(2.12) merely by a single integral (instead of two), whose integrand is known.

Three situations are worth mentioning as an illustration.

(i) The actual domain Ω consists of the half-space with possible detraction of bounded regions (typically, cavities in geotechnical applications, fig.8.1). Then Mindlin's fundamental solution with zero tractions on the plane bounding Ω and $\Omega^* \cap \Omega$ permits to confine the intervening unknowns to Γ_3 and Γ_4 over the traction-free portions Γ_1 of that plane. Mindlin's solution differs from Kelvin's by nonsingular "complementary" terms, so that all considerations of the preceding Sections hold true unaltered.

(ii) In both linear and elastic-plastic fracture mechanics two-dimensional problems, computational gains (no discretization on the crack edges) can be achieved by using a fundamental solution which explicitly accounts for the presence of a traction-free straight crack in the interior of the elastic plane Ω^* [30].

(iii) In this context, the extreme case is represented by the fundamental solution over Ω with homogeneous boundary conditions, i.e. by the Green's function of the problem. Clearly, the adoption of this function would reduce the b.v. problem to quadratures, but would shift the difficulty to the determination of the Green's function itself.

(b) For two-dimensional problems, say in the plane $x_1\, x_2$, integral equations hold formally analogous to the preceding ones, but written on the basis of a 2D fundamental solution. This can be generated from a three-dimensional one (e.g. Kelvin's or Mindlin's) by referring to a uniform distribution of forces along a straight line parallel to the x_3

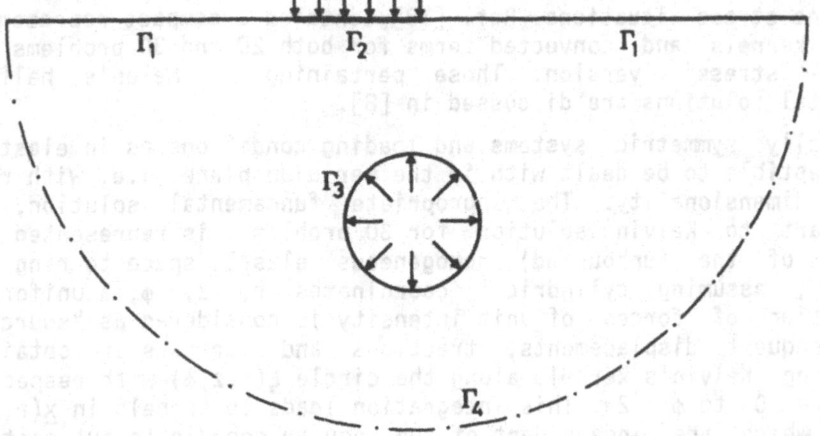

Fig. 8.1

axis, and by integrating over $-\infty < \xi_3 < \infty$ (thus lowering the singularities at $\underline{\xi} = \underline{x}$ of the kernels). The kernel $u^*_{\alpha i}$ for displacements contains the function ln r and, hence, ranges from $-\infty$ to $+\infty$, being zero at an arbitrary distance r and unbounded for $r \to \infty$ (the latter circumstance is consistent with the unbounded nature of the total load acting on the elastic space in this case). The above mentioned arbitrariness has been recently shown to imply the possibility of ill-conditioning in the solution of elastic 2D problems and possible remedies have been suggested, see e.g. [9]. Clearly, these remarks and proposed provisions apply unaltered to 2D inelastic analysis but cannot be discussed here. As for unbounded domains, a peculiarity of 2D formulations is that the two boundary integrals over the fictitious circle cancel each other but do not vanish individually as its radius tends to infinity, see e.g. [8].

Focusing now on the domain integral accounting for the inelastic strains, it is worth noting that the reciprocal work addend pertaining to x_3 axis vanishes in the "initial stress" version (2.10), even in plane strain cases ($\varepsilon^*_{\alpha 33}\ \sigma^p_{33} = 0$ as $\varepsilon^*_{\alpha 33} = 0$). This represents a slight formal advantage over the "initial strain" version, like (2.9), since $\sigma^*_{\alpha 33}\ \varepsilon^p_{33} \neq 0$ in plane strain situations (unless ε^p_{33} and ε^e_{33} are assumed to be zero individually as a nonrigorous but customary hypothesis for plane strains). The kernels and the convected term of the integral equations (2.4)(2.12) for two dimensional problems are given in [8] for both the "initial stress" and "initial strain" formulations and for plane strain and plane stress situations. Ref. [18] contains a compact representation of the kernels and convected terms for both 2D and 3D problems in the "initial stress" version. Those pertaining to Melan's half-plane fundamental solutions are discussed in [8].

(c) Axially symmetric systems and loading conditions as in elasticity, are susceptible to be dealt with in the meridian plane, i.e. with reduced domain dimensionality. The appropriate fundamental solution, as a counterpart to Kelvin solution for 3D problems, is represented by the response of the (unbounded) homogeneous elastic space to ring loads. Precisely, assuming cylindrical coordinates r, z, ϕ, a uniform ring distribution of forces of unit intensity is considered as "source" and the consequent displacements, tractions and stresses are obtained by integrating Kelvin's kernels along the circle $\underline{\xi}(r,z,\phi)$ with respect to ϕ from $\theta = 0$ to $\phi = 2\pi$. This integration leads to kernels in $\underline{x}(r,z)$ and $\underline{\xi}(r,z)$, which are independent of the angular coordinate but containing elliptic integrals. The presence of these functions makes the analytical integrations over singularities impossible even over elements with constant interpolation functions. For detailed discussions see e.g. [42] in elasticity and [43] [44] in plasticity.

(d) Axially symmetric systems under general boundary conditions in elasticity by BE and FE alike can be analyzed by expanding in Fourier series both external actions and unknown response fields. Thus one benefits of the familiar decomposition (due to the orthogonality of trigonometric functions) into as many reduced problems as Fourier harmonics, each one independent of the angular coordinate, and concerning Fourier coefficients of the physical variables. This decoupling rests basically on linearity and superposition of effects; in fact all applications available so far in the BE literature concern elasticity and potential problems, see e.g. [42]. However, if the unknown plastic strains are expanded (truncated) Fourier series with respect to ϕ, the structural response to external actions and inelastic strains can still be advantageously expressed as a superposition of harmonics but the nonlinear constitution inevitably implies coupling. Such procedure of inelastic analysis has been applied in the FE context [45], not yet, to the writers' knowledge, in the BE context, though no significant difference is expected in this respect.

(e) Reflective symmetry combined with either symmetry or antisymmetry of external actions leads to the customary reduction of the domain to one half, with suitable boundary conditions on the simmetry plane (or axis in 2D). A less traditional alternative, typical of the BE method, consists of taking into account the symmetry relation among nodal variables in the formulation for the whole domain. This permits to avoid discretization over the symmetry plane (or axis).

(f) Many engineering systems and structures are rotationally repetitive or "cyclically symmetric". Specifically, they can be conceived as formed by a number, say N, of segments which are geometrically and physically identical, but rotated from each other by $2\pi/N$ radiants around an axis. Successive rotations generate from any point a set of N corresponding points. A special one-to-one ("rotational") transformation applied to vectors of scalar quantities pertaining to corresponding points in the above sense, can be used to decompose the linear analysis of cyclically symmetric systems under general loading into N' reduced problems N' being $\frac{1}{2}N + 1$ for even N, $\frac{1}{2}(N+1)$ for odd N. The maximum size of these turns out to be the size of the overall original problem and is given by the double of the number of unknowns on the boundary of a single segment without involving interfaces with adjacent segments. The esclusion of discretization over interface boundaries pointed out in ref.[46] is peculiar of BE approaches (not common to FE) and is advantageous inasmuch it alleviates both number of unknowns and approximation errors. Such gain is not achieved when dealing with cyclically repetitive loading without recourse to rotational transformation of variables. The extension from elastic and potential problems to inelastic analysis, discussed in [47], suffers from limitations similar to those mentioned at (d) for axisymmetry.

When each segment exhibits reflective symmetry, i.e. in the presence of cyclic and reflective symmetries combined ("dihedrical" symmetry) the decomposition process can be further enhanced. It is worth noting that cyclic symmetry at least in principle contains as special cases for N = 2 and $N \to \infty$ the reflective and axial symmetry, respectively.

(g) The homogeneity hypothesis assumed in what precedes, though confined to the elastic properties, is quite restrictive in practice. Zonewise homogeneous solids and structures can be dealt with by costructing the integral equation for each (elastically) homogeneous zone separately, and by enforcing matching conditions (equilibrium and compatibility) at interface nodes. The zoning procedure is equally applicable to elastic and inelastic analysis, with similar advantages (banded matrix) and drawbacks (more unknowns), see e.g. [6][8].

(h) Layered systems, such as stratified soils, exhibit besides a zoned domain a chain-like pattern, which can ideally be exploited in the BE context, since the interface unknowns normally play a central role in the analysis and are the primary variables in the integral equation approach. In elasticity an elegant method resting on BE discretization and transfer matrices has been developed and succesfully applied [48]. Some intrinsic limitations of this approach were recently pointed out and alternative "ad hoc" BE procedures have been proposed and shown to avoid those drawbacks and provide some computational gains [49][50]. The potential usefulness of such special-purpose procedures in inelastic analysis of layered systems has still to be investigated.

(k) If the elastic properties exhibit orthotropy or general anisotropy, the fundamental solutions become more complex accordingly and have been presented in a recent paper by Cruse dealing with 2D elastic plastic fracture mechanics problems [30].

(j) When dealing with boundaries extended to infinity, one can profitably adopt infinite BE, with interpolation functions suitably matching the expected asymptotic behaviours of the solution. This can avoid the introduction of fictitious boundaries preserving the peculiar accuracy of the BE approach in both elastic and plastic analysis.

8.2 Alternative formulations

The by now traditional direct inelastic BE method has been considered so far together with some mostly recent variants dictated by peculiar features of the system to analyze. More general modifications of the integral equation approach have been proposed in the literature and will be concisely outlined below.

(a) It is well known [6][8] that in the presence of inelastic strains, Navier equations with their relevant boundary conditions still hold true, formally unchanged, provided that suitable pseudo body forces and pseudo boundary tractions be defined to account for such strains. Precisely, the above equations can be used to determine the real displacement field due to given loads and imposed strains, if the original static quantities b_i, p_i are replaced by the fictitious ones $\bar{b}_i$, $\bar{p}_i$. These are defined as follows, keeping in mind the definition of "initial stresses" eq.(2.10b):

$$\bar{b}_i \equiv b_i - \sigma^p_{ij,j} = b_i - 2G(\varepsilon^p_{ij,j} + \frac{\nu}{1-2\nu}\varepsilon^p_{kk,j}) \tag{8.1a}$$

$$\bar{p}_i \equiv p_i + \sigma^p_{ij} n_j = p_i + 2G(\varepsilon^p_{ij} n_j + \frac{\nu}{1-2\nu}\varepsilon^p_{kk} n_i) \tag{8.1b}$$

The possibility of embodying plastic strains into pseudo body forces and tractions, above noticed in the differential formulation, carries through to the expression of Betti theorem and to the integral equations which flow from it as described in Sec.2. The boundary equation for displacements contains the plastic strains in a modified domain integral and in a new boundary integral and reads:

$$c_{\alpha\beta}(\underline{\xi})\, u_\beta(\underline{\xi}) + \int_\Gamma p^*_{\alpha i}(\underline{\xi},\underline{x})\, u_i(\underline{x})\, d\Gamma = \int_\Gamma u^*_{\alpha i}(\underline{\xi},\underline{x})\, p_i(\underline{x})\, d\Gamma + \tag{8.2}$$

$$+ \int_\Omega u^*_{\alpha i}(\underline{\xi},\underline{x})\, b_i(\underline{x})\, d\Omega - \int_\Omega u^*_{\alpha i}(\underline{\xi},\underline{x})\sigma^p_{ij,j}(\underline{x})\, d\Omega + \int_\Gamma u^*_{\alpha i}(\underline{\xi},\underline{x})\, \sigma^p_{ij}(\underline{x})\, n_j\, d\Gamma$$

It is worth noting that both the new integrals can be evaluated in the usual sense. The kernel of the former has a weaker singularity than its counterpart in eq.(2.9); hence, the differentiation with respect to ξ, leading to the integral equation for stresses (or strains), gives rise to no convected term and to a domain integral containing plastic strains to be still evaluated in the usual sense. Thus the counterpart of eq.(2.12) reads:

$$\sigma_{\alpha\beta}(\underline{\xi}) = \int_\Gamma u^*_{\alpha\beta i}(\underline{\xi},\underline{x})\ p_i(\underline{x})\ d\Gamma - \int_\Gamma p^*_{\alpha\beta i}(\underline{\xi},\underline{x})\ u_i(\underline{x})\ d\Gamma +$$

$$+ \int_\Omega u^*_{\alpha\beta i}(\underline{\xi},\underline{x})\ b_i(\underline{x})\ d\Omega - \int_\Omega u^*_{\alpha\beta i}(\underline{\xi},\underline{x})\ \sigma^p_{ij,j}(\underline{x})\ d\Omega + \qquad (8.3)$$

$$+ \int_\Gamma u^*_{\alpha\beta i}(\underline{\xi},\underline{x})\ \sigma^p_{ij}(\underline{x})\ n_j\ d\Omega$$

Eq.(8.2) could also be generated from eq.(2.9) integrating by parts the domain integral for the plastic strains by interpreting the kernel as differential factor in its integral. This formulation, put forward by Mukherjee [7] and Telles [8], has been successfully applied by the former to inelastic torsion problems. An unresolved, difficult to assess, trade-off arises between the elimination of the domain singularity on one hand and on the other hand the need for accurate evaluation of plastic strains on the boundary, in view of the new integral, and over the domain in view of the plastic strain derivatives.

(b) A boundary element formulation was recently proposed in elasticity [15] with the purpose to obtain more accurate values for stresses and strains at points near and on the boundary Γ. Its adoption in the BE solution of elastic-plastic problems, although straightforward, has not been numerically tested yet, but appears fruitful. In fact, plastic strains often develop near the boundary; hence, an accurate evaluation of stresses and strains at "strain points" near or on the boundary turns out to be crucial for an effective enforcement of the inelastic constitutive laws. In what follows the basic ideas of the new formulation are outlined with reference to simply connected domains only. Further details can be found in [15] where multiply connected regions are also dealt with.

The modified integral equation stems from Somigliana identity eq.(2.4), through the following two steps:

(i) the original kernel $p^*_{\alpha i}(\underline{\xi},\underline{x})$ is shown to be amenable to the form:

$$p^*_{\alpha i}(\underline{\xi},\underline{x}) = \frac{d}{ds}\ \{\frac{1}{4\pi(1-\nu)}\ [2(1-\nu)\phi\delta_{ij} + e_{jk}\ r_{,i}\, r_{,k} +$$

$$+ (1-2\nu)e_{ij}\ \ln r]\} = \frac{d}{ds} w^*_{\alpha i}(\underline{\xi},\underline{\xi}',\underline{x}) \qquad (8.4)$$

where (see fig. 8.2) $\underline{\xi}'$ is a "reference point", chosen as intersection between the boundary Γ and the x_1 axis through the load point $\underline{\xi}$; s and ϕ are defined in fig.8.2; e_{ij} is the permutation symbol ($e_{12} = 1$ $e_{21} = -1$ $e_{11} = e_{22} = 0$); $w^*_{\alpha i}$ represents a new kernel with lower order singularity.

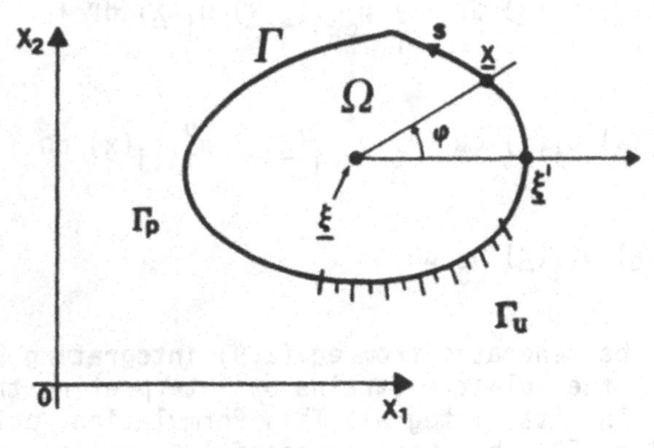

Fig. 8.2

(ii) using eq.(8.4) the integral term on the l.h.s. of eq.(2.4) can be processed through integration by parts giving:

$$\int_\Gamma p^*_{\alpha i}(\underline{\xi},\underline{x})\, u_i(\underline{x})\, d\Gamma = -u_\alpha(\underline{\xi}') + \int_\Gamma w^*_{\alpha i}(\underline{\xi},\underline{\xi}',\underline{x}) \frac{du_i}{ds}\, d\Gamma \qquad (8.5)$$

On this basis, the integral equation (2.4) for internal points is re-cast into the form:

$$u_\alpha(\underline{\xi}) - u_\alpha(\underline{\xi}') = \int_\Gamma u^*_{\alpha i}(\underline{\xi},\underline{x})\, p_i(\underline{x})\, d\Gamma + \qquad (8.6)$$
$$- \int_\Gamma w^*_{\alpha i}(\underline{\xi},\underline{\xi}',\underline{x}) \frac{du_i}{ds}(\underline{x})\, d\Gamma + \int_\Gamma \sigma^*_{\alpha ij}(\underline{\xi},\underline{x})\, \varepsilon^p_{ij}(\underline{x})\, d\Omega$$

For $\underline{\xi}$ on Γ the resulting integral equation, obtained by modifying eq.(2.9) in the same way as above, formally coincides with eq.(8.6) up to the l.h.s. which vanishes in the present case as $\underline{\xi} = \underline{\xi}'$. Thus, tractions and tangential derivatives of the displacements are primary variables in this formulation. The boundary conditions on Γ_u can be fed into the integral equation directly in terms of tangential derivatives of displacements. In the BE discretization, one can adopt as usual, e.g. straight elements with linear interpolation for p_i and $\frac{du_i}{ds}$. Ghosh et al. [15] recommend to deal with jumps in tractions and/or in tangential derivatives of displacements placing nodes away from the points of discontinuity.

By solving the boundary equation system, supplemented by auxiliary conditions of the type (8.7) when Γ_u is not simply connected (but Ω is so), one determines all the nodal unknowns $p_i(\underline{\xi}^k)$ for the nodes on Γ_u and $\frac{du_i}{ds}(\underline{\xi}^k)$ for those on Γ_p. Once the modelled distribution of $\frac{du_i}{ds}$ along Γ is known, the boundary displacements, if needed, can be determined by applying equations of the type:

$$u_i(\underline{\xi}^k) - u_i(\underline{\xi}^{k-1}) = \int_{\underline{\xi}^{k-1}}^{\underline{\xi}^k} \frac{du_i}{ds}(\underline{x})\, d\Gamma \qquad (i = 1, 2) \qquad (8.7)$$

where nodes $\underline{\xi}^k$ are numbered through increasing integer values of superscript k, in counterclockwise direction. Repeated use of eq.(8.7), starting from a constrained node, determines in sequence the values of the displacement components at the nodes (and at any other boundary point).

For a multiply connected region problem, the displacement components appear explicitly in the relevant boundary integral equations together with p_i and $\frac{du_i}{ds}$, see ref [15]. Therefore in this case equations of type (8.7), written for pairs of adjacent nodal points $\underline{\xi}^k$, $\underline{\xi}^{k-1}$ lying on the same outer or inner boundary, are to be coupled with the integral equations in order to determine the boundary solution; and this, of course, rises the computational cost.

At boundary solution level, the advantage of the new formulation over the standard one, is that the accuracy of the stresses on Γ, still computed via Cauchy equations and Hooke's law from the tractions and the tangential elongation, is greatly enhanced since the latter is no longer obtained through numerical differentiation (as in the standard method) but directly in terms of $\frac{du_i}{ds}$.

Acknowledgement

Certain parts of this study were carried out, and some of the surveyed literature was gathered, in the framework of a EEC-supported study contract (F.R.C.C.; W.G.2), which is gratefully acknowledged.

9. REFERENCES

1. Martin, J.B.: Plasticity, The MIT Press, Cambridge, 1985.
2. Cohn, M.Z. and Maier, G., Engineering Plasticity by Mathematical Programming, Proc. NATO Advanced Study Institute, University of Waterloo, Pergamon Press, New York, 1979.
3. Swedlow, J.L. and Cruse, T.A., Formulation of boundary integral equations for three-dimensional elasto-plastic flow, Int. J. Solids and Struct., 7(1971), 1673-1682.
4. Riccardella, P.C., A implementation of the boundary-integral technique for planar problems in elasticity and elastoplasticity, Report No. SM 73-10 Department of Mechanical engineering, Carnegie Mellon University, Pittsburg, PA, 1973.
5. Mendelson, A. and Albers, L.U., Application of boundary integral equations to elastoplastic problems, in: T.A. Cruse, F.J. Rizzo (Eds.), Boundary Integral Equation Method: Computational Applications in Applied Mechanics, ASME, New York, pp. 47-84, 1975.
6. Banerjee, P.K. and Butterfield, R.: Boundary elements methods in Engineering Science, McGraw-Hill, London, 1981.
7. Mukherjee, S., Boundary Element Methods in Creep and Fracture, Appl. Science Publishers, London, 1982.
8. Brebbia, C.A., Telles, J.C.F. and Wrobel, L.C.: Boundary Element Techniques, Springer-Verlag, Berlin, 1984.
9. Kuhn, G., Boundary element techniques in elastostatics and linear fracture mechanics, in Finite Element and Boundary Element Techniques, from Mathematical and Engineering Point of View, (Eds. Stein, E. and Wendland, W.L.) CISM, Springer-Verlag, Berlin, 1988.
10. Chandra, A. and Mukherjee, S., Boundary element formulations for large strain-large deformation problems of viscoplasticity, Int. J. Solids Strctures, 20(1984), 41-53.
11. Danson, D.J., A boundary element formulation of problems in linear isotropic elasticity with body forces, in Proc. 3rd Int. Conf. on Boundary Element Methods (Ed. Brebbia, C.A.), Springer-Verlag, Berlin, 1981, 105-122.
12. Gurtin, E.M., The linear theory of elasticity, in: Encyclopedia of Physics (Ed. Flugge, S.), Vol. VIa/2, Springer-Verlag, Berlin, 1972.
13. Bui, H.D., Some remarks about the formulation of three-dimensional thermoelastoplastic problems by integral equations, Int. J. Solids and Struct., 14(1978), 935-939.
14. Kellog, O.D., Foundation of Potential Theory, Dover Publication, New York, 1953.
15. Ghosh, N., Rajiyah, H., Ghosh, S. and Mukherjee, S., A new boundary element method formulation for linear elasticity, J. Appl. Mechanics, 53(1986), 69-76.
16. Sutton, M.A., Liu, C.H., Dickerson, J.R. and McNeill, S.R., The two-dimensional boundary integral equation method in elasticity with a consistent boundary formulation, Engineering Analysis, 3(1986), 79-84.
17. Manolis, G.D: and Banerjee, P.K., Conforming versus non-conforming boundary elements in three-dimensional elastostatics, Int. J. Num. Meth. Engng., 23(1986), 1885-1904.

18.Banerjee, P.K. and Raveendra, S.T., Advanced boundary element analysis of two- and three-dimensional problems of elasto-plasticity, Int. J. Num. Methods in Engng., 23, pp. 985-1002, 1986.
19.Mustoe, G.G.W., Advanced integration schemes over boundary elements and volume cells for two- and three-dimensional non-linear analysis, in: Developments in BE methods - 3 (Eds. Banerjee, P.K. and Mukherjee S.), Elsevier Appl. Sci. Publ., London, 1984, 213-270.
20.Banerjee, P.K. and Davies, T.C.: Advanced implementation of boundary element methods for three-dimensional problems of elastoplasticity and viscoplasticity, in: Developments in BE Methods-3, (Eds. Banerjee, P.J., Mukherjee, S.), Elsevier Appl. Sci. Publ., London, 1984, 1-26.
21.Han, G.M., Li, H.B. and Mang, H.A., A new method for evaluating singular integrals in stress analysis of solids by the direct boundary element method: Int. J. Num. Meth. in Engrg., 21(1985), 2071-2098.
22.Martin, J.B., An internal variable approach to the formulation of finite element problems in plasticity, in: Physical Nonlinearities in Structural Analysis (Eds. Hult J. and Lemaitre J.), Springer-Verlag, Berlin, 1981, 165-176.
23.Maier, G., "Linear" flow-laws of elastoplasticity: a unified general approach, Rendic. Acc. Naz. Lincei, Serie VIII, Vol. XLVIII - 5 (1969), 266-276.
24.Maier, G., A matrix structural theory of piecewise-linear plasticity with interacting yield planes, Meccanica, 5(1970), 55.
25.Cathie, D.N. and Banerjee, P.K., Boundary element methods for plasticity and creep including a viscoplastic approach, Res. Mechanica, 4(1982), 3-22.
26.Ortiz, M. and Popov, E.P., Accuracy and stability of integration algorithms for elastoplastic consitutive equations, Int. J. Num. Meth. Engng., 21(1985), 1561-1675.
27.Simo, J.C. and Taylor, R.L., A return mapping algorithm for plane stress elastoplasticity, Int. J. Num. Meth. Engng., 22(1986), 649-670.
28.Aravas, N., On the numerical integration of a class of pressure dependent plasticity models, 1987, to appear.
29.Martin, J.B., Reddy, B.D., Griffin, T.B. and Bird, W.W., Applications of mathematical programming concepts to incremental elastic-plastic analysis, Engineering Structural 1987, to appear.
30.Cruse, T.A. and Polch, E.Z., Elastoplastic BIE analysis of cracked plates and related problems, Part. 1, Int. J. Num. Meth. Engng., 23(3), (1986), 429-438.
31.Maier, G., On elastoplastic analysis by boundary elements, Mechanics Research Communications, 10(1983), 45-52.
32.Maier, G. and Polizzotto C.: A Galerkin approach to boundary elements elastoplastic analysis, Computer Methods in Applied Mech. and Engng., 60(1987), 175-194.
33.Tullberg, O. and Bolteus, L., A critical study of different boundary element stiffness matrices, Proc. Int. Conf. on Boundary Elements (Ed. Brebbia, C.A.), Springer-Verlag, 1982, 621-649.
34.Crouch. S.L. and Starfield, A.M., Boundary Element Methods in Solid Mechanics, G. Allen & Unwin, London, 1983.
35.Polizzotto, C., An energy approach to the boundary element method, 1987, to appear.

36.Maier, G. and Novati, G., Elastic plastic boundary element analysis as a linear complementarity problem, Appl. Math. Modelling, 7(1983), 74-82.
37.Maier, G., Munro, J. and Lloyd Smith, D., Mathematical programming applications to engineering plastic analysis Appl. Mech. Rev., 35(1982), 1631-1643; update, Appl. Meth. Update 1986, ASME, 1986, 377-383.
38.Maier, G. and Polizzotto C.: A boundary element approach to limit analysis, in: Proc. 5th Int. Conf. on Boundary Elements, (Eds. Brebbia, C.A., Futagami, T., Tanaka, M.), Hiroshima, Springer-Verlag, (1983), 551-556.
39.Maier, G. and Polizzotto C., On shakedown analysis by boundary elements, in: Verba Volant, Scripta Manent, Ch. Massonet Anniversary Volume, Liège, 265-277, 1984.
40.Maier, G. and Nappi, A., On bounding post-shakedown quantities by the boundary element method, Engineering Analysis, 1(1984), 223-229.
41.Polizzotto, C., A BEM approach to the bounding techniques, in: Proc. 7th Int. Conf. on Boundary Elements (Eds. Brebbia, C.A. and Maier, G.), Springer-Verlag, Berlin, 1985, 13-114.
42.Mayr, M., Drexler, W. and Kuhn, G., A semianalytical boundary integral approach for axisymmetric elastic bodies with arbitrary boundary conditions, Int. J. Solids Structures, 16(1980), 863-871.
43.Cathie, D.N. and Banerjee, P.K., Numerical solutions in axisymmetric elastoplasticity, in: Innovative numerical analysis for the applied engineering sciences (Eds. Shaw, R. et al.), University of Virginia Press, Charlottesville, VA, 331-340, 1980.
44.Telles, J.C.F., A boundary element formulation for axisymmetric plasticity, in Proc. 5th Int. Conf. on Boundary Elements (Eds. Brebbia, C.A., Futagami, T. and Tanaka, M.), Hiroshima, Springer-Verlag, 1983, 577-589.
45.Winnicki, L.A. and Zienkiewicz, O.C., Plastic (or visco-plastic) behaviour of axisymmetric bodies subjected to non-symmetric loading: semi-analytical finite element solution, Int. J. Num. Meth. Engng., 14(1979), 1399-1412.
46.Maier, G., Novati, G. and Perreira, P., Boundary element analysis of rotationally symmetric systems under general boundary conditions, Civil Engineering Systems, 1(1983), 42-49.
47.Maier, G., Novati, G. and G. Perreira, P., On Boundary element elastic and inelastic analysis in the presence of cyclic symmetry, in: Proc. 6th Int. Conf. on Boundary Element Methods in Engineering (Ed. Brebbia, C.), Springer-Verlag, Berlin 1984, 19-41.
48.Banerjee, P.K. and Butterfield, R.: Boundary element methods in geomechanics, in Finite Elements in Geomechanics (Ed. Gudehus, G.), Wiley, New York, 1977, 529-570.
49.Maier, G. and Novati, G., On boundary element-transfer matrix analysis of layered elastic systems, Engineering Analysis, 3(1986), 208-216.
50.Maier, G. and Novati G., Boundary element elastic analysis of layered soils by a successive stiffness method, Int. J. Num. Meth. Geomech., (1987), to appear.

ON ASYMPTOTIC ERROR ESTIMATES FOR COMBINED BEM AND FEM

W.L. Wendland
Universität Stuttgart, Stuttgart, F.R.G.

Dedicated to Prof. Dr. Dr. h.c. mult. Wolfgang Haack on the occasion of his 85th birthday.

Abstract

As can be seen from several recent engineering applications, the coupling of finite elements with boundary elements plays an increasingly important role in computational methods. This paper presents a survey on the corresponding current mathematical analysis in the framework of asymptotic convergence and error estimates. The presented analysis covers classical coupling with boundary element Galerkin methods as for the Laplacian and corresponding variants with faster mesh refinement of the boundary elements on the coupling boundary or - alternatively - of the finite elements which already apply to a large class of elliptic boundary value problems. Further we present recent symmetric Galerkin formulations for general strongly elliptic boundary value problems in accordance with the Hellinger-Reissner principle. For two-dimensional problems we present some new results for boundary element collocation on the coupling boundary.

1980 AMS-MOS Mathematics subject classifications: 65N30; 65N45; 65R99.

Introduction

As is well known, finite element methods are now in common use due to their high flexibility and their applicability also to nonhomogeneous physical properties as well as to nonlinear problems. However, for curved and complicated boundaries as well as for exterior problems their accuracy and corresponding modifications are far less efficient. In particular, in the latter cases the boundary element techniques prove to be very efficient since these reduce the original boundary value problem in $\mathbf{R}^n$ to the $(n-1)$-dimensional compact boundary manifold incorporating also its geometry. However, unmodified boundary element methods need an explicit fundamental solution of the differential equations and, hence, require essentially homogeneous physical properties and linear laws.
Therefore it seems to be most efficient to combine the two methods and to exploit the features of each due to their corresponding superiority, i.e., a "Marriage à la mode. The best of both worlds". (Bettess, Kelly, Zienkiewicz [10],[11].)

At present, there are basically two different concepts for this combination. One consists of subdividing the original domain $\Omega \in \mathbf{R}^n$, $n = 2$ or 3, into a finite number of subregions and using in each of them either FEM or BEM, where the latter lives on all boundaries of the subregions.
The second concept uses BEM for the modelling of special finite element functions within a mixed FEM formulation [80].
Here we analyse some of the numerical procedures within the first concept which we can find in various recent applications. The domain Ω of the corresponding boundary value problem will be divided into two subdomains Ω_1, Ω_2, where $\Gamma_c = \partial\Omega_1$ and $\Gamma = \partial\Omega$ are separated. We will formulate our analysis for the case of a bounded domain Ω (see Figure 1). For exterior infinite domains Ω (Figure 2), the corresponding boundary integral equations usually become even simpler and our results are also valid, correspondingly. Very often, as in substructuring techniques, the two subdomains have a common boundary part Γ_c which connects two points on Γ (Figure 3). Since our analysis will apply to this situation only for the symmetric Galerkin formulation in Sections 1.5 and 4 we will indicate the corresponding results but otherwise omit this important case which in general still has not been rigorously analysed yet.

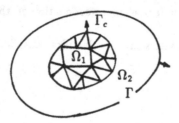

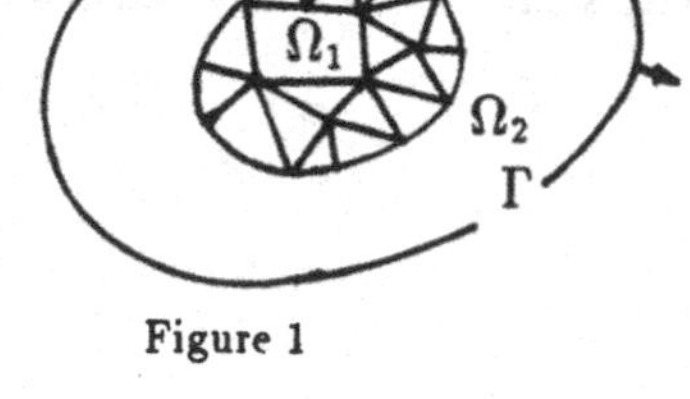

Figure 1

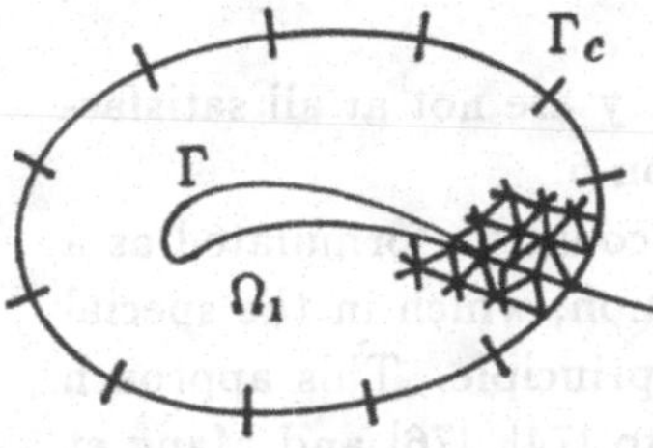

Figure 2

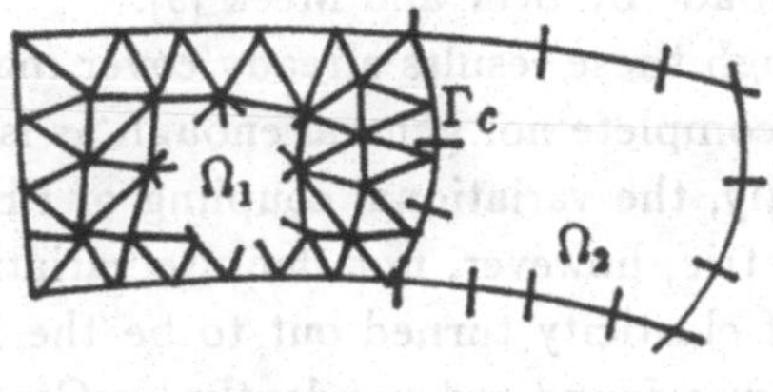

Figure 3

Since the asymptotic analysis of coupling methods will rest on results for purely FEM and purely BEM, correspondingly, we first give a more or less unifying short survey of asymptotic error analysis in Sections 1 and 2. For conforming FEM applied to variational problems, this error analysis is now standard knowledge, see Babuška and Aziz [8]. In a similar framework, for BEM based on Galerkin's schemes, the asymptotic error estimates are also well established by Nedelec [69] and Hsiao, Stephan and the author [50], [85]. As we will see, these are well suited for the analysis of the combines FEM-BEM based on the variational formulation and Galerkin's procedures. In practical codes, however, the BEM is mostly used as a collocation scheme. Corresponding error estimates for BEM collocations is now essentially settled for two-dimensional problems (see Section 2). In three-dimensional problems, however, BEM collocation was only analysed yet for the very special Fredholm boundary integral equations of the second kind with smoothing operators excluding e.g. the application to elasticity [90].

Asymptotic error estimates for the <u>coupled</u> FEM-BEM methods go back to Brezzi, Johnson and Nedelec [16], [17], [52] who considered the exterior Dirichlet problem for the Laplacian. The method is based on a direct variational formulation yielding an unsymmetric bilinear form. This approach is commonly used. We present the corresponding asymptotic error analysis of the Galerkin scheme for some slightly more general equations. However, we need to have a rather strong assumption for the coupling operator between FEM and BEM-Galerkin, which here must be a compact perturbation of the identity. This assumption is violated for many practical problems as e.g. in elasticity. Therefore, we require instead a more careful coupling boundary and a faster mesh refinement of the boundary elements than of the finite elements. This idea was developed by Brezzi and Johnson [16], and it turns out that it still works when the above mentioned compactness property (of the Laplacian case) is violated. It also allows the rigorous justification of corresponding unsymmetrically coupled FEM-collocation BEM in two dimensions as will be shown in Section 3.3. The different rates of refinements can also be used the other way around: Convergence can be shown if the BEM refinement is faster than the FEM refinement. Such additional requirements for the general case are in agreement with numerical experi-

ences made by Beer and Meek [9].
Although these results already cover many applications, they are not at all satisfactorily complete nor general enough as is indicated in Section 5.
Recently, the variational coupling of FEM-BEM Galerkin could be formulated as a symmetric, however, non-definite variational bilinear equation, which in the special case of elasticity turned out to be the Hellinger-Reissner principle. This approach which was found independently by Costabel [25], Polizzotto [74], [76] and Mang et al. [56] yields asymptotically convergent new FEM-BEM Galerkin methods. If in the coupling boundary integral equation the fundamental solution is replaced by the Green function of the finite element domain, then the coupling boundary element equation degenerates to a representation formula and the new formulation becomes the known method of reduction to the boundary by Feng Kang [37]. Using the boundary potentials in Ω_2, the new method can also be interpreted as a hybrid FEM and the corresponding mixed formulation is the same as our symmetric formulation. This result is also new and due to Grannell [40].

All of our analysis is based on the variational formulation of boundary integral equations and their characterization by strongly elliptic variational forms as considered by Costabel and the author in [32]. This approach with strong ellipticity generalizes the principle of minimal energy to general elliptic boundary value problems. A different approach to a unifying concept including FEM-BEM coupling was proposed by Herrera [44]. Without strong ellipticity, however, convergence results with this concept can only be rigorously justified for least squares methods, which we do not consider here.
The lecture notes are organized as follows. In Chapter 1 we present the basic ideas of asymptotic error analysis based on the coercive variational formulations. In Section 1.1 these are formulated for strongly elliptic boundary value problems. Section 1.2 is devoted to the variational formulation of boundary integral equations. In Section 1.3 we present the basic variational formulation and boundary integral equations underlying unsymmetric coupling of FEM and BEM. Section 1.4 contains the recent symmetric variational coupling.
In Chapter 2 we collect the asymptotic error analyses of conforming FEM methods, of FEM Galerkin and, for two-dimensional problems, of naive BEM collocation. In Chapter 3 we present asymptotic error analyses for FEM-BEM coupling in nonsymmetric formulation. Section 3.1 present FEM-BEM Galerkin and the results from [52]. In Section 3.2 we consider faster BEM mesh refinement providing asymptotic convergence for corresponding FEM-BEM Galerkin and with faster FEM mesh refinement and in Section 3.5 the corresponding result for FEM-BEM collocation. This last situation corresponds to exterior flow problems (Figure 2).
Chapter 4 is devoted to the symmetric Galerkin coupling of FEM and BEM.
In Chapter 5 we try to give a summary of rigorous results, which are currently avail-

able. Up to now, many of the coupling techniques are not justified yet, and hopefully, these lecture notes will stimulate the interest in and the numerical analysis of these very powerful engineering computational methods.

Acknowlegdements:
The author is very much indebted to M. Costabel and J. C. Nedelec for their helpful suggestions and discussions. He also wants to thank F. Brezzi and C. Johnson who have brought several references to his attention. The author wants to express his gratitude to Ms. Tabbert, Ms. Wagner and Mr. Witt, who typed the manuscripts, and to Mr. Schmitz and Mr. Berger, who were patiently proofreading.

1 Error Analysis for Variational Equations

In this section we shall consider the abstract principles underlying asymptotic error bounds for FEM, BEM and their coupling. These are stability and consistency based on the concept of abstract strong V-ellipticity [86] and duality arguments [70],[71].

Let $\mathbf{a}(u, v)$ be a given continuous bilinear form on a Hilbert space H with scalar product $(\cdot,\cdot)$ and norm $\|\cdot\|$. Let $V \in H$ be a given closed subspace of H. Further, let $H_{\pm}^{\{T\}}$ be two given additional Hilbert spaces with dense continuous imbeddings $H_{+}^{\{T\}} \subset H \subset H_{-}^{\{T\}}$ such that the scalar product $(\cdot,\cdot)$ extends from $H \times H$ to a continuous duality on $H_{+}^{\{T\}} \times H_{-}^{\{T\}}$,

$$(u, v) \leq c\|u\|_{H_{+}^{\{T\}}}\|v\|_{H_{-}^{\{T\}}} \tag{1.1}$$

to which also $\mathbf{a}(\cdot,\cdot)$ can be extended continuously. On V let us consider the variational equations with a continuous linear fucntional $\ell(\cdot)$ given on H.

Find $u \in V$ *such that*

$$\mathbf{a}(u, v) = \ell(v) \text{ for all } v \in V. \tag{1.2}$$

For the approximation of (1.2) let $V_h \subset H_{+}^{\{T\}}$ be the trial spaces, a sequence of approximating finite dimensional subspaces, and $T_h \subset H_{-}^{\{T\}}$ be a corresponding sequence of finite dimensional test spaces, here with $\dim V_h = M = \dim T_h$. Many of the approximation methods for (1.2) can be formulated in the form of a Petrov-Galerkin method:
Find $u_h \in V_h$ *such that*

$$\mathbf{a}(u_h, \phi_h) = \ell(\phi_h) \text{ for all } \phi_h \in T_h. \tag{1.3}$$

As is well known, stability of the method together with approximation properties of $V_h \in V$ implies convergence.

Theorem 1.1 [7]: *If there exists a positive* γ *such that the inequality*

$$\sup_{\phi_h \in \mathcal{T}_h} |\mathbf{a}(v_h, \phi_h)| / \|\phi_h\|_{H_+^{\{T\}}} \geq \gamma \|v_h\|_{H_-^{\{T\}}} \tag{1.4}$$

holds for all $v_h \in V_h$ *then* (1.3) *is uniquely solvable and there exists a constant* c *such that there holds the error estimate,*

$$\|u - u_h\|_{H_+^{\{T\}}} \leq c \inf_{v_h \in V_h} \|u - v_h\|_{H_+^{\{T\}}}.$$

(In case $V_h = \mathcal{T}_h$ see also [19]). The stability estimate (1.4) is often called the BBL condition, i.e. Babuška-Brezzi-Ladyzenskaya condition.[1]

Proof: (i) Let $\{\mu_j\}_{j=1}^M$ denote a basis of V_h and $\{\lambda_k\}_{k=1}^M$ a basis of $\mathcal{T}_h$. Then with

$$u_h = \sum_{j=1}^{M} \gamma_j \mu_j, \quad \gamma_j \in \mathbf{R} \,(\text{or } \mathbf{C}),$$

the Equations (1.3) are equivalent to quadratic system of linear equations

$$\sum_{j=1}^{M} \mathbf{a}(\mu_j, \lambda_k)\gamma_j = \boldsymbol{\ell}(\lambda_k), \quad k = 1, \ldots, M \tag{1.5}$$

for $\gamma_1, \ldots, \gamma_M$. For the homogeneous equations (1.5) with $\boldsymbol{\ell}(\lambda_k) = 0$ we have $\mathbf{a}(u_h, \phi_h) = 0$ and (1.4) implies $u_h = 0$ which yields vanishing of $\gamma_1, \ldots, \gamma_M$. Hence, the linear system (1.5) is uniquely solvable. Moreover, (1.4) also implies with (1.2) and (1.3) an a-priori bound for u_h:

$$\|u_h\|_{H_+^{\{T\}}} \leq \frac{1}{\gamma} \sup_{0 \neq \phi_h \in \mathcal{T}_h} \{|\mathbf{a}(u_h, \phi_h)| \,/\, \|\phi_h\|_{H_-^{\{T\}}}\} \leq c\|u\|_{H_+^{\{T\}}} \tag{1.6}$$

where c is independent of h or u.

(ii) Solving (1.3) defines a family of uniformly bounded linear mappings

$$u \mapsto u_h =: U_h(u) \quad \text{with} \quad \|U_h\|_{H_+^{\{T\}}, H_+^{\{T\}}} leqc \leq c.$$

Moreover, for $u \in V_h$ the unique solvability of (1.3) implies $u = u_h$, i.e., U_h is a family of projections,

$$U_h(v_h) = v_h \quad \text{for every} \quad v_h \in V_h.$$

[1]Here and in the sequel c will denote a generic constant being independent of h, u, u_h but having actual different values due to the context.

Hence,

$$\|u - u_h\|_{H_+^{\{T\}}} = \|(I - U_h)(u - v_h)\|_{H_+^{\{T\}}} \leq (1+c)\|u - v_h\|_{H_+^{\{T\}}}$$

for every $u_h \in V_h$, which implies (1.5). □

Similarly to [86], p. 135 we define:

$\mathbf{a}(u,v)$ is V-coercive if there exists a positive constant γ_0 and a compact bilinear form $\mathbf{c}(u,v)$ on $H \times H$ such that for all $v \in V$ there holds

$$Re(\mathbf{a}(v,\ v) + \mathbf{c}(v,\ v)) \geq \gamma_0 \|v\|^2 . \tag{1.7}$$

Let us consider conforming Bubnov-Galerkin methods where $V_h = T_h \subset V$ for V-coercive bilinear forms, and suppose that for every $v \in V$ there exists an approximating family $v_h \in V_h$ such that

$$\lim_{h \to 0} \|v - v_h\| = 0 . \tag{1.8}$$

Mikhlin proved for this case the following theorem.

Theorem 1.2 [45], [67], [87]: *Let* $\mathbf{a}(u,v)$ *be V-coercive and let* (1.2) *be uniquely solvable. Then the Bubnov-Galerkin method is stable for almost all* $h > 0$, *i.e. there exists a positive* h_o *such that the equations* (1.5) *with* $V_h = T_h \subset V$ *are uniquely solvable for every* h *with* $0 < h \leq h_o$ *and there holds* (1.4) *with* $H = H_+^{\{T\}} = H_-^{\{T\}}$. *Hence,* (1.5) *yields*

$$\|u - u_h\| \leq c \inf_{v_h \in V_h} \|u - v_h\| .$$

Proof [45], [87]: (i) Since (1.3) is equivalent to (1.5), non-uniqueness for a sequence $h' \to 0$ would imply the existence of a corresponding sequence of eigensolutions $\overset{\circ}{u}_{h'} \in V_{h'}$ with

$$\mathbf{a}(\overset{\circ}{u}_{h'},\ v_{h'}) = 0 \text{ for all } v_{h'} \in V_{h'} \text{ and } \|\ \overset{\circ}{u}_{h'}\ \| = 1. \tag{1.9}$$

Hence, there existed a weakly convergent subsequence

$$\overset{\circ}{u}_{h''} \rightharpoonup \overset{\circ}{u} \quad \text{in } V$$

for which

$$\mathbf{c}(\overset{\circ}{u}_{h''} - \overset{\circ}{u},\ \overset{\circ}{u}_{h''} - \overset{\circ}{u}) \to 0$$

for $h'' \to 0$ due to the compactness of $\mathbf{c}$. Now (1.7) and (1.9) would imply

$$\begin{aligned}\|\ \overset{\circ}{u} - \overset{\circ}{u}_{h''}\ \|^2 &\leq \tfrac{1}{\gamma_0} Re\{\mathbf{a}(\overset{\circ}{u} - \overset{\circ}{u}_{h''},\ \overset{\circ}{u} - \overset{\circ}{u}_{h''}) + \mathbf{c}(\overset{\circ}{u}_{h''} - \overset{\circ}{u},\ \overset{\circ}{u}_{h''} - \overset{\circ}{u})\} \\ &\leq \tfrac{1}{\gamma_0} Re\{\mathbf{a}(\overset{\circ}{u},\ \overset{\circ}{u} - \overset{\circ}{u}_{h''}) - \mathbf{a}(\overset{\circ}{u}_{h''},\ \overset{\circ}{u})\} + o(1) \\ &\leq \tfrac{1}{\gamma_0} Re\{-\mathbf{a}(\overset{\circ}{u}_{h''}, \overset{\circ}{u})\} + o(1) .\end{aligned}$$

Let $w_{h''}$ be a sequence which approximates $\mathring{u}$ due to (1.8), then $a(\mathring{u}_{h''}, w_{h''}) = 0$ because of (1.9) and

$$\| \mathring{u} - \mathring{u}_{h''} \|^2 \leq \frac{1}{\gamma_0} Re\, \mathbf{a}(\mathring{u}_{h''}, w_{h''} - \mathring{u}) + o(1)$$

$$\leq c \| w_{h''} - \mathring{u} \| + o(1) \to 0$$

for $h'' \to 0$. Hence, $\mathring{u} = \lim_{h'' \to 0} \mathring{u}_{h''}$ in V , $\| \mathring{u} \| = 1$ and

$$\mathbf{a}(\mathring{u}, v) = \lim_{h'' \to 0} \mathbf{a}\left(\mathring{u}_{h''}, v_{h''}\right) = 0 ,$$

where $v_{h''} \to v$ in V with any $v \in V$.

Hence, $\mathring{u}$ would be an eigensolution to (1.2) contradicting our assumption. Consequently, there is some $h_0 > 0$ such that the equations (1.3) are uniquely solvable for any h with $0 < h \leq h_0$.

(ii) Now, from (i) we have that for the given ℓ and the corresponding solution u of (1.2), the equations

$$\mathbf{a}(u_h - u, \phi_h) = 0 \quad \text{for all} \quad \phi_h \in V_h = \mathcal{T}_h$$

are uniquely solvable for all $0 < h \leq h_0$. If (1.4) were not true then to u_h there would exist an unbounded subsequence $u_{h'}$ with $\|u_{h'}\| \to \infty$. Hence, for

$$\tilde{u}_{h'} := u_{h'}/\|u_{h'}\|, \quad \|\tilde{u}_{h'}\| = 1 ,$$

we have

$$\mathbf{a}(\tilde{u}_{h'} - u/\|u_{h'}\|, v_{h'}) = 0 \quad \text{for all} \quad v_{h'} \in V_{h'} .$$

Again, we could find a weakly convergent subsequence

$$\tilde{u}_{h''} \rightharpoonup \mathring{u}$$

and

$$|\mathbf{a}(\tilde{u}_{h''}, v_{h''})| \leq \|v_{h''}\| o(1) \quad for \quad h'' \to 0 . \tag{1.10}$$

In exactly the same manner as in case (i) we would find

$$\| \mathring{u} - \tilde{u}_{h''} \|^2 \leq \frac{1}{\gamma_o} |\mathbf{a}(\tilde{u}_{h''}, \mathring{u})| + o(1)$$

and with $v_{h''} \to \mathring{u}$ and (1.10), $\| \mathring{u} - \tilde{u}_{h''} \| \to 0$ for $h'' \to 0$. Then, again with (1.10), we could find $\mathring{u}$ to be an eigensolution of (1.2) in the contrary to our assumption. Hence, (1.4) cannot be violated. □

Now let us exploit the duality between $H_+^{\{T\}}$ and $H_-^{\{T\}}$ with respect to $(\cdot,\cdot)$ in H, $V_h \subset H_+^{\{T\}} \subseteq H \subseteq H_-^{\{T\}}$. To this end let $w \in V$ be the variational solution of the adjoint problem to any given ϕ :

$$\mathbf{a}(v,\, w) = (v,\, \phi) \quad \text{for all} \quad v \in V\,. \tag{1.11}$$

Now we require the

<u>regularity assumption:</u> *For given* $\phi \in H_+^{\{T\}}$, *the solution* w *of* (1.11) *also belongs to* $H_+^{\{T\}}$ *and satisfies the a-priori estimate*

$$\|w\|_{H_+^{\{T\}}} \leq c\,\|\phi\|_{H_+^{\{T\}}}\,. \tag{1.12}$$

In FEM and BEM applications, $H_+^{\{T\}}$ will consist of smoother functions than H and the quantity

$$E(H_+^{\{T\}}\,,\, V_h) := \sup_{\|v\|_{H_+^{\{T\}}} \leq 1} \ \inf_{v_h \in V_h} \|v - v_h\| \tag{1.13}$$

will tend to zero with some order of h. This property will yield super approximation results for the Galerkin solutions.

<u>Theorem 1.3 [5], [71]:</u> *Let in addition to the assumptions of Theorem 1.2 also the regularity assumption* (1.12) *be valid. Then we have the super approximation result*

$$\|u - u_h\|_{H_-^{\{T\}}} \leq c E(H_+^{\{T\}}\,,\, V_h) \cdot \inf_{v_h \in V_h} \|u - v_h\|\,. \tag{1.14}$$

<u>Remark:</u> For BEM see also [50]. If the solution of the adjoint problem in the dual space cannot be obtained one still gets (1.14) under slightly weaker assumptions [31].
<u>Proof:</u> By duality,

$$\|u - u_h\|_{H_-^{\{T\}}} \leq c \sup_{\|\phi\|_{H_+^{\{T\}}} \leq 1} |(u - u_h, \phi)|$$

$$\leq c \sup_{\|\phi\|_{H_+^{\{T\}}} \leq 1} |\mathbf{a}(u - u_h,\, w)|$$

where w is the solution of (1.11) to any chosen ϕ. Now use the orthoganility

$$\mathbf{a}(u - u_h,\, v_h) = 0 \quad \text{for every} \quad v_h \in V_h$$

to obtain

$$\|u - u_h\|_{H_-^{\{T\}}} \leq c \sup_{\|\phi\|_{H_+^{\{T\}}} \leq 1} \inf_{v_h \in V_h} |\mathbf{a}(u - u_h, w - v_h)|$$

$$\leq c \sup_{\|\phi\|_{H_+^{\{T\}}} \leq 1} \inf_{v_h \in V_h} \|u - u_h\| \, \|w - v_h\|$$

$$\leq c \, \|u - u_h\| \sup_{\|\phi\|_{H_+^{\{T\}}} \leq 1} E(H_+^{\{T\}}, V_h) \, \|w\|_{H_+^{\{T\}}} .$$

Inequality (1.12) and Theorem 1.2 yield the desired estimate (1.14). □

Examples:

1.1 The variational form of elliptic boundary value problems [57]

Let $\Omega \in \mathbf{R}^n$ denote a smoothly bounded compact domain with boundary Γ. Let L denote a strongly elliptic partial differential operator of order $2m$ in divergence form or a system of such,

$$L\phi = \sum_{|j| \leq m} \sum_{|k| \leq m} (-1)^{|j|} (\partial_x)^j \left(a_{jk}(x)(\partial_x)^k \phi\right), \tag{1.15}$$

with sufficiently smooth coefficients $a_{jk}(x)$, where $j = (j_1, \dots, j_n)$ and $k = (k_1, \dots, k_n)$ denote multiindices, $|j| = j_1 + \dots + j_n$, $0 \leq j_\ell, k_\ell \in \mathbf{Z}$, $(\partial_x)^j = \Pi_{\ell=1}^n (\partial_x)^{j_\ell}$. Let $B\gamma = \begin{pmatrix} B_1 \\ B_2 \end{pmatrix} \gamma$ be a system of regular boundary differential operators, where B_1 denotes the essential and B_2 the natural boundary conditions associated with a regular strongly elliptic boundary value problem

$$Lu = f \text{ in } \Omega \text{ and}$$
$$B\gamma u = g = 0 \text{ on } \Gamma \tag{1.16}$$

where $\gamma u = (u, \frac{\partial}{\partial n} u, \dots, (\frac{\partial}{\partial n})^{2m-1} u)^T |_\Gamma$ denotes the column of Cauchy data at Γ. Here

$$H = H^m(\Omega) \text{ and}$$

$$V = \{v \in H^m(\Omega) \text{ satisfying the essential boundary conditions } B_1 \gamma v = 0 \text{ on } \Gamma\} .$$

The bilinear form **a** is given by the Dirichlet bilinear form

$$\mathbf{a}(u,\, v) = \sum_{|j|,|k|\le m} \int_\Omega a_{jk}(x)(\partial_x^k u)(\partial_x^j \bar{v})\, dx\, . \tag{1.17}$$

V-coercivity (1.7) of $\mathbf{a}(\cdot,\cdot)$ now corresponds to the Gårding inequality [86]. The dualities are defined by the scales of Sobolev spaces

$$H_+^{\{T\}} = H^{m\pm T}(\Omega)\ ,\ H = H^m(\Omega),$$

and the a-priori estimate (1.12) corresponds to the regularity results for the regular strongly elliptic adjoint boundary value problem [57]. Here, (1.12) is often called the shift theorem.

As standard examples consider the Dirichlet and Neumann problems of homogeneous isotropic ideal elasticity for the vector-valued displacement $\mathbf{u}$ satisfying the Navier system

$$L\mathbf{u} = \mu\Delta\mathbf{u} + (\lambda+\mu)\text{grad div}\mathbf{u} = \mathbf{f}\, . \tag{1.18}$$

The homogeneous Dirichlet conditions are given by

$$B\gamma\mathbf{u} = B_1\gamma\mathbf{u} = \mathbf{u}|_\Gamma = \mathbf{g} = \mathbf{0}\, , \tag{1.19}$$

these are essential boundary conditions. Here, $m = 1$, $n = 2$ or 3, $\mu > 0$ and $\lambda > -\frac{2}{n}\mu$. The latter are the Lamé constants. The bilinear form **a** is given by

$$\mathbf{a}(\mathbf{u},\, \mathbf{v}) =$$

$$\int_\Omega \{((\gamma + \mu - \frac{n-2}{n}\mu)\,\text{div}\,\mathbf{u}\,\text{div}\,\mathbf{v} \ + \tfrac{1}{2}\mu\sum_{j\neq k}(u_{j/k}+u_{k/j})(v_{j/k}+v_{k/j}) + \frac{\mu}{n}\sum_{j\neq k}(u_{j/k}-u_{k/j})(v_{j/k}-v_{k/j})\}dx \tag{1.20}$$

where $u_{j/k} = \partial_{x_k} u_j$ with u_j the j-th component of $\mathbf{u}$. In the homogeneous Neumann problem for (1.18), the tractions on the boundary are given by

$$B\gamma\mathbf{u} = B_2\gamma\mathbf{u} = T(\mathbf{u})|_\Gamma = \lambda\mathbf{n}\,\text{div}\,\mathbf{u} + 2\mu\frac{\partial\mathbf{u}}{\partial n} + \mu\mathbf{n}\times\text{curl}\,\mathbf{u}|_\Gamma = \mathbf{g}\, . \tag{1.21}$$

For $\mathbf{g} = \mathbf{0}$, this is a natural boundary condition belonging to (1.20). For the interior traction problem (1.18), (1.21) one finds that

$$\underset{\sim}{m}(x)\omega = \begin{Bmatrix} 1 & 0 & -x_2 & 0 & 0 & x_3 \\ 0 & 1 & x_1 & 0 & -x_3 & 0 \\ 0 & 0 & 0 & 1 & x_2 & -x_1 \end{Bmatrix} \begin{pmatrix} \omega_1 \\ \vdots \\ \omega_6 \end{pmatrix}$$

are eigensolutions. Therefore, here the given forces $\mathbf{f}$ in (1.18) need to satisfy the equilibrium conditions of vanishing total forces and total momentum, and for a unique solution we require side conditions

$$\int_\Gamma \underset{\sim}{m}^T \mathbf{u}\, ds = \mathbf{0}. \tag{1.22}$$

1.2 The variational form of boundary integral equations [24], [32]

Let us assume that to L in (1.16) there exists the Green tensor $\underset{\sim}{G}(x, y)$ for $\mathbf{R}^n$, i.e. the fundamental solution satisfying radiation conditions. Let us consider (1.16) in the interior Ω_+ or the exterior Ω_-,

$$\begin{aligned} Lu &= f \text{ in } \Omega_\pm \text{ and} \\ B\gamma u &= g \text{ on } \Gamma . \end{aligned} \tag{1.23}$$

For the exterior problem we need the radiation conditions at ∞ to have unique solvability [51], [32]. Let us decompose L along the boundary Γ with respect to tangential and normal derivatives via

$$L\phi = \sum_{j=0}^{2m} L_j \left(\frac{\partial}{\partial n}\right)^j \phi$$

where the L_j are appropriately defined tangential differential operators. The Green representation formula for L reads as

$$u_\pm(x) = \int_{\Omega_\pm} \underset{\sim}{G}(x, y) f(y)\, dy \tag{1.24}$$

$$\mp \sum_{\substack{k+\ell+1\le 2m \\ k,\ell\ge 0}} \int_\Gamma \left(\left(\frac{\partial'}{\partial n(y)}\right)^k \underset{\sim}{G}(x, y)\right) L_{k+\ell+1}(y) \left(\frac{\partial}{\partial n(y)}\right)^\ell u_\pm(y)\, dS(y)$$

for $x \in \Omega_\pm$

and the application of γ to (1.24) provides a relation between the Cauchy data on Γ. Here $\frac{\partial'}{\partial n}$ denotes the operator adjoint to $\frac{\partial}{\partial n}$. For incorporating the boundary conditions let us add to B the complementing boundary operators S such that the square matrix of tangential operators

$$M = \begin{pmatrix} B \\ S \end{pmatrix}$$

becomes invertible. Then the Cauchy data γu in (1.24) can be expressed by g and an additional boundary function w on Γ,

$$M\gamma u = \begin{pmatrix} g \\ w \end{pmatrix} \quad \text{and} \quad \gamma u = M^{-1} \begin{pmatrix} g \\ w \end{pmatrix}, \tag{1.25}$$

where u and also w are different for Ω_+ and Ω_- , respectively.

Inserting (1.25) into (1.24) gives

$$\begin{aligned} u_\pm(x) &= \int_{\Omega_\pm} \underset{\sim}{G}(x, y) f(y)\, dy \\ &\mp \sum_{\substack{k+\ell+1\le 2m\\ k,\ell\ge 0}} \int_\Gamma \left(\left(\frac{\partial'}{\partial n(y)}\right)^k \underset{\sim}{G}(x, y)\right) L_{k+\ell+1}(y)(M^{-1})_\ell \begin{pmatrix} g\\ w\end{pmatrix} dS(y) \end{aligned} \tag{1.26}$$

where $(M^{-1})_\ell$ denotes the ℓ-th row of M^{-1} .

Hence, $u(x)$ is known in Ω_+ or Ω_- if the corresponding boundary function w is known on Γ . The application of M to (1.26) (which involves the jump relations for potentials) then gives on Γ

$$\begin{pmatrix} g\\ w\end{pmatrix}(x) = \sum_{j=0}^{2m-1} M_j \left(\frac{\partial}{\partial n(x)}\right)^j \int_{\Omega_\pm} \underset{\sim}{G}(x, y) f(y)\, dy \tag{1.27}$$

$$\mp \sum_{j=0}^{2m-1} \sum_{\substack{k+\ell+1\le 2m\\ k,\ell\ge 0}} M_j \left(\frac{\partial}{\partial n(x)}\right)^j \int_\Gamma \left(\left(\frac{\partial'}{\partial n(y)}\right)^k \underset{\sim}{G}(x, y)\right) L_{k+\ell+1}(M^{-1})_\ell \begin{pmatrix} g\\ w\end{pmatrix} dS(y),$$

where M_j denotes the j-th column of M . This is an overdetermined system of two equations on Γ for the yet unknown w . If we choose the first of the equations in (1.27) we obtain the so called boundary integral equation of the first kind [32] for w on Γ ,

$$\begin{aligned} \tfrac{1}{2} Aw(x) &:= \\ &- \sum_{\substack{k+\ell+1\le 2m\\ k,\ell\ge 0}} B\gamma_\pm \int_\Gamma \left(\left(\frac{\partial'}{\partial n(y)}\right)^k \underset{\sim}{G}(x, y)\right) L_{k+\ell+1}(y) \left(M^{-1}(y)\right)_\ell \begin{pmatrix} 0\\ w(y)\end{pmatrix} dS(y) \\ &= \tfrac{1}{2} F(x) \\ &:= \pm g(x) + \sum_{\substack{k+\ell+1\le 2m\\ k,\ell\ge 0}} B\gamma_\pm \int_\Gamma \left(\left(\frac{\partial'}{\partial n(y)}\right)^k \underset{\sim}{G}(x, y)\right) L_{k+\ell+1} \left(M^{-1}\right)_\ell \begin{pmatrix} g\\ 0\end{pmatrix} dS(y) \\ &\mp B\gamma_\pm \int_{\Omega_\pm} \underset{\sim}{G}(x, y) f(y)\, dy \quad \text{for } x\in\Gamma . \end{aligned} \tag{1.28}$$

If we choose the second of the two equations in (1.27) we obtain the so called boundary integral equation of the second kind [32] for w on Γ ,

$$\begin{aligned}\tfrac{1}{2}\underset{\sim}{A}w(x) \;:=&\\ w(x) \;\pm& \sum_{\substack{k+\ell+1\leq 2m\\ k,\ell\geq 0}} S\gamma_{\pm}\int_{\Gamma}\left(\left(\frac{\partial'}{\partial n(y)}\right)^{k}\underset{\sim}{G}(x,\,y)\right)L_{k+\ell+1}(y)\left(M^{-1}(y)\right)_{\ell}\begin{pmatrix}0\\ w(y)\end{pmatrix}dS(y)\end{aligned}$$

$$\begin{aligned}=\tfrac{1}{2}\underset{\sim}{F}(x) \;:=\pm& \sum_{\substack{k+\ell+1\leq 2m\\ k,\ell\geq 0}} S\gamma_{\pm}\int_{\Gamma}\left(\left(\frac{\partial'}{\partial n(y)}\right)^{k}\underset{\sim}{G}(x,\,y)\right)L_{k+\ell+1}(y)\left(M^{-1}\right)_{\ell}\begin{pmatrix}g\\ 0\end{pmatrix}dS(y)\\ &+S\gamma_{\pm}\int_{\Gamma}\underset{\sim}{G}(x,\,y)\,f(y)dy\,,\quad x\in\Gamma\,.\end{aligned} \tag{1.29}$$

Both types of equations are used in boundary element methods, alternatively. Their mapping properties, however, are rather different and, moreover, the V-coercivity can be obtained for (1.28) from energy principles whereas for (1.29) one gets strong ellipticity only for smooth Γ under additional stronger assumptions.

For the variational form of these equations let us first consider (1.28) Here the associated bilinear form is defined by

$$\mathbf{a}(w,\,v) := \langle Aw,\,v\rangle_{L_2(\Gamma)} := \int_{\Gamma}(Aw(x))\,\bar{v}(x)\,dS(x)\,, \tag{1.30}$$

the linear functional by

$$\ell(v) := \langle F,\,v\rangle_{L_2(\Gamma)}$$

and the Hilbert spaces by

$$H = H^{\alpha}(\Gamma) := \prod_{j=0}^{m-1} H^{\alpha_j}(\Gamma) \quad \text{where} \quad \alpha_j = \mu_j - m + \frac{1}{2}$$

and where μ_j with $0 \leq \mu_j \leq 2m-1$, $\mu_0 < \mu_1 < \ldots < \mu_{n-1}$, are the orders associated with B , namely

$$ord\,(B_{jk}) = \mu_j - k\,,\; B = ((B_{jk}))_{\substack{j=0,\ldots,\,m-1\\ k=0,\ldots,\,2m-1}}\,. \tag{1.31}$$

Mostly $V = H$ but sometimes $H = V \oplus V_1$ where V_1 is a finite dimensional space orthogonal to V which is characterized by appropriate linear orthogonality conditions like (1.22).

Theorem 1.4 [32] : *Let the original boundary value problem (1.23) provide an energy bilinear form* $\Phi(u,v)$ *having the following properties:*

For every compact $K \subset \mathbf{R}^n$ *there exists a constant* C_k *such that for all* $u,\, v \in C_0^\infty(K)$

$$|\Phi(u, v)| \le C_k \|u\|_{H^m(\mathbf{R}^n)} \|v\|_{H^m(\mathbf{R}^n)} .$$

For distributions u *with* $u|_{\Omega_\pm} \in C^\infty(\overline{\Omega}_\pm)$ *and compact support in* K, Φ *satisfies*

$$Re\, \Phi(u, u) = Re \left\{ \int_{\Omega_+ \cup \Omega_-} (Lu) \cdot \bar{u}\, dx \right.$$

$$\left. + \int_\Gamma \{(B\gamma u_+) \cdot (S\gamma \bar{u}_+) - (B\gamma u_-) \cdot (S\gamma \bar{u}_-)\}\, dS \right\}.$$

Moreover, there exist positive constants λ, ε *and* $c \ge 0$ *such that for* u *satisfying in addition* $B\gamma u_+ = B\gamma u_-$ *the Gårding inequality holds:*

$$Re\, \Phi(u, u) \ge \gamma \left\{ \|u\|_{H^m(\Omega_+)} + \|u\|_{H^m(\Omega_-)} \right\}^2 - c\|u\|^2_{H^{m-\epsilon}} .$$

Then the integral operator of the first kind A *in (1.28) is* $H^\alpha(\Gamma)$*- coercive on the boundary* Γ .

Here, the dualities are defined by the scales of Sobolev spaces on Γ ,

$$H_\pm^{\{\tau\}} = \prod_{j=0}^{m-1} H^{\alpha_j \pm \tau}(\Gamma) .$$

This approach is even valid for non-smooth Γ [27] and is closely related to the principle of virtual work.

For the boundary integral equations of the second kind (1.29), the corresponding variational form needs to be defined by

$$\underset{\sim}{\mathbf{a}}(w, v) := \sum_{j,k=0}^{m-1} \langle \underset{\sim}{A}_{jk} w_k, v_j \rangle_{H^{\mu_j - \mu_0}(\Gamma)} , \tag{1.32}$$

and the linear functional by

$$\underset{\sim}{\boldsymbol{\ell}}(v) := \sum_{j=0}^{m-1} \langle \underset{\sim}{\mathbf{F}}_j, v_j \rangle_{H^{\mu_j - \mu_0}(\Gamma)} .$$

The Hilbert spaces here are chosen as

$$H = H^\alpha(\Gamma) = \prod_{j=0}^{m-1} H^{\alpha_j}(\Gamma) \quad \text{where} \quad \alpha_j = \mu_j - \mu_0$$

and

$$H_{\pm}^{\{\tau\}} := \prod_{j=0}^{m-1} H^{\alpha_j \pm \tau}(\Gamma)\,,$$

correspondingly.

Note, that the simple L_2-scalar product on Γ in (1.32) can only be obtained for $m = 1$ (see Remark 3.15 in [32]). In general, for the operator $\underset{\sim}{A}$ in (1.29) and the associated bilinear form (1.32), only a weaker modification of the H^α-coercivity in the form of the so called *strong ellipticity* can be shown for sufficiently smooth Γ with the help of the principal symbol of $\underset{\sim}{A}$ in the framework of pseudo-differential operators (see [92], [73]). For non-smooth Γ, for $\underset{\sim}{A}$, strong ellipticity and, hence H^α-coercivity in general is not valid anymore [39]. Also, a simple and general relation between the energy of the two fields u_+, u_- and (1.32) is not known.

For the special example of the displacement or Dirichlet problem (1.18), (1.19) (with $g \neq 0$, too) we have the $(H^{-1/2}(\Gamma))^n$-coercive

Fredholm boundary integral equation of the first kind

for (1.28) with weakly singular kernel,

$$\mathbf{A}\mathbf{t}(x) = 2\int_{\Gamma\setminus\{x\}} \underset{\sim}{G}(x,\, y)\mathbf{t}(y)dS_y = \underset{\sim}{\mathbf{F}}(x) \quad \text{for } x \in \Gamma\,, \tag{1.33}$$

where

$$\underset{\sim}{G}(y,\, x) = c_o\left(\left(\gamma(x,\, y)\delta_{jk} + \frac{\lambda+\mu}{\lambda+3\mu}(x_j - y_j)(x_k - y_k)\,/\,|x-y|^n\right)\right), \tag{1.34}$$

$$c_0 = (\lambda + 3\mu)\,/\,(4\pi(n-1)(\lambda+2\mu)), \quad \gamma(x,\, y) = \begin{cases} -\log|x-y| & \text{for } n = 2,\\ 1\,/\,|x-y| & \text{for } n = 3,\end{cases}$$

and where δ_{jk} denotes the Kronecker symbol (see [51]). Here $\alpha = -\frac{1}{2}$ and

$$V = \left\{\mathbf{v} \in \left(H^{-\frac{1}{2}}(\Gamma)\right)^n \text{ with } \int_\Gamma \underset{\sim}{m}^T(x)\mathbf{v}(x)dS(x) = 0\right\}. \tag{1.35}$$

These equations are still valid for Lipschitz boundaries Γ [27].

For the same boundary value problem, we may take also the

Cauchy singular integral equation

of the second kind (1.29), [51],

$$\underset{\sim}{A}\mathbf{t}(x) := (I \pm K_c')\mathbf{t}(x) = \mathbf{t}(x) \mp 2\int_{\Gamma\setminus\{x\}} \left(T_{(x)}\underset{\sim}{G}(x,y)\right)\mathbf{t}(y)dS(y)$$

$$= \underset{\sim}{\mathbf{F}}(x) \quad \text{for } x \in \Gamma . \tag{1.36}$$

Here $\alpha = 0$ (and with the appropriate radiation condition in Ω_-)

$$V = \left\{\mathbf{t} \in (L_2(\Gamma))^n \quad \text{with} \quad \int_\Gamma \underset{\sim}{m}^T(x)\mathbf{t}(x)dS(x) = \mathbf{0}\right\} . \tag{1.37}$$

For smooth Γ, $\underset{\sim}{A}$ here is V-coercive [51].

Note, that for (1.33) we have $\mathbf{t} \in \left(H^{-\frac{1}{2}}(\Gamma)\right)^n$, which is the trace space of boundary tractions belonging to elastic fields of finite energy whereas for (1.36) we consider solutions $\mathbf{t} \in (L_2(\Gamma))^n$ which are smoother than necessary for a sensible physical setting. Moreover, for two-dimensional domains with corners, the V-coercivity in $(L_2(\Gamma))^2$ gets lost [39].

For the traction problem, also called the Neumann problem (1.18), (1.21) (with $g \neq 0$, too) the integral equation of the first kind (1.28) now takes the form

$$A\mathbf{u}(x) = -2T_{(x)}\int_{\Gamma\setminus\{x\}} \left(T_{(y)}\underset{\sim}{G}(x,y)\right)^T \mathbf{u}(y)\,dS(y) \tag{1.38}$$

This is a

hypersingular integral equation

with $\alpha = \frac{1}{2}$ and

$$V = \left\{\mathbf{v} \in \left(H^{\frac{1}{2}}(\Gamma)\right)^n \quad \text{with} \quad \int_\Gamma \underset{\sim}{m}^T(x)\mathbf{v}(x)dS(x) = \mathbf{0}\right\} , \tag{1.39}$$

which is V-coercive and which is also valid for Lipschitz boundaries (see [51] and [27]. The explicit kernel can be found in [15] pp. 187 ff. and 191.)

The corresponding integral equation of the second kind (1.29) again is a

Cauchy singular integral equation ,

$$\underset{\sim}{A}\mathbf{u}(x) := (I \mp K_c)\,\mathbf{u}(x) = \mathbf{u}(x) \mp \int_{\Gamma\setminus\{x\}} \left(T_{(y)}\underset{\sim}{G}(x,\,y)\right)^T \mathbf{u}(y)dS(y) \tag{1.40}$$

$$= \underset{\sim}{\mathbf{F}}(x), \quad \text{for } x \in \Gamma\,.$$

Again, $\alpha = 0$ and V is given by (1.37). As for (1.36), $\underset{\sim}{A}$ is V-coercive for smooth Γ and this property gets lost on domains with corners. Note that K_c and K_c' are adjoint to each other in $(L_2)^n$. Whereas in (1.38) $\mathbf{u} \in \left(H^{\frac{1}{2}}(\Gamma)\right)^n$ corresponds again to the trace space of elastic fields of finite energy, in (1.40) $\underset{\sim}{A} \in (L_2)^n$ is weaker allowing elastic fields of infinite energy which are physically irrelevant.

For general boundary integral equations underlying BEM as well as for the second kind equations (1.29), V-coercivity, or the slightly more general strong ellipticity of the associated bilinear forms must be shown via the calculus of pseudo-differential operators separately. For most stationary elliptic problems from applications it turns out that A on Γ defines a strongly elliptic pseudo-differential operator of order 2α and its strong ellipticity can be seen from positive definiteness of its principal symbol. The latter can be found from the Fourier transformed operator A (see [92] and references given there).

1.3 The variational form of unsymmetrically coupled boundary integral equations and domain equations [91]

Let us consider the general boundary value problem (1.23) in $\Omega = \Omega_1 \cup \Omega_2 \cap \Gamma_c$ where $\partial\Omega_1 = \Gamma_c$ is the coupling boundary, $\overline{\Omega_1} = \Omega_1 \cap \Gamma_c \subset \Omega$ and $\Omega_2 = \Omega\setminus\left(\overline{\Omega_1}\right)$ is the ring domain (see Figure 1).

Let us assume that the bilinear form (1.17) associated with L in Ω_1 admits the first Green formula with a pair of complementary boundary operators B_c, S_c on Γ_c providing (1.25) on Γ_c , let

$$a_1(u,\,v) := \sum_{|j|,|k|\le m} \int_{\Omega_1} a_{jk}(x)\left(\partial_x^k u\right)\left(\partial_x^j \overline{v}\right) dx$$

$$= \int_{\Omega_1} f\overline{v}\,dx + \int_{\Gamma_c} (S_c\gamma u)\cdot\left(\overline{B_c\gamma v}\right) dS \tag{1.41}$$

with v an arbitrary C^∞-test function in Ω_1 .

According to the subdomains Ω_1, Ω_2 we also split the desired solution u into

$$u_{(1)} = u|_{\Gamma_1} \quad \text{and} \quad u_{(2)} = u|_{\Gamma_2} \quad \text{along } \Gamma_c$$

and couple both projections by

$$\gamma_1 u_{(1)} = \gamma_2 u_{(2)} \text{ on } \Gamma_c \text{ or } B_c\gamma_1 u_{(1)} = B_c\gamma_2 u_{(2)}\,, \quad S_c\gamma_1 u_{(1)} = S_c\gamma_2 u_{(2)}\,.$$

Further we introduce

$$\Lambda = S_c\gamma u \quad \text{on} \quad \Gamma_c\,. \tag{1.42}$$

If Λ were given then $u \in H^m(\Omega_1)$ would satisfy

$$a_1(u,\, v) = (f,\, v)_{L_2(\Omega_1)} + \langle \Lambda,\, B_c\gamma v\rangle_{L_2(\Gamma_c)} \tag{1.43}$$

for all test functions $v \in C^\infty\left(\overline{\Omega_1}\right)$.

In Ω_2 we represent the solution via boundary potentials assuming for brevity that $f|_{\Omega_2} = 0$,

$$u(x) = \Big\{ \sum_{\substack{k+\ell+1\le 2m\\ k,\ell\ge 0}} \int_{\Gamma_c} \left(\left(\frac{\partial'}{\partial n(y)}\right)^k \underset{\sim}{G}(x,\, y)\right) L^c_{k+\ell+1}(y)\left(M_c^{-1}\right)_\ell \begin{pmatrix} B_c\gamma u\\ \Lambda\end{pmatrix} dS(y)$$

$$-\int_\Gamma \left(\left(\frac{\partial'}{\partial n(y)}\right)^k \underset{\sim}{G}(x,\, y)\right) L_{k+\ell+1}(y)\left(M^{-1}\right)_\ell \begin{pmatrix} g\\ w\end{pmatrix} dS(y)\Big\},\quad x \in \Omega_2\,. \tag{1.44}$$

For finding Λ on Γ_c and w on Γ we proceed as in Section 1.2 and get the two boundary integral equations

$$A_c\Lambda(x) := -\sum_{\substack{k+\ell+1\le 2m\\ k,\ell\ge 0}} 2B_c\gamma_2 \int_{\Gamma_c} \left(\left(\frac{\partial'}{\partial n(y)}\right)^k \underset{\sim}{G}(x,\, y)\right) L^c_{k+\ell+1}\left(M_c^{-1}(y)\right)_\ell \begin{pmatrix} 0\\ \Lambda\end{pmatrix} dS(y)$$

$$= -B_c\gamma u$$

$$+\left\{ \sum_{\substack{k+\ell+1\le 2m\\ k,\ell\ge 0}} 2B_c\gamma_2 \int_{\Gamma_c} \left(\left(\frac{\partial'}{\partial n(y)}\right)^k \underset{\sim}{G}(x,\, y)\right) L^c_{k+\ell+1}\left(M_c^{-1}\right)_\ell \begin{pmatrix} B_c\gamma u\\ 0\end{pmatrix} dS(y) - B_c\gamma u\right\}$$

$$+2\sum_{\substack{k+\ell+1\le 2m\\ k,\ell\ge 0}} B_c\gamma_2|_{\Gamma_c} \int_{\Gamma} \left(\left(\frac{\partial'}{\partial n(y)}\right)^k \underset{\sim}{G}(x,\, y)\right) L_{k+\ell+1}\left(M^{-1}\right)_\ell \begin{pmatrix} g\\ w\end{pmatrix} dS(y) \tag{1.45}$$

$$=: -B_c\gamma u + K_c\left(B_c\gamma u\right) + C_{12}w + K_{12}g \quad \text{for} \quad x \in \Gamma_c\,,$$

$$Aw(x) := -\sum_{\substack{k+\ell+1\le 2m\\ k,\ell\ge 0}} 2B\gamma_2|_\Gamma \int_\Gamma \left(\left(\frac{\partial'}{\partial n(y)}\right)^k \underset{\sim}{G}(x,\,y)\right) L_{k+\ell+1}\left(M^{-1}\right)_\ell \begin{pmatrix} 0\\ w\end{pmatrix} dS(y)$$

$$= -\sum_{\substack{k+\ell+1\le 2m\\ k,\ell\ge 0}} 2B\gamma_2|_\Gamma \int_{\Gamma_c} \left(\left(\frac{\partial'}{\partial n(y)}\right)^k \underset{\sim}{G}(x,\,y)\right) L^c_{k+\ell+1}y)\left(M_c^{-1}(y)\right)_\ell \begin{pmatrix} B_c\gamma u\\ \Lambda\end{pmatrix} dS(y)$$

$$+2g(x) + \sum_{\substack{k+\ell+1\le 2m\\ k,\ell\ge 0}} 2B\gamma_2|_\Gamma \int_\Gamma \left(\left(\frac{\partial'}{\partial n(y)}\right)^k \underset{\sim}{G}(x,\,y)\right) L_{k+\ell+1}\left(M^{-1}\right)_\ell \begin{pmatrix} g\\ 0\end{pmatrix} dS(y) \tag{1.46}$$

$$=: -C_{20}u - C_{21}\Lambda + g + K_{22}g \quad \text{for} \quad x \in \Gamma\,.$$

Now, if we write also (1.45) and (1.46) in variational form, the three equations (1.43), (1.45), (1.46) can be written as the following variational equations:

$$\textit{Find } \mathcal{U} = (u,\,\Lambda,\,w) \in \mathcal{H} = H^m(\Omega_1) \times H^{\alpha c}(\Gamma_c) \times H^{\alpha}(\Gamma) \textit{ such that}$$
$$\mathcal{A}(\mathcal{U},\,\mathcal{Y}) = \mathcal{L}(\mathcal{Y}) \quad \textit{for all} \quad \mathcal{Y} = (v,\,\phi,\,z) \in \mathcal{H}\,. \tag{1.47}$$

Here

$$\begin{aligned}\mathcal{A}(\mathcal{U},\,\mathcal{Y}) := a_1(u,\,v) \quad &+\langle\phi,\,A_c\Lambda\rangle_{L_2(\Gamma_c)} + \langle z,\,Aw\rangle_{L_2(\Gamma)}\\ &+\langle\phi,\,B_c\gamma u\rangle_{L_2(\Gamma_c)} - \langle\Lambda,\,B_c\gamma v\rangle_{L_2(\Gamma_c)} - \langle\phi,\,K_c\left(B_c\gamma u\right)\rangle_{L_2(\Gamma_c)}\\ &-\langle\phi,\,C_{12}w\rangle_{L_2(\Gamma_c)} + \langle z,\,C_{20}u\rangle_{L_2(\Gamma)} + \langle z,\,C_{21}\Lambda\rangle_{L_2(\Gamma)}\end{aligned} \tag{1.48}$$

and

$$\mathcal{L}(\mathcal{Y}) = (f,\,v)_{L_2(\Omega_1)} + \langle\phi,\,K_{12}g\rangle_{L_2(\Gamma)} + \langle z,\,g + K_{22}g\rangle_{L_2(\Gamma)}\,.$$

Remarks: In flow problems, scattering problems and many further applications, the domain $\Omega_2 = \Omega_-$ is an exterior domain up to infinity where the boundary condition on Γ is to be replaced by an appropriate radiation condition at infinity (see Figure 2). Here we need to incorporate the radiation condition into (1.44), and equation (1.46) can be scipped whereas (1.45) simplifies with $C_{12} = 0$ and $K_{12} = 0$. Correspondingly, the third, seventh, eighth and ninth terms in (1.48) are to be cancelled.

In general, however, Ω_1 will become a ring domain and (1.41) will be modified according to additional conditions on the additional boundary parts of Ω_1 not belonging to Γ_c . Since the principal steps of our analysis will remain valid we omit the details of the corresponding modification.

In (1.45), on the coupling boundary Γ_c , we have chosen the first kind integral equation corresponding to (1.28). In many applications also the second kind equation (1.29) for Λ on Γ_c is chosen instead of (1.45). Correspondingly, one often finds (1.29) on Γ for w instead of (1.46). From the existence of the solution to the original boundary value problem (1.23) in Ω it follows that the above variational problem (1.47) also has a unique solution provided (1.23) is uniquely solvable. However, $\mathcal{A}$ becomes H-coercive only if the operator $K_c : H^m(\Omega) \to H^{-\alpha_c}(\Gamma_c)$ is compact which is true only for a special class of problems, as will be seen in Section 3.

In our special examples of the Dirichlet and Neumann problems in elasticity we have the following special operators.

Example: Coupled formulation in elasticity

$$a_1(\mathbf{u}, \mathbf{v}) \text{ is given by (1.20) for } \mathbf{u}, \mathbf{v} \in \left(H^1(\overline{\Omega}_1)\right)^n ,$$

$$\begin{aligned} B_c\gamma\mathbf{v} &:= \mathbf{v}|_{\Gamma_c} \text{ and } \boldsymbol{\Lambda} = T(\mathbf{u})|_{\Gamma_c} \text{ with (1.30)} ,\\ A_c\boldsymbol{\Lambda} &:= 2\int_{\Gamma_c\setminus\{x\}} \underset{\sim}{G}(x, y)\boldsymbol{\Lambda}(y)dS(y) \text{ as in (1.33)},\\ K_c\mathbf{u} &:= -2\int_{\Gamma_c\setminus\{x\}} \left(T_{(y)}(\underset{\sim}{G}(x, y))\right)^T \mathbf{u}(y)dS(y) . \end{aligned} \tag{1.49}$$

	Dirichlet problem (1.19)	Neumann problem (1.21)
$A\mathbf{w} =$	given by (1.33) for $\mathbf{w} \in \left(H^{-\frac{1}{2}}(\Gamma)\right)^n$	given by (1.38) for $\mathbf{w} \in \left(H^{\frac{1}{2}}(\Gamma)\right)^n$
$C_{12}\mathbf{w}$	$-2\int_\Gamma \underset{\sim}{G}(x, y)\mathbf{w}(y)dS(y)$, $x \in \Gamma_c$	$2\int_\Gamma \left(T_{(y)}(\underset{\sim}{G}(x, y))\right)^T \mathbf{w}(y)dS(y), x \in \Gamma_c$
$C_{20}\mathbf{u} =$	$-2\int_{\Gamma_c} \left(T_{(y)}(\underset{\sim}{G}(x, y))\right)^T \mathbf{u}(y)dS(y)$, $x \in \Gamma$	$2T_{(x)}\int_{\Gamma_c} \left(T_{(y)}(\underset{\sim}{G}(x, y))\right)^T \mathbf{u}(y)dS(y)$, $x \in \Gamma$
$C_{21}\boldsymbol{\Lambda} =$	$2\int_{\Gamma_c} \underset{\sim}{G}(x, y)\boldsymbol{\Lambda}(y)dS(y)$, $x \in \Gamma$	$-2\int_{\Gamma_c} T_{(x)}(\underset{\sim}{G}(x, y))\boldsymbol{\Lambda}(y)dS(y)$, $x \in \Gamma$

The operator (1.49) is a classical Cauchy principal value singular integral operator with Cauchy kernel in case $n = 2$ and Giraud-Mikhlin kernel in case $n = 3$ [66], p. 392 ff. Hence, in the important case of elasticity K_c is <u>not</u> compact and therefore the coupled system does not provide the decisive V-coercivity (1.7) for $\mathcal{A}$ anymore.

1.4 The variational form of symmetrically coupled boundary integral and domain equations [25]

Although the variational formulation (1.47), (1.48) seems to be rather natural in the case of elasticity, it does not correspond to one of the basic mechanical principles for the complete fields in $\Omega_1 \cup \Omega_2$. Recently, Costabel [25],[26], Polizzotto [74], [76] and Mang et. al. [56] independently proposed a variational formulation, which corresponds to the Hellinger-Reissner principle of the complete fields and which will also provide convergence and error analysis for the corresponding BEM-FEM treatment. (In [41] this approach again is exemplified for potential and elasticity problems). Moreover, this method also generalizes to transmission problems [28], [29], [30] and to coupling of fluids and elastic bodies [46].

In Section 1.3 we used on the coupling boundary Γ_c only equation (1.45) corresponding to the boundary integral equation of the first kind (1.28). Here we also use (1.29) appropriately which reads on Γ_c as

$$\Lambda(x) + K_c'\Lambda(x) + D_c(B_c\gamma u) = C_{12}w + K_{12}g(x), x \in \Gamma_c\,, \tag{1.50}$$

where

$$K_c'\Lambda = \left\{ \sum_{\substack{k+\ell+1\leq 2m\\ k,\ell\geq 0}} 2S_c\gamma_2 \int_{\Gamma_c\setminus\{x\}} \left(\left(\frac{\partial'}{\partial n(y)}\right)^k \underset{\sim}{G}(x,\,y)\right) L^c_{k+\ell+1}\left(M_c^{-1}\right)_\ell \begin{pmatrix} o \\ \Lambda \end{pmatrix} dS(y) - \Lambda(x) \right\} \tag{1.51}$$

turns out to be the adjoint operator to K_c in (1.45), and where D_c is given by

$$D_c(B_c\gamma u) := \tag{1.52}$$

$$-2 \sum_{\substack{k+\ell+1\leq 2m\\ k,\ell\geq 0}} S_c\gamma_2 \int_{\Gamma_c\setminus\{x\}} \left(\left(\frac{\partial'}{\partial n(y)}\right)^k \underset{\sim}{G}(x,\,y)\right) L^c_{k+\ell+1}\left(M_c^{-1}\right)_\ell \begin{pmatrix} B_c\gamma u \\ o \end{pmatrix} dS(y)$$

corresponding to the first kind boundary integral operator of the complementary boundary condition $S_c\gamma$ on Γ_c. The remaining operators for $x \in \Gamma_c$ are given by

$$C'_{12}w(x) := - \sum_{\substack{k+\ell+1\leq 2m \\ k,\ell\geq 0}} 2S_c\gamma_2 \int_\Gamma \left(\left(\frac{\partial'}{\partial n(y)} \right)^k \underset{\sim}{G}(x,\, y) \right) L_{k+\ell+1} \left(M^{-1} \right)_\ell \begin{pmatrix} o \\ w \end{pmatrix} dS(y)\,,$$

$$K'_{12}g(x) := - \sum_{\substack{k+\ell+1\leq 2m \\ k,\ell\geq 0}} 2S_c\gamma_2 \int_\Gamma \left(\left(\frac{\partial'}{\partial n(y)} \right)^k \underset{\sim}{G}(x,\, y) \right) L_{k+\ell+1} \left(M^{-1} \right)_\ell \begin{pmatrix} g \\ o \end{pmatrix} dS(y)\,.$$

Note that in the last two operators the integration is taken over Γ whereas the observation point $x \in \Gamma_c$ and here Γ and Γ_c are separated. Therefore, these operators are *compact*.

Now the variational formulation is given by (1.47):

$$\begin{aligned} &\textit{Find } \mathcal{U} = (u,\, \Lambda,\, w) \in \mathcal{H} = H^m(\Omega_1) \times H^{\alpha_c}(\Gamma_c) \times H^{\alpha}(\Gamma) \\ &\textit{such that } \mathcal{A}(\mathcal{U},\, \mathcal{Y}) = \mathcal{L}(\mathcal{Y}) \textit{ for all } \mathcal{Y} = (v,\, \phi,\, z) \in \mathcal{H}\,. \end{aligned} \tag{1.53}$$

For the symmetric formulation we define

$$\begin{aligned} \mathcal{A}(\mathcal{U},\mathcal{Y}) := 2a_1(u,\, v) \;+ \int_{\Gamma_c} \{ &D_c(B_c\gamma u)\cdot(B_c\gamma v) - (A_c\Lambda)\cdot\varphi \\ &-\Lambda\cdot(B_c\gamma v) \;-\; (B_c\gamma u)\cdot\varphi \\ &+(K'_c\Lambda)\cdot(B_c\gamma v) + (K_c(B_c\gamma u))\cdot\varphi \\ &+(C'_{20}w)\cdot(B_c\gamma v) + (C_{12}w)\cdot\varphi\}ds \\ &+\int_\Gamma \{(Aw)\cdot z + (C_{20}u)\cdot z + (C_{21}\Lambda\cdot z\}dS \end{aligned} \tag{1.54}$$

and

$$\mathcal{L}(\mathcal{Y}) = 2(f,v)_{L_2(\Omega_1)} - \langle K'_{12}g,(B_c\gamma u)\rangle_{L_2(\Gamma_c)} - \langle K_{12}g,\varphi\rangle_{L_2(\Gamma_c)} + 2\langle g + K_{22}g, z\rangle_{L_2(\Gamma_c)}\,.$$

Note that (1.54) is symmetric with

$$\mathcal{A}(\mathcal{U},\, \mathcal{Y}) = \mathcal{A}(\mathcal{Y},\, \mathcal{U})$$

which corresponds to the symmetry of virtual work. However, the bilinear equations (1.54) are equivalent to the system

$$\begin{aligned} 2a_1(u,v) \; &+\langle\{D_c(B_c\gamma u) - \Lambda + K_c'\Lambda + C_{20}'w\}\,,\, B_c\gamma v\rangle_{L_2(\Gamma_c)} \\ &+\langle\{Aw + C_{20}u + C_{21}\Lambda\}\,, z\rangle_{L_2(\Gamma)} \\ &= 2(f,v)_{L_2(\Omega_1)} - \langle K_{12}'g, B_c\gamma v\rangle_{L_2(\Gamma_c)} + 2\langle g + K_{22}g, z\rangle_{L_2(\Gamma)}, \end{aligned}$$

$$\langle\{A_c\Lambda + (B_c\gamma u) - K_2(B_c\gamma u) + C_{12}w\}\,,\varphi\rangle_{L_2(\Gamma_c)} = \langle K_{12}g,\varphi\rangle_{L_2(\Gamma_c)} \tag{1.55}$$

which can also be written as

$$\mathcal{A}(\mathcal{U},\mathcal{Y}) = \mathcal{L}(\mathcal{Y}) \quad \text{for all} \quad \mathcal{Y}\in\mathcal{H} \tag{1.56}$$

with a different bilinear form

$$\begin{aligned} \mathcal{A}(\mathcal{U},\mathcal{Y}) = 2a_1(u,v) \; &+\langle\big\{D_c(B_c\gamma u) - \Lambda + K_c'\Lambda + C_{20}'w\big\}\,, B_c\gamma v\rangle_{L_2(\Gamma_c)} \\ &+\langle A_c\Lambda + B_c\gamma u - K_2\gamma u + C_{12}w,\varphi\rangle_{L_2(\Gamma_c)} \\ &-\langle\{Aw + C_{20}u + C_{21}\Lambda\}\,, z\rangle_{L_2(\Gamma)} \end{aligned} \tag{1.57}$$

and

$$\begin{aligned} \mathcal{L}(\mathcal{Y}) \; &= 2(f,v)_{L_2(\Omega_1)} + \langle K_{12}g,\varphi\rangle_{L_2(\Gamma_c)} \\ &-\langle K_{12}'g, (B_c\gamma v)\rangle_{L_2(\Gamma_c)} + 2\langle(g + K_{22}g), z\rangle_{L_2(\Gamma)} \end{aligned}$$

(For this formulation see also [46].)

Theorem 1.5 [25]: *If the assumptions of Theorem 1.4 are satisfied, then the bilinear form* $\mathcal{A}(\mathcal{U},\mathcal{W})$ (1.57) *is* $\mathcal{H}$*-coercive, i.e. there exists* $\gamma_o > 0$ *and a compact bilinear form* $\mathcal{C}(\mathcal{U},\mathcal{W})$ *on* $\mathcal{H}\times\mathcal{H}$ *such that for all* $\mathcal{Y}\in\mathcal{H}$ *there holds*

$$Re(\mathcal{A}(\mathcal{Y},\,\mathcal{Y}) + \mathcal{C}(\mathcal{Y},\,\mathcal{Y})) \geq \gamma_0\|\mathcal{Y}\|^2_{\mathcal{H}}.$$

This Theorem follows immediately from the individual Gårding inequalities of the four terms in (1.57)

$$2a_1(v,v),\;\; \langle D_c(B_c\gamma v),(B_c\gamma v)\rangle_{L_2(\Gamma_c)},\;\; \langle A_c\varphi,\varphi\rangle_{L_2(\Gamma_c)},\;\; \langle Az,z\rangle_{L_2(\Gamma)}$$

on $H^1(\Omega_1)$, $H^1(\Omega_1)$, $H^{\alpha_c}(\Gamma_c)$, $H^{\alpha}(\Gamma)$, respectively, and the compactness of C'_{20}, C_{20}, C_{12} and C_{21} which have C^{∞}-kernels.

Remarks:

i) Feng Kang and Yu [37], [38], [94], [95], [96] use in (1.28) and (1.50) the Green function associated with the Dirchlet condition $B_c\lambda u = \gamma u$ and the boundary Γ_c. Then Λ can be expressed explicitly whereas the corresponding coupling equations correspond to (1.43).

ii) Grannell [40] gives a mixed variational formulation with a_1 in Ω_1 and boundary potentials in Ω_2 generated by densities on Γ_c. If these are chosen as in (1.44) then (1.53) and the mixed formulation coincide. With Brezzi's coerciveness proof for mixed methods Grannell also again obtains Theorem 1.5.

iii) Since for the integral equation of the first kind (1.28) and second order differential equations, i.e. $m = 1$, the $H^{\alpha}(\Gamma)$-coercivity, Theorem 1.4 remains valid for Lipschitz domains , the results in [24], [27] provide the Gårding inequality in Theorem 1.5 also for the coupling of substructures as in Figure 3 [26]. However, the shift theorem now holds only for a small interval of Sobolev space indices.

2 Survey on the Asymptotic Error Estimates for FEM and for BEM

Finite element domain methods nowadays are very well established, and most of their analyses has been rather completely developed, too. Here we shall collect some basic results concerning conforming methods for V-coercive problems. Most of them can be carried over to Galerkin boundary element methods. However, the more popular collocation boundary element methods require a more elaborate analysis which is by no means sufficiently complete yet except in the two-dimensional case.

2.1 Conforming finite element domain methods [8]

Here let us consider elliptic boundary value problems in variational form as in Example 1.1. Following [8], let us assume that we are dealing with a conforming $S_h^{k,r}$-system of FEM spaces, where k, $r \in \mathbf{N}_o$, which provides the

Approximation property [8] Theorem 4.12.: *Let* $\ell, s \in \mathbf{R}$ *satisfy* $\ell \leq r < k$,

$\ell \le s \le k$. *Then to every* $u \in H^s(\Omega)$ *there exists a sequence* $u_h \in S_h^{k,r}$ *such that*

$$\|u - u_h\|_{H^\ell(\Omega)} \le c_{\ell,s} h^{s-\ell} \|u\|_{H^s(\Omega)}. \tag{2.1}$$

These elements are defined subject to corresponding families of triangulations or quadrangulations of the given domain Ω . Here, h denotes the parameter of mesh-width of the family. E.g. (2.1) is satisfied if the elements are piecewise polynomials containing complete polynomials of order $k-1$ on each triangle or quadrangle, respectively. r denotes the smoothness, i.e. $S_h^{k,r} \subset H^r(\Omega)$. Note, that (2.1) implies for $\ell \le r < k, \ell \le s \le k$ the approximation property

$$E(H_+^{\{s-\ell\}};\ S_h^{k,r}) \ \le\ ch^{s-\ell} \ \text{ in } H^\ell(\Omega), \tag{2.2}$$

if $H_+^{\{0\}} = H^\ell(\Omega)$ and $H_+^{\{s-\ell\}} = H^s(\Omega)$ in (1.13).

Helfrich showed in [43] even the stronger approximation property: *There exists a family* u_h *satisfying* (2.1) *which is independent of* ℓ .

For conforming FEM we set

$$V_h = S_h^{k,r} \bigcap V \subset H^m(\Omega), \tag{2.3}$$

which implies $m \le r$. Moreover, the condition (2.3) also implies that we require V_h to satisfy the homogeneous boundary conditions associated with V . The latter yield rather tedious precautions for the FEM at curved boundaries. The standard FEM then reads, *solve*

$$\sum_{j=1}^{M} a(\mu_j, \mu_\ell)\gamma_j = \ell(\mu_\ell), \qquad \ell = 1, \ldots, M \tag{2.4}$$

for the coefficients γ_j *defining the approximate solution*

$$u_h = \sum_{j=1}^{M} \gamma_j \mu_j$$

according to (1.6) and $\mathcal{T}_h = V_h$. With the above approximation properties we obtain

Theorem 2.1: *Let* a *be* V*-coercive, let* (1.2) *associated with* (1.15) - (1.17) *be uniquely solvable and let the shift Theorem* (1.12) *be valid with* $H_\pm^{\{\tau\}} = H^{m \pm \tau}(\Omega)$. *Then there exists* $h_o > 0$ *such that* (2.4) *is uniquely solvable for every* $0 < h \le h_o$ *and the* FEM *Galerkin method converges with optimal order,*

$$\|u - u_h\|_{H^\ell(\Omega)} \le ch^{s-\ell} \|u\|_{H^s(\Omega)}, \tag{2.5}$$

where $m-k \le \ell \le m \le s \le k,\ m \le r < k$.

Proof: The proof is straightforward by taking $V = H = H^m$, $\tau = 0$ in Theorem 1.2 to find $h_o > 0$ and unique solvability of (2.4) for $0 < h \le h_o$ Moreover, (1.4) is satisfied with $V_h = S_h^{k,r} \cap V = \mathcal{T}_h$. Hence, Theorem 1.1 with (2.1) and $H^{\{o\}} = H^m$ yields (2.5) with $\ell = m$. Then take $H = H^m(\Omega)$, $H_-^{\{m-\ell\}} = H^\ell(\Omega)$ and $H_+^{\{m-\ell\}} = H^{2m-\ell}(\Omega)$ in Theorem 1.3.. Then, (2.1) and (2.2) yield (2.5). □

Very often the family of triangulations defining the finite element spaces satisfies additional regularity properties which imply that there holds the inverse assumption: *For* $s \le \ell \le r < k$ *there holds for all* $v_h \in S_h^{k,r}$,

$$\|v_h\|_{H^\ell(\Omega)} \le c h^{s-\ell} \|v_h\|_{H^s(\Omega)}. \tag{2.6}$$

Corollary 2.2: *If, in addition to the assumptions in Theorem 2.1 also the inverse assumption* (2.6) *holds then the optimal order estimate* (2.5) *is valid for* $m \le \ell \le s \le k$ *and* $\ell \le r$, *too.*

2.2 Conforming boundary element Galerkin methods

In this section we follow [50], [69] [85]. For the BEM let us assume that the boundary Γ is given by a finite number of parameter representations, each defined on a corresponding domain in $\mathbf{R}^{n-1}$, $n = 2$ or 3 . On every parameter domain we now define a family of triangulations or quadrangulations, respectively, and subject to those we are choosing an $S_h^{k,r}$-system of finite elements (as in Section 2.1 on Ω , respectively). The parameter representation of Γ maps each of the polygonal nets from the parameter domains onto a corresponding curved net on Γ . Correspondingly, the finite elements are transplanted from the parameter domains onto Γ . For this construction and the corresponding computations, one needs the parameter representations explicitly, which is desirable for $n = 2$ - here Γ is just a curve . In general, however, and in practical computations, Γ is also often replaced by Lagrangian spline or finite element interpolations Γ_h . Then the trial and test spaces on Γ_h are defined by the compositions of the finite elements in the parameter domain with those representing Γ_h i.e. by generalized isoparametric elements as in shell theory [22], [68].

2.2.1 Two-dimensional problems

If Γ is one simple closed smooth curve then we always can choose a global regular 1 -periodic parameter representation,

$$\Gamma : y = Z(t) = Z(t+1) \quad \textit{for} \quad t \in \mathbf{R}. \tag{2.7}$$

Now all functions on Γ may be seen as 1-periodic functions depending on t . Correspondingly, the representation (2.7) generates boundary integral equations with 1-periodic functions. If Γ consists of a finite system of mutually nonintersecting smooth closed curves, then a representation like (2.7) holds for every boundary component, correspondingly, and every function on Γ can be seen as a vector valued 1-periodic function whose j-th component would correspond to the component Γ_j of Γ . Eventually, we will find a system of boundary integral equations with everything 1-periodic.

On the unit-interval we choose a family of partitions,

$$0 = t_o < t_1 < \ldots < t_N = 1,$$

which is extended to $\mathbf{R}$ via 1-periodicity, $h := max\{t_{\ell+1} - t_\ell\}$. Here $S_h^{k,r}$ is the family of 1-periodic piecewise polynomials of order $k-1$ subject to the above partitions which are $r-1$-times continuously differentiable. The basis $\{\mu_j\}$ is chosen with smallest possible supports (B-splines). With (2.7) these splines are defined on Γ , too. (see Figure 4.)

If Γ is also to be approximated then we take a $S_h^{\kappa-1,1}$ Lagrangian system with

$$\Gamma_h : y = Z_h(t), \quad \text{where} \quad Z_h(t_\ell) = Z(t_\ell) \, , \; \ell = 1, \ldots, N. \tag{2.8}$$

As in [68] we require

$$r \leq k-1 \leq \kappa \quad \text{and} \quad 0 \leq r \leq 1. \tag{2.9}$$

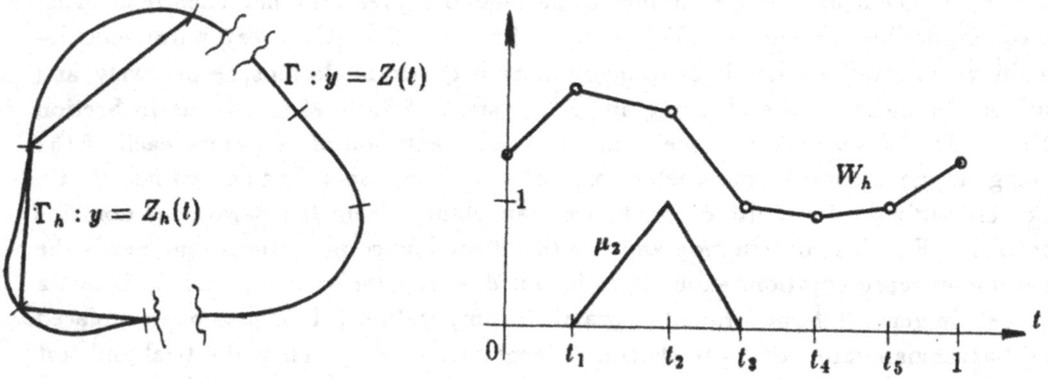

Figure 4: Piecewise linear splines without and with polygonal approximation.

2.2.2 Three-dimensional problems

For $n = 3$, Γ is a sufficiently smooth closed regular two-dimensional surface which can be represented by a finite covering of locally regular parameter representations

$$y = Z_j(t) \quad \text{for} \quad t \in U_j \subset \mathbf{R}^2, \; j = 1, \ldots, L,$$

where the boundaries ∂U_j can be chosen to be polygonal. Then each of the U_j will be partitioned by a family of triangulations Π_h or of quadrangulations on which we define $S_h^{k,r}$ to be a k,r-system on Γ .

If also Γ is to be approximated then we take a $S_h^{\kappa+1,1}$ continuous Lagrangian system on Π_h with respect to a unisolvent point set in every triangle such that the interpolation will be unique. Again,

$$\Gamma_h : y = Z_{jh}(t), \quad Z_{jh} \in S_h^{\kappa+1,1}, \tag{2.10}$$

defines a piecewise smooth continuous interpolating surface of Γ . Composing the finite elements with (2.10) we find the proposed "shell" elements.

2.3 Asymptotic convergence

The above boundary elements have approximation properties on the boundary Γ corresponding to (2.1). However, for $n = 2$ we deal with one-dimensional periodic spline functions for which (2.1) on Γ holds in a slightly larger scale of Sobolev spaces.

Approximation property:

For $-\infty < \ell \leq s \leq k$ *there holds*

$$\|u - u_h\|_{H^\ell(\Gamma)} \leq c_{\ell,s} h^{s-\ell} \|u\|_{H^s(\Gamma)}$$

provided for

$$\begin{aligned} n = 2: \ & \ell < r + 1/2 && \textit{without and}. \\ & \ell < \min\{r, 1\} + 1/2 && \textit{with boundary approximation}, \\ n = 3: \ & \ell \leq r && \textit{without and} \\ & \ell \leq \min\{r, 1\} && \textit{with boundary approximation}. \end{aligned} \tag{2.11}$$

Theorem 2.3: *Let* A *be a* $H^\alpha(\Gamma)$*-coercive and injective pseudo-differential operator of order* 2α *on* Γ *. Then there exists* $h_o > 0$ *such that for all h with* $0 < h \leq h_o$ *the* BEM-*Galerkin equations are uniquely solvable and the corresponding solutions converge with optimal order,*

$$\|u - u_h\|_{H^\ell(\Gamma)} \leq c h^{s-\ell} \|u\|_{H^s(\Gamma)} \tag{2.12}$$

provided $2\alpha - k \le \ell \le \alpha \le s \le k$ *and for*

$$n = 2: \quad \alpha < r + 1/2 \quad \textit{without or}$$
$$\alpha < \min\{r, 1\} + 1/2 \quad \textit{with boundary approximation},$$

$$n = 3: \quad \alpha \le r \quad \textit{without or}$$
$$\alpha \le \min\{r, 1\} \quad \textit{with boundary approximation}.$$

Proof: For the approximation of the boundary integral equations on the boundary Γ we proceed here similar to the proof for the finite element domain method in Section 2.1. Here, however, function spaces and the operators are to be considered on Γ. There now all the conditions of Theorem 1.2 are satisfied with $H = H^{\alpha}(\Gamma)$, there exists $h_o > 0$ such that for all h with $0 < h \le h_o$, the Galerkin equations are uniquely solvable. With the approximation property and (2.11), Theorem 1.1 can also be applied yielding the proposed estimate (2.12) with $\ell = \alpha$. Then, with

$$H_+^{\{\tau\}} = H^{\alpha+\tau}(\Gamma),\ H_-^{\{\tau\}} = H^{\alpha-\tau}(\Gamma)$$

and the shift theorem for psdeudo-differential operators [89] pp. 60ff., the Aubin-Nitsche duality argument, i.e. Theorem 1.3 can be applied and yields the remaining cases with of (2.12) $2\alpha - k \le \ell \le \alpha$. □

As for the FEM, one has often regular boundary element families available providing the

Inverse assumption: *For* $s \le \ell < r + 1/2 < k$ *in case* $n = 2$ *or* $s \le \ell \le r < k$ *in case* $n = 3$ *there holds for all* $v_h \in S_h^{k,r}(\Gamma)$,

$$\|v_h\|_{H^\ell(\Gamma)} \le ch^{s-\ell}\|v_h\|_{H^s(\Gamma)}. \tag{2.13}$$

Corollary 2.4: *If, in addition to the assumptions in* Theorem 2.3, *also the inverse assumption* (2.13) *holds then the optimal order estimate* (2.12) *is valid for* $\alpha \le \ell \le s \le k$ *and* ℓ *satisfying* (2.11), *too.*

2.4 Naive boundary element collocation methods for two-dimensional problems.

Almost all codes of engineering BEM [61] are based on collocation methods. Here the test space T_h in (1.3) is the span of Dirac functionals associated with M appropriately chosen collocation points $x_1, \ldots, x_M \in \Gamma$ or Γ_h, respectively. With these $v \in T_h$ in (1.29), the BEM collocation equations read as follows:

Find $\gamma_1, \ldots, \gamma_M$ *by solving the quadratic system of linear equations*

$$\sum_{j=1}^{M} (A\mu_j)(x_\ell)\gamma_j = F(x_\ell), \quad \ell = 1, \ldots, M. \tag{2.14}$$

In the naive collocation method, the points x_ℓ are chosen as an unisolvent set for the interpolation in $S_h^{k,r}$. Error estimates for the collocation of general boundary integral equations in BEM, however, yet are available only in the case $n = 2$ and with the "smoothest" splines, where

$$k = r + 1.$$

The naive choice of collocation points for the polynomial splines of degree r means on the curve Γ

$$x_\ell = \begin{cases} Z(t_\ell) & \text{for } r \text{ odd,} \\ Z((t_{\ell+1} + t_\ell)/2) & \text{for } r \text{ even,} \quad \ell = 1, \ldots, M = N. \end{cases} \tag{2.15}$$

If also Γ is approximated by Γ_h then choose Z_h instead of Z in (2.15).

Here, one has the following optimal order convergence results.

Theorem 2.5: *Assume that to A there exists a smooth matrix valued function* $\Theta(x)$ *on* Γ *such that* ΘA *is a* $H^\alpha(\Gamma)$*-coercive injective pseudo-differential operator of order* 2α *on* Γ . *Then, for the naive collocation there exists* $h_o > 0$ *such that for all h with* $0 < h \leq h_o$ *the collocation equations* (2.14) *with* (2.15) *are uniquely solvable and the corresponding solutions satisfy*

$$\|u - u_h\|_{H^\ell(\Gamma)} \leq c h^{s-\ell} \|u\|_{H^s(\Gamma)} \tag{2.16}$$

provided one of the following conditions is satisfied,

i) r *is odd,* $2\alpha < r$ *and* $2\alpha \leq \ell \leq \alpha + k/2 \leq s \leq k$ *(see* [1]*), or* (2.17)

ii) the family of partitions is quasiuniform, $2\alpha < r, 2\alpha \leq \ell < r + 1/2$, $\ell \leq s \leq k$, $\alpha + k/2 \leq s$ *(see* [2], [79] *and the references given there)* (2.18)
or

iii) the family of partitions is quasiuniform, r *is even* $r \leq 2\alpha < r + 1/2$ *and* $\ell < s$ *or* $2\alpha + 1/2 < \ell = s$ *(see* [2]*).* (2.19)

Remarks to the proof: For all three cases, the proof can be reduced to showing (1.4) in appropriate spaces. In the cases ii) and iii) this reduction hinges on explicit Fourier series expansions and, hence, is omitted here. Note that multiplication of the collocation equations (2.14) by the regular matrices $\Theta(x_\ell)$ on both sides will not change the method. Thus, we can restrict the proof to the case $\Theta(x) = 1$.

In case i) r is odd. then $(r+1)/2 = k/2$ -times integration by parts of (1.29) implies that the collocation equations (2.14) at the break points are equivalent to

$$\langle (A - (J - J_\Delta)\,A)\,(u_h - u)\,,\, v_h\rangle_{H^{k/2}(\Gamma)} = 0 \tag{2.20}$$

for all $v_h \in V_h = S_h^{k,r}(\Gamma)$, where

$$Jw := \int_0^1 w(t)dt \text{ and } J_\Delta w := \sum_{\ell=1}^{M} w\,(t_\ell)\,(t_{\ell+1} - t_{\ell-1})\,/2\,. \tag{2.21}$$

Hence, with

$$\mathbf{a}(u,\, v) := \langle Au,\, v\rangle_{H^{k/2}(\Gamma)} \quad \text{on } V = H = H_{\pm}^{\{o\}} = H^{\alpha+k/2}(\Gamma) \tag{2.22}$$

we can write (2.20) as the modified variational problem:

Find $u_h \in V_h$ *such that*

$$\mathbf{a}\,(u_h,\, v_h) = \mathbf{a}\,(u + A^{-1}\,(J - J_\Delta)\,A\,(u_h - u)\,, v_h) \quad \textit{for all } v_h \in V_h\,.$$

$H^\alpha(\Gamma)$-coerciveness of the pseudo-differential operator A implies the V-coerciveness of the above bilinear form $\mathbf{a}$ defined by (2.22). Moreover, with

$$H_{\pm}^{\{\tau\}} = H^{\alpha \pm \tau + k/2}(\Gamma)$$

all the assumptions of Theorems 1.2 and 1.3 are fulfilled by implying with (2.11) the estimate

$$\|u_h - u - A^{-1}\,(J - J_\Delta)\,A\,(u_h - u)\,\|_{H^\ell(\Gamma)}$$

$$\leq ch^{s-\ell}\left\{\|u\|_{H^s(\Gamma)} + \|A^{-1}\,(J - J_\Delta)\,A\,(u_h - u)\,\|_{H^s(\Gamma)}\right\}.$$

Since

$$\|A^{-1}\,(J - J_\Delta)\,A\,(u_h - u)\,\|_{H^s(\Gamma)} \leq c|\,(J - J_\Delta)\,A\,(u_h - u)\,|$$

we find

$$\|u_h - u\|_{H^\ell(\Gamma)} - c|\,(J - J_\Delta)\,A\,(u_h - u)\,| \leq ch^{s-\ell}\|u\|_{H^s(\Gamma)}\,. \tag{2.23}$$

Now we set first $\ell = 2\alpha + 1 \le s$ and use the convergence of the trapezoidal rule,

$$\begin{aligned} |(J - J_\Delta)A(u_h - u)| &\le ch\|A(u_h - u)\|_{H^1(\Gamma)} \\ &\le ch\|u_h - u\|_{H^{2\alpha+1}(\Gamma)}. \end{aligned} \tag{2.24}$$

Then, for h small enough the desired estimate (2.16) follows with $\ell = 2\alpha + 1 \le s$ from

$$\|u_h - u\|_{H^{2\alpha+1}(\Gamma)}(1 - ch) \le ch^{s-2\alpha-1}\|u\|_{H^s(\Gamma)}. \tag{2.25}$$

For the remaining case $2\alpha \le \ell < 2\alpha + 1$ we again get (2.23), and with (2.24) and (2.25) we find

$$\begin{aligned} \|u_h - u\|_{H^\ell(\Gamma)} &\le ch^{s-\ell}\|u\|_{H^s(\Gamma)} + ch\|u_h - u\|_{H^{2\alpha+1}(\Gamma)} \\ &\le \left(ch^{s-\ell} + ch^{s-2\alpha}\right)\|u\|_{H^s(\Gamma)} \end{aligned}$$

which implies (2.16) due to $s - 2\alpha \ge s - \ell$. □

3 The Coupling of BEM and FEM in the Nonsymmetric Form

As we have seen in Section 1.3, the coupled domain and boundary integral methods can be formulated as a system with one variational equation (1.43) in Ω_1 which is coupled with the two boundary integral equations (1.45) on Γ_c and (1.46) on Γ . This type of coupling is mostly used in the applications and only recently one finds the symmetric coupling in variational form which will be analysed in the next chapter. If the nonsymmetric coupling equations are written in variational form then the whole system becomes the variational equation (1.47).

If the boundary integral equations (1.45), (1.46) can be solved exactly then a finite element method for (1.43) can be seen as a variational method in Ω_1 alone whose trial and test functions are extended via (1.44) by corresponding solutions of (1.45), (1.46). The exact solution of the boundary integral equation on Γ_c is e. g. possible if $\underset{\sim}{G}(x, y)$ is chosen as the Green function of L subject to the boundary conditions $B_c\gamma\underset{\sim}{G}|_{x\in\Gamma_c} = 0$ which is practical for special domains Ω_1 like the unit ball in $\mathbf{R}^n$ or a half space. Methods of this type have been used in [21], [65], [84] and also analysed in [37], [38], [94], [95], [96]. This method still converges if the solution of the boundary integral equations on the trial space is approximated far more accurately than the FEM variational method in Ω_1 . For equation (1.45) on Γ_c e. g. this can be accomplished by using a spectral method which can be found in [66], [14], [36], [59] or global approximation as in [97]. If we use BEM on Γ_c , this idea would correspond to a faster mesh refinement for the BEM than for the FEM in Ω_1 . In Sections 3.2 and 3.3 we will see that such a method will indeed converge for the largest class of problems and with Galerkin BEM as well as, for $n = 2$, with BEM collocation.

Conversely, if the equation (1.43) is solved exactly and the boundary integral equation (1.45) on Γ_c and (1.46) on Γ are treated with a BEM then this method can be considered as a modified BEM whose convergence can be justified for strongly elliptic boundary integral equations. The convergence remains valid if the FEM approximation in Ω_1 is done with a faster mesh refinement in the domain Ω_1 than on Γ_c . This case will be considered in Section 3.4.

In practical problems, however, the BEM mesh on Γ_c is defined by the adjacent finite elements on Ω_1 and, hence, both mesh sizes are the same [56], [62]. The convergence of such a method follows directly from the variational formulation (1.54) if the bilinear form $\mathcal{A}$ is $\mathcal{H}$-coercive. This can be shown for problems of classical potential theory and boundary value problems whose principal part of the differential operator L is also the Laplacian. In this case, the refinements of FEM and BEM can be chosen arbitrarily as will be seen in Section 3.1.

For the coupled FEM-BEM method let us choose a conforming $S_h^{k,r}(\Omega_1)$ system of finite elements without boundary conditions, i. e.

$$V_h = S_h^{k,r}(\Omega_1) \subset V = H^m(\Omega_1) \tag{3.1}$$

and let the coupling boundary condition $B_c\gamma$ e. g. be the Dirichlet condidition, i. e.

$$B_c\gamma u\big|_{\Gamma_c} = \left(u, \frac{\partial}{\partial n}u, \dots, \left(\frac{\partial}{\partial n}\right)^{m-1} u\right)^T \Big|_{\Gamma_c} \tag{3.2}$$

and $B_c = ((\delta_{ij}))$ for $i = 0, \dots , m-1$ and $j = 0, \dots , 2m-1$, provided the Dirichlet problem for L is uniquely solvable.

On Γ_c let us approximate equation (1.45) for Λ with a BEM method subject to a $S_{\tilde{h}}^{\tilde{k},\tilde{r}}(\Gamma)$ boundary element system and equation (1.46) for w on Γ by a $S_{\approx h}^{\approx k, \approx r}(\Gamma)$ system, correspondingly.

3.1 Essentially identical coupling [17], [52]

The coupling between equations (1.43) and (1.45) is furnished by the operator

$$B_c\gamma u\,\big|_{\Gamma_c} - K_c\,(B_c\gamma u) \quad \text{on } \Gamma_c\,. \tag{3.3}$$

In (1.43), K_c on the boundary is missing and $\mathcal{H}$-coerciveness of $\mathcal{A}$ in (1.47) can only be expected if K_c is a compact mapping which will be the essential restriction in this section. Following Brezzi, Nedelec and Johnson who have considered the Dirichlet problem for the Laplacian $L = \Delta$ in (1.23), we will show optimal order convergence if the equation (1.43) is approximated by usual FEM and the equations (1.45) and (1.46) are treated by the BEM-Galerkin method as in Section 2.2, i. e.

$$\begin{aligned} &\textit{find } \mathcal{U}_h = \left(u_h,\, \lambda_{\tilde h},\, w_{\tilde{\tilde h}}\right) \in \mathcal{V}_h := S_h^{k,r}(\Omega_1) \times S_{\tilde h}^{\tilde k,\tilde r}(\Gamma_c) \times S_{\tilde{\tilde h}}^{\tilde{\tilde k},\tilde{\tilde r}}(\Gamma) \textit{ such that} \\ &\mathcal{A}(\mathcal{U}_h,\, \mathcal{Y}_h) = \mathcal{L}\,(\mathcal{Y}_h) \text{ it for all } \mathcal{Y}_h \in \mathcal{V}_h\,. \end{aligned} \tag{3.4}$$

Theorem 3.1: *Let* $K_c\,(B_c\gamma u) : H^m(\Omega_1) \to \prod_{j=0}^{m-1} H^{m - \mu_{cj} - 1/2}(\Gamma_c)$ *in (1.48) be a compact mapping and let (1.47) be uniquely solvable. Let the family* $\mathcal{V}_h$ *satisfy the following restrictions:*

$$\begin{aligned} & m \le r < k && \textit{for } S_h^{k,r}(\Omega_1)\,, \\ & -\tilde k \le \alpha_{c_j} \le \tilde r && \textit{for } S_{\tilde h}^{\tilde k,\tilde r}(\Gamma_c)\,, \quad \alpha_{c_j} = m - 1/2 - \mu_{c_j}\,, \quad j = 0, \ldots, m, \\ & -\tilde{\tilde k} \le \alpha_j \le \tilde{\tilde r} && \textit{for } S_{\tilde{\tilde h}}^{\tilde{\tilde k},\tilde{\tilde r}}(\Gamma)\,, \quad \alpha_j = m - 1/2 - \mu_j\,, \quad j = 0, \ldots, m. \end{aligned} \tag{3.5}$$

Then there exists $h_o > 0$ *such that for all* $0 < h,\, \tilde h,\, \tilde{\tilde h} \le h_o$ *the FEM-BEM Galerkin equations (3.4) are uniquely solvable. Moreover, we have asymptotic convergence of optimal order, i. e.*

$$\begin{aligned} &\|u_h - u\|_{H^{m-\ell}(\Omega_1)} + \|\Lambda_{\tilde h} - \Lambda\|_{H^{\alpha_c - \tilde\ell}(\Gamma_c)} + \|w_{\tilde{\tilde h}} - w\|_{H^{\alpha - \tilde{\tilde\ell}}(\Gamma)} \\ &\le c\left\{ h^{s+\ell}\|u\|_{H^{m+s}(\Omega_1)} + \tilde h^{\tilde s + \tilde\ell}\|\lambda\|_{H^{\alpha_c + \tilde s}(\Gamma_c)} + \tilde{\tilde h}^{\tilde{\tilde s} + \tilde{\tilde \ell}}\|w\|_{H^{\alpha + \tilde{\tilde s}}(\Gamma)} \right\} \end{aligned} \tag{3.6}$$

for $s,\, \ell \in [0,\, k],\ \tilde s,\, \tilde\ell \in [0,\, \tilde k],\ \tilde{\tilde s},\, \tilde{\tilde\ell} \in [0,\, \tilde{\tilde k}]$.

Proof: With $\mathcal{A}(\cdot,\cdot)$ given by (1.48) we find for $\mathcal{Y} = \mathcal{U} \in \mathcal{V} = \mathcal{H}$,

$$\mathcal{A}(\mathcal{Y},\, \mathcal{Y}) = a_1(v,\, v) \;\; +\langle \phi,\, A_c\phi\rangle_{L_2(\Gamma_c)} + \langle z,\, Az\rangle_{L_2(\Gamma)} - \langle \phi,\, K_cB_c\gamma v + C_{12}z\rangle_{L_2(\Gamma_c)}$$

$$-\langle z, C_{20}v + C_{21}\phi\rangle_{L_2(\Gamma)}\,.$$

The first three terms on the right hand side provide the Gårding inequality separately due to Sections 1.1 and 1.2 whereas all the remaining bilinear forms are compact, since $K_cB_c\gamma$ is assumed to be compact and since C_{12}, C_{20} and C_{21} are defined by integral operators with C^∞-kernels.

Hence, $\mathcal{A}(\mathcal{Y},\, \mathcal{V})$ is $\mathcal{H}$-coercive with $\mathcal{H} = H^m(\Omega_1) \times H^{\alpha_c}(\Gamma_c) \times H^{\alpha}(\Gamma)$ and Theorem 1.2 implies the stability estimate (1.4) as well as quasi-optimal convergence (1.5) of the method for $H^{\{o\}}_{\pm} = \mathcal{H}$ and for $h > 0$ small enough. Next, observe that with $H^{\{\tau\}}_{\pm} = H^{m\,\pm\,\tau}(\Omega_1) \times H^{\alpha_c\,\pm\,\tau}(\Gamma_c) \times H^{\alpha\,\pm\,\tau}(\Gamma)$ and the regularity results for strongly elliptic boundary value problems [57], the regularity assumption (1.12) holds for $\mathcal{A}$ in these spaces. Hence, the duality arguments, Theorem 1.3, give the desired proposition (3.6). □

Examples: For the

Dirichlet problem of the Laplacian,

(1.23) takes the form

$$\begin{aligned} -\Delta u &= f \quad \text{in} \quad \Omega \;\text{ and} \\ u|_\Gamma &= g \quad \text{on} \quad \Gamma. \end{aligned} \tag{3.7}$$

Here

$$B_c\gamma u = u_{|\Gamma_c} \;\text{ and }\; \Lambda = \frac{\partial u}{\partial n}_{|\Gamma_c} = S_c\gamma u \;\text{ on }\; \Gamma_c \;\text{ and }\; w = \frac{\partial u}{\partial n}_{|\Gamma} \;\text{ on }\; \Gamma,$$

$$a_1(u,v) = \int_{\Omega_1} \nabla u \cdot \nabla v \; dx,$$

$$2(n-1)\pi \underset{\sim}{G}(x,y) = \gamma(x,y) = \begin{cases} -\log|x-y| & \text{for } n = 2, \\ \;\;1/|x-y| & \text{for } n = 3, \end{cases}$$

$$
\begin{aligned}
A_c\Lambda &= \tfrac{1}{(n-1)\pi}\int_{\Gamma_c}\gamma(x,y)\Lambda(y)dS(y) && \text{on } \Gamma_c,\\
Aw &= \tfrac{1}{(n-1)\pi}\int_{\Gamma_c}\gamma(x,y)w(y)dS(y) && \text{on } \Gamma,\\
K_cB_c\gamma u &= \tfrac{1}{(n-1)\pi}\int_{\Gamma_c\setminus\{x\}}\left(\frac{\partial}{\partial n(y)}\gamma(x,y)\right)u(y)dS(y) && \text{on } \Gamma_c,\\
C_{12}w &= -\tfrac{1}{(n-1)\pi}\int_{\Gamma}\gamma(x,y)w(y)dS(y) && \text{for } x\in\Gamma_c,\\
C_{20}u &= \tfrac{1}{(n-1)\pi}\int_{\Gamma_c}\left(\frac{\partial}{\partial n(y)}\gamma(x,y)\right)u(y)dS(y) && \text{for } x\in\Gamma,\\
C_{21}\Lambda &= -\tfrac{1}{(n-1)\pi}\int_{\Gamma_c}\gamma(x,y)\Lambda(y)dS(y) && \text{for } x\in\Gamma.
\end{aligned}
\tag{3.8}
$$

The operator $K_cB\gamma$ is defined by the classical double layer potential operator whose kernel $\frac{\partial}{\partial n(y)}\gamma(x,y)$ is for $n=2$ a $C^{\kappa-2}$ - function if $\Gamma\in C^{\kappa}$ (See [78] Nr. 81) whereas for $n=3$ this kernel is weakly singular defining a pseudo-differential operator of order -1 [90] which for $\Gamma\in C^{\infty}$ is compact as a mapping of any $H^s(\Gamma)$ into itself. Hence, here all the assumptions of Theorem 3.1 are fulfilled.

For the Neumann problem one finds essentially the same formulation except the definition of A, which now corresponds to (1.38) or (1.40) and the operators C_{12}, C_{20}, C_{21}, which are defined differently but still have C^{∞} kernels for Γ_c, $\Gamma\in C^{\infty}$.

Although the compactness property for $K_cB_c\gamma$ is rather restrictive, there are several applications which allow a FEM-BEM coupling with compact $K_cB_c\gamma$. With the Laplacian we find applications in [90], [64], for ideal flows in [93], for heat conduction in [72] and for electrical fields in [18]. For the Helmholtz equation, $K_cB_c\gamma$ differs from (3.6) by smooth or weakly singular integral operators (see e.g. Equations (1.4), (1.5) in [90]) and therefore Theorem 3.1 applies to the ocean and water wave problems [12] and acoustics [65]. The same properties are valid in viscous flows for the Stokes problem (where $K_cB_c\gamma = w$, Equation (2.5) in Hebeker [42]) and in [47], [48] and [49] and also in electromagnetic fields [83] (where $K_cB_c\gamma$ corresponds to L, L', Equation (4.43), in Colton and Kress [23])(see also [60]). However, the details for these problems are still to be worked out yet.

This approach with compact $K_cB_c\gamma$ can even be generalized to more general finite element methods as to the mixed FEM for Stokes flows by Sequeira [81], [82].

3.2 The nonsymmetric coupling of Galerkin methods with finer BEM mesh [16]

As we have seen in Section 1.3 in the example of elasticity, the operator $K_c B_c \gamma$ in (1.45) is not compact. Hence, for an important class of applications as in [3], [4], [9], [10], [11], [13], [33], [34], [35], [53], [54], [55], [58], [88] the assumptions of Theorem 3.1 cannot be verified. In these cases we can still prove convergence of optimal order if we impose additional inverse assumptions on the coupling boundary either for the traces $B_c\gamma v_h|_{\Gamma_c}$ or for the boundary elements on Γ_c and also a faster mesh refinement of the boundary elements than of the domain finite elements V_h in Ω_1 or a faster mesh refinement of the domain finite elements in Ω_1 than of the boundary elements on Γ_c, respectively. For symmetric coupling such additonal assumptions will not be necessary. Let us begin with a faster refinement of the boundary elements. Following [16], here we can show:

Theorem 3.2: *For the coupled FEM-BEM with* (1.48) *let the equation* (1.47) *be uniquely solvable and let the family* V_h *satisfy* (3.5). *Furthermore, let*

$$\tilde{h} = o(h) \quad \text{for} \quad h \to 0 \tag{3.9}$$

and require the inverse assumption for the traces of v_h *on* Γ_c ,

$$\|B_c\gamma v_h\|_{H^{\sigma-\alpha_c}(\Gamma_c)} \le ch^{-\sigma}\|v_h\|_{H^m(\Omega_1)} \tag{3.10}$$

with some $\sigma > 0$ *where* $\alpha_{cj} = m - \mu_{cj} - \frac{1}{2}$. *Then there exists* $h_o > 0$ *such that for all* $0 < h \le h_o$, $0 < \tilde{\tilde{h}} \le h_o$ *the FEM-BEM Galerkin equations* (3.4) *are uniquely solvable. Moreover, we have asymptotic convergence of optimal order, i.e. again* (3.6) .

Proof: For the proof we exploit the idea of solving the boundary integral equations exactly which corresponds to solving the additional boundary value problem in Ω_2 ,

$$\begin{aligned} L\tilde{u} &= 0 \ \text{in} \ \Omega_2 , \\ B_c\gamma\tilde{u} &= \tilde{g} \ \text{on} \ \Gamma_c \ \text{and} \ B\gamma\tilde{u} = g \ \text{on} \ \Gamma . \end{aligned} \tag{3.11}$$

Correspondingly, we introduce the finite dimensional space

$$\tilde{V}_h := \{\tilde{v}_h | \ \tilde{v}_{h|\Omega_1} = v_h \in S_h^{k,r}(\Omega_1)$$

$$\text{and } \tilde{v}_h \text{ solves (3.11) in } \Omega_2 \text{ with } \tilde{g} = B_c\gamma\tilde{v}_h \text{ and } g = 0\}$$

which is a subspace, $\tilde{V}_h \subset V \subset H^m(\Omega)$, for the variational formulation with (1.17) associated with the original boundary value problem (1.16) and $g = 0$. Hence, let us consider

$$\tilde{u} = \overset{\circ}{\tilde{u}}_h + u^*$$

where u^* is the solution of (3.11) in Ω_2 with $\tilde{g} = 0$, $u^*|_{\Omega_1} = 0$ and solve the auxiliary problem:

Find $\overset{\circ}{\tilde{u}}_h \in \tilde{V}_h$ *satisfying*

$$a_1(\overset{\circ}{\tilde{u}}_h,\, \tilde{v}_h) + a_2(\overset{\circ}{\tilde{u}}_h,\, \tilde{v}_h) - a_1(u,\, \tilde{v}_h) + a_2(u^* - u,\, \tilde{v}_h) = 0 \tag{3.12}$$

$$\textit{for all } \tilde{v}_h \in \tilde{V}_h$$

where a_i denote the Dirichlet bilinear forms (1.17) associated with the domains Ω_i for $i = 1$ and 2.
Note that (3.12) also corresponds to finding $\tilde{u}_h \in u^* + \tilde{V}_h$ satisfying

$$\mathbf{a}(\tilde{u}_h - u, \tilde{v}_h) := a_1(\tilde{u}_h - u,\, \tilde{v}_h) + a_2(\tilde{u}_h - u,\, \tilde{v}_h) = 0$$

for all $\tilde{v}_h \in \tilde{V}_h$. This again is a variational problem as analysed in Section 1.1. Hence, for (3.12) we have Theorems 1.1 -1.3 available yielding for $s \le k$

$$\|\tilde{u}_h - u\|_{H^{m-\ell}(\Omega_1)} + \|\tilde{u}_h - u\|_{H^{m-\ell}(\Omega_2)} \le ch^{s+\ell}\Big\{\|u\|_{H^s(\Omega_1)} + \|u\|_{H^s(\Omega_2)}\Big\}. \tag{3.13}$$

The auxiliary solution, on the other hand, satisfies the system of equations

$$a_1(\tilde{u}_h - u,\, v_h) - \langle\tilde{\Lambda} - \Lambda,\, B_c\gamma v_h\rangle_{L_2(\Gamma_c)} = 0 \;\text{ for all }\; v_h \in S_h^{k,r}(\Omega_1)\,,$$

$$A_c\tilde{\Lambda} = -B_c\gamma\tilde{u}_h + K_cB_c\gamma\tilde{u}_h + C_{12}\tilde{w} \;\text{ on }\; \Gamma_c \;\text{ and} \tag{3.14}$$

$$A\tilde{w} = -C_{20}\tilde{u}_h - C_{21}\tilde{\Lambda} \;\text{ on }\; \Gamma\,,$$

where

$$\tilde{\Lambda} = S_c\gamma\tilde{u}_h \;\text{ from }\; \Omega_2 \;\text{ on }\; \Gamma_c \;\text{ and }\; \tilde{w} = S\gamma\tilde{u}_h \;\text{ on }\; \Gamma.$$

If we insert (3.4) then we see that the system

$$a_1(\tilde{u}_h - u_h,\, v_h) - \langle\tilde{\Lambda} - \Lambda_{\tilde{h}},\, B_c\gamma v_h\rangle_{L_2(\Gamma_c)} = 0, \tag{3.15}$$

$$\begin{aligned}\langle\varphi_{\tilde{h}}, A_c(\tilde{\Lambda} - \Lambda_{\tilde{h}})\rangle_{L_2(\Gamma_c)} &= \langle\varphi_{\tilde{h}}, B_c\gamma(u_h - \tilde{u}_h) + K_cB_c\gamma(\tilde{u}_h - u_h)\\ &\quad + C_{12}(\tilde{w} - w_{\tilde{\tilde{h}}})\rangle_{L_2(\Gamma_c)},\end{aligned} \tag{3.16}$$

$$\langle \varphi_{\tilde{\tilde{h}}}, A(\tilde{w} - w_{\tilde{\tilde{h}}}) \rangle_{L_2(\Gamma)} = \langle \varphi_{\tilde{\tilde{h}}}, C_{20}(u_h - \tilde{u}_h) + C_{21}(\Lambda_{\tilde{h}} - \tilde{\Lambda}) \rangle_{L_2(\Gamma)} \tag{3.17}$$

will be satisfied by any Galerkin solution of (3.4).

If we consider the equations (3.16) and (3.17) separately, then we find from Theorems 2.1 and 2.3 and with (3.10) the estimates

$$\begin{aligned} &\|\Lambda_{\tilde{h}} - \tilde{\Lambda} - A_c^{-1}\left\{B_c\gamma(\tilde{u}_h - u_h) - K_cB_c\gamma(\tilde{u}_h - u_h) - C_{12}(w_{\tilde{\tilde{h}}} - \tilde{w})\right\}\|_{H^{\alpha_c}(\Gamma_c)} \\ &\qquad \le c\tilde{h}\left\{\|B_c\gamma(\tilde{u}_h - u_h)\|_{H^{1-\alpha_c}(\Gamma_c)} + \|w_{\tilde{\tilde{h}}} - \tilde{w}\|_{H^\alpha(\Gamma)}\right\} \\ &\qquad \le o(1)\|\tilde{u}_h - u_h\|_{H^m(\Omega_1)} + c\tilde{h}\|w_{\tilde{\tilde{h}}} - \tilde{w}\|_{H^\alpha(\Gamma)}, \\ &\|w_{\tilde{\tilde{h}}} - \tilde{w} - A^{-1}\{C_{20}(\tilde{u}_h - u_h) + C_{21}(\tilde{\Lambda} - \Lambda_{\tilde{h}})\|_{H^\alpha(\Gamma)} \\ &\qquad \le c\,\tilde{\tilde{h}}\left\{\|\tilde{u}_h - u_h\|_{H^m(\Omega_1)} + \|\tilde{\Lambda} - \Lambda_{\tilde{h}}\|_{H^{\alpha_c}(\Gamma_c)}\right\}. \end{aligned} \tag{3.18}$$

In order to show (3.6) we first prove stability, i.e. (3.6) for $\ell = \tilde{\ell} = \tilde{\tilde{\ell}} = s = \tilde{s} = \tilde{\tilde{s}} = 0$ proceeding as in the proof of Theorem 1.2. If now (3.6) were not true then we could find a subsequence $(u_{h'}, \Lambda_{\tilde{h}'}, w_{\tilde{\tilde{h}}'})$ whose norms tend to infinity. Then introduce

$$(\mathring{u}_{h'}, \mathring{\Lambda}_{\tilde{h}'}, \mathring{w}_{\tilde{\tilde{h}}'}) := (u_{h'}, \Lambda_{\tilde{h}'}, w_{\tilde{\tilde{h}}'}) / \left\{\|u_{h'}\|_{H^m(\Omega_1)} + \|\Lambda_{\tilde{h}'}\|_{H^{\alpha_c}(\Gamma_c)} + \|w_{\tilde{\tilde{h}}'}\|_{H^\alpha(\Gamma)}\right\}$$

and a subsequence will converge weakly in $H^m(\Omega_1) \times H^{\alpha_c}(\Gamma_c) \times H^\alpha(\Gamma)$,

$$(\mathring{u}_{h''}, \mathring{\Lambda}_{\tilde{h}''}, \mathring{w}_{\tilde{\tilde{h}}''}) \rightharpoonup (\mathring{u}, \mathring{\Lambda}, \mathring{w}).$$

Dividing (3.15) - (3.17) by

$$\left\{\|u_{h''}\|_{H^m(\Omega_1)} + \|\Lambda_{\tilde{h}''}\|_{H^{\alpha_c}(\Gamma_c)} + \|w_{\tilde{\tilde{h}}''}\|_{H^\alpha(\Gamma)}\right\} \to \infty$$

yields with (3.13) for the subsequence $(\mathring{u}_{h''}, \mathring{\Lambda}_{\tilde{h}''}, \mathring{w}_{\tilde{\tilde{h}}''})$ strong limits

$$\begin{aligned} &\|\mathring{\Lambda}_{\tilde{h}''} - A_c^{-1}\left\{-B_c\gamma\,\mathring{u}_{h''} + K_cB_c\gamma\,\mathring{u}_{h''} + C_{12}\,\mathring{w}_{\tilde{\tilde{h}}''}\right\}\|_{H^{\alpha_c}(\Gamma_c)} \to 0, \\ &\|\mathring{w}_{\tilde{\tilde{h}}''} + A^{-1}\left\{C_{20}\,\mathring{u}_{h''} + C_{21}\,\mathring{\Lambda}_{\tilde{h}''}\right\}\|_{H^\alpha(\Gamma)} \to 0. \end{aligned} \tag{3.19}$$

Hence, $(\overset{\circ}{u}, \overset{\circ}{\Lambda}, \overset{\circ}{w})$ satisfies the homogeneous system of equations,

$$a_1(\overset{\circ}{u}, v) - \langle \overset{\circ}{\Lambda}, B_c\gamma v\rangle_{L_2(\Gamma_c)} = 0 \quad \text{for all} \quad v \in H^m(\Omega_1),$$

$$A_c \overset{\circ}{\Lambda} + B_c\gamma \overset{\circ}{u} - K_cB_c\gamma \overset{\circ}{u} - C_{12} \overset{\circ}{w} = 0,$$

$$A \overset{\circ}{w} + C_{20} \overset{\circ}{u} + C_{21} \overset{\circ}{\Lambda} = 0.$$

This implies with Fredholm's alternative,

$$\overset{\circ}{u}= 0, \quad \overset{\circ}{\Lambda}= 0, \quad \overset{\circ}{w}= 0.$$

Since the operators C_{20}, C_{21} and C_{12} are completely continuous, the weak convergence to zero implies strong convergence, $C_{21}\, \overset{\circ}{u}_{h''} \to 0$, $C_{21}\, \overset{\circ}{\Lambda}_{\tilde{h}''} \to 0$ and $C_{12}\, \overset{\circ}{w}_{\approx\atop h''} \to 0$ and (3.19) yield

$$\overset{\circ}{w}_{\approx\atop h''} \to 0 \text{ in } H^{\alpha}(\Gamma).$$

To $\overset{\circ}{u}_{h''}$ let us construct the extension to Ω_2 via (3.11), i.e. $\overset{\tilde{\circ}}{u}_{h''} \in \tilde{V}_{h''}$. Then the representation formula (1.44) yields for $\lambda_{h''} := S_c\gamma\, \overset{\tilde{\circ}}{u}_{h''}$ the integral equation on Γ_c,

$$\lambda_{h''} = A_c^{-1}\{-B_c\gamma\, \overset{\circ}{u}_{h''} + K_cB_c\gamma\, \overset{\circ}{u}_{h''}\},$$

and (3.19) implies

$$\| \overset{\circ}{\Lambda}_{\tilde{h}''} - \lambda_{h''}\|_{H^{\alpha_c}(\Gamma_c)} \to 0.$$

Therefore, the application of the Green formula (1.44) in Ω_2 yields from (3.15)

$$a_1(\overset{\circ}{u}_{h''}, v_{h''}) + a_2(\overset{\tilde{\circ}}{u}_{h''}, \tilde{v}_{h''}) \to 0$$

uniformly for all $\|v_{h''}\|_{H^m(\Omega_1)} \leq 1$. Since a_1 and a_2 are both coercive and $\overset{\tilde{\circ}}{u}_{h''} \to 0$ in $H^m(\Omega_1)$ and in $H^m(\Omega_2)$, the choice $\tilde{v}_{h''} = \overset{\tilde{\circ}}{u}_{h''}$ implies

$$\| \overset{\circ}{u}_{h''} \|_{H^m(\Omega_1)} \to 0.$$

Then (3.19) finally gives

$$\| \overset{\circ}{\Lambda}_{\tilde{h}''} \|_{H^{\alpha}(\Gamma_c)} \to 0$$

and the strong convergence $(\mathring{u}_{h''},\ \mathring{\Lambda}_{\tilde{h}''},\ \mathring{w}_{\approx \atop h''}) \to 0$ is in contradiction to

$$\| \mathring{u}_{h''} \|_{H^m(\Omega_1)} + \| \mathring{\Lambda}_{\tilde{h}''} \|_{H^{\alpha_c}(\Gamma_c)} + \| \mathring{w}_{h''} \|_{H^{\alpha}(\Gamma)} = 1 .$$

As a consequence, we obtain the stability estimate

$$\|u_h\|_{H^m(\Omega_1)} + \|\Lambda_{\tilde{h}}\|_{H^{\alpha_c}(\Gamma_c)} + \|w_{\approx \atop h}\|_{H^{\alpha}(\Gamma)}$$

$$\leq c \left\{ \|u\|_{H^m(\Omega_1)} + \|\Lambda\|_{H^{\alpha_c}(\Gamma_c)} + \|w\|_{H^{\alpha}(\Gamma)} \right\}$$

provided $0 < h, \overset{\approx}{h} \leq h_o$ with an appropriately chosen $h_o > 0$. This stability estimate yields (1.4) for (3.4) . Then Theorems 1.1 and 1.3 yield the desired estimate (3.6). □

Remarks:

i) Note that for (3.18) we have used that the adjoint boundary value problem

$$Lv = 0 \ \text{ in } \ \Omega_1, \ \ S_c \gamma v = 0 \ \text{ on } \ \Gamma_c$$

admits only the trivial solution $v = 0$. If this problem has a nontrivial solution then with standard arguments for a perturbed uniquely solvable problem we can proceed in the same manner.

ii) The proof rests on the auxiliary problem (3.11). Therefore Theorem 3.2 remains valid if the boundary element equations (1.45) and /or (1.46) and their variational forms in (1.48) are replaced by the corresponding boundary integral equations of the second kind (1.29), respectively, provided $\underset{\sim}{A}_c$ and/or $\underset{\sim}{A}$ are strongly elliptic in the respective Sobolev spaces.

3.3 FEM-BEM collocation for $n = 2$ with finer BEM mesh

Since the accuracy of the solutions to (3.15) - (3.17) has been estimated separately we can apply the results for point collocation in Section 2.3, too. This yields the following method.

Find u_h, $\Lambda_{\tilde h}$, $w_{\tilde{\tilde h}}$ *such that*

$$\begin{aligned}
a_1(u_h, v_h) &= (f, v_h)_{L_2(\Omega_1)} + \langle \Lambda_{\tilde h}, B_c\gamma v_h\rangle_{L_2(\Gamma_c)} ,\\
(A_c\Lambda_{\tilde h})(x_\ell) &= (-B_c\gamma u_h + K_c B_c\gamma u_h + C_{12} w_{\tilde{\tilde h}} + K_{12} g)(x_\ell) \quad \text{for } x_\ell \in \Gamma_C ,\\
&\qquad \ell = 1, \dots, M_C ,\\
(Aw_{\tilde{\tilde h}})(\tilde x_\ell) &= (-C_{20}u_h - C_{21}\Lambda_{\tilde h} + g + K_{22}g)(\tilde x_\ell) \qquad \text{for } \tilde x_\ell \in \Gamma ,\ \ell = 1, \dots, M ,
\end{aligned} \tag{3.20}$$

where $M_C = \dim S_{\tilde h}^{\tilde k, \tilde r}(\Gamma_C)$, $M = \dim S_{\tilde{\tilde h}}^{\tilde{\tilde k}, \tilde{\tilde r}}(\Gamma)$, $\tilde r + 1 = \tilde k$, $\tilde{\tilde r} + 1 = \tilde{\tilde k}$ *and the collocation points* x_ℓ *and* $\tilde x_\ell$ *are chosen according to naive collocation viz.* (2.15), *respectively.*

Theorem 3.3: *For the coupled FEM-BEM collocation* (3.20) *let* (1.47) *be uniquely solvable and let* $m \le r \le k$ *for* $S_h^{k,r}(\Omega_1)$ *and one of the conditions* (2.17) *or* (2.18) *or* (2.19) *be satisfied with* α_c *on* Γ_c *for* $S_{\tilde h}^{\tilde k, \tilde r}(\Gamma_c)$ *and with* α *on* Γ *for* $S_{\tilde{\tilde h}}^{\tilde{\tilde k}, \tilde{\tilde r}}(\Gamma)$, *respectively. Furthermore let* (3.9) *and* (3.10) *be satisfied.*
Then there exists $h_o > 0$ *such that for all* $0 < h, \tilde{\tilde h} \le h_o$ *the FEM-BEM equations* (3.20) *are uniquely solvable, and we have asymptotic convergence of optimal order, i.e.*

$$\begin{aligned}
&\|u_h - u\|_{H^{m-\ell}(\Omega_1)} + \|\Lambda_{\tilde h} - \Lambda\|_{H^{\tilde\ell}(\Gamma_c)} + \|w_{\tilde{\tilde h}} - w\|_{H^{\tilde{\tilde \ell}}(\Gamma)}\\
&\le c\left\{ h^{s+\ell}\|u\|_{H^{m+s}(\Omega_1)} + \tilde h^{\tilde s - \tilde \ell}\|\Lambda\|_{H^{\tilde s}(\Gamma_c)} + \tilde{\tilde h}^{\tilde{\tilde s} - \tilde{\tilde \ell}}\|w\|_{H^{\tilde{\tilde s}}(\Gamma)}\right\}
\end{aligned} \tag{3.21}$$

where $s \in [0, k], 2\alpha_c \le \tilde\ell \le \alpha_c + \tilde k/2 \le \tilde s \le \tilde k$, $2\alpha \le \tilde{\tilde \ell} \le \alpha + \tilde{\tilde k}/2 \le \tilde{\tilde s}, \ell = \alpha_c - \tilde \ell$.

Proof: For the proof replace in (3.16), (3.17) the bilinear forms by corresponding collocation equations as in (3.20). Then we show stability of the method in $H^{m-\ell}(\Omega_1) \times H^{\tilde\ell}(\Gamma_c) \times H^{\tilde{\tilde\ell}}(\Gamma)$ with similar arguments as in the proof of Theorem 3.2. If the collocation equations in (3.20) are considered separately, then Theorem 2.5 provides with (3.9) and (3.10) the estimates

$$\|\Lambda_{\tilde{h}} - \Lambda - A_c^{-1}\left\{B_c\gamma(\tilde{u}_h - u_h) - K_cB_c\gamma(\tilde{u}_h - u_h) - C_{21}(w_{\tilde{\tilde{h}}} - \tilde{w})\right\}\|_{H^{\tilde{\ell}}(\Gamma_c)}$$

$$\leq c\tilde{h}\left\{\|B_c\gamma(\tilde{u}_h - u_h)\|_{H^{\tilde{\ell}+1}(\Gamma_c)} + \|w_{\tilde{\tilde{h}}} - \tilde{w}\|_{H^{\tilde{\tilde{\ell}}}(\Gamma)}\right\}$$

$$\leq o(1)\|B_c\gamma(\tilde{u}_h - u_h)\|_{H^{\tilde{\ell}}(\Gamma_c)} + c\tilde{h}\|w_{\tilde{\tilde{h}}} - \tilde{w}\|_{H^{\tilde{\tilde{\ell}}}(\Gamma)},$$

$$\|w_{\tilde{\tilde{h}}} - \tilde{w} - A^{-1}\left\{C_{20}(\tilde{u}_h - u_h) + C_{21}(\tilde{\Lambda} - \Lambda_{\tilde{h}})\right\}\|_{H^{\tilde{\tilde{\ell}}}(\Gamma)}$$

$$\leq c\,\tilde{\tilde{h}}\left\{\|\tilde{w}\|_{H^{\tilde{\tilde{\ell}}+1}(\Gamma)} + \|\tilde{u}_h - u_h\|_{H^{m-\ell}(\Omega_c)} + \|\tilde{\Lambda} - \Lambda_{\tilde{h}}\|_{H^{\tilde{\ell}}(\Gamma_c)}\right\}.$$

If the method were unstable, then we could find a sequence as in the proof of Theorem 3.2,

$$(\mathring{u}_{h''}, \mathring{\Lambda}_{\tilde{h}''}, \mathring{w}_{\tilde{\tilde{h}}''}) \to (0,0,0) \text{ in } H^{m-\ell}(\Gamma_1) \times H^{\tilde{\ell}}(\Gamma_c) \times H^{\tilde{\tilde{\ell}}}(\Gamma)$$

with

$$\|\mathring{u}_{h''}\|_{H^{m-\ell}(\Gamma_1)} + \|\mathring{\Lambda}_{\tilde{h}''}\|_{H^{\tilde{\ell}}(\Gamma_c)} + \|\mathring{w}_{\tilde{\tilde{h}}''}\|_{H^{\tilde{\tilde{\ell}}}(\Gamma)} = 1.$$

The above inequalities yield again with

$$C_{12}\,\mathring{w}_{\tilde{\tilde{h}}''} \to 0 \text{ in } H^{\tilde{\ell}}(\Gamma_c),\ C_{20}\,\mathring{u}_{h''} \to 0 \text{ and } C_{21}\,\mathring{\Lambda}_{\tilde{h}''} \to 0 \text{ in } H^{\tilde{\tilde{\ell}}}(\Gamma),$$

$$\|\mathring{\Lambda}_{\tilde{h}''} + A_c^{-1}\left\{B_c\gamma\,\mathring{u}_{h''} - K_cB_c\gamma\,\mathring{u}_{h''}\right\}\|_{H^{\tilde{\ell}}(\Gamma_c)} \to 0,$$

$$\|\mathring{w}_{\tilde{\tilde{h}}''}\|_{H^{\tilde{\tilde{\ell}}}(\Gamma)} \to 0.$$

With $\lambda_{h''}$ as in the proof of Theorem 3.2 we find

$$\|\mathring{\Lambda}_{\tilde{h}''} - \lambda_{h''}\|_{H^{\tilde{\ell}}(\Gamma_c)} \to 0 \text{ for } h'' \to 0 \text{ and (3.9)},$$

which implies

$$a_1(\mathring{u}_{h''}, v_{h''}) + a_2(\mathring{\tilde{u}}_{h''}, \tilde{v}_{h''}) \to 0$$

for $h'' \to 0$, uniformly for all $v_{h''}$ with $\|v_{h''}\|_{H^{m+\ell}(\Gamma_1)} \leq 1$. For $a_1 + a_2$ we have a Gårding inequality in $H^m(\Omega_1 \overset{\bullet}{\cup} \Omega_2)$ where for $\mathring{u}_{h''} \to 0$ the compact terms converge to zero uniformly for the above $\tilde{v}_{h''}$. Hence,

$$0 = \lim_{h'' \to 0} \sup_{\|v_{h''}\|_{H^{m+\ell}(\Omega_1)} \leq 1} |a_1(\mathring{u}_{h''}, v_{h''}) + a_2(\mathring{u}_{h''}, \tilde{v}_{h''})|$$

$$\geq \gamma_0 \lim_{h'' \to 0} \| \mathring{u}_{h''} \|_{H^{m-\ell}(\Omega_1)} .$$

This convergence $\| \mathring{u}_{h''} \|_{H^{m-\ell}(\Omega_1)} \to 0$ implies with the above inequality also

$$\| \mathring{\Lambda}_{\tilde{h}''} \|_{H^{\tilde{\ell}}(\Gamma_c)} \to 0.$$

Hence, we arrive at a contradiction to

$$\| \mathring{u}_{h''} \|_{H^{m-\ell}(\Gamma_c)} + \| \mathring{\Lambda}_{\tilde{h}''} \|_{H^{\tilde{\ell}}(\Gamma_c)} + \| \mathring{w}_{\tilde{\tilde{\ell}}} \|_{H^{\tilde{\tilde{\ell}}}(\Gamma)} = 1,$$

and (3.20) must be stable, i.e. there exist $h_0 > 0$ and a constant c such that

$$\|u_h\|_{H^{m-\ell}(\Omega_1)} + \|\Lambda_{\tilde{h}}\|_{H^{\tilde{\ell}}(\Gamma_c)} + \|w_{\tilde{\tilde{h}}}\|_{H^{\tilde{\tilde{\ell}}}(\Gamma)}$$

$$\leq c \left\{ \|u\|_{H^{m+\ell}(\Omega_1)} + \|\Lambda\|_{H^{\tilde{\ell}}(\Gamma_c)} + \|w\|_{H^{\tilde{\tilde{\ell}}}(\Gamma)} \right\}$$

for all $\tilde{h}, \tilde{\tilde{h}} \leq h_0$.
For the remaining part of the proof one may proceed as in Section 1.1. We omit the details. □

Remark:
Again, Theorem 3.3 also holds if the boundary element collocation equations (3.20) are replaced on Γ_c and/or on Γ by the corresponding equations defined with the boundary integral equations of the second kind (1.29), respectively, provided $\underset{\sim}{A}c$ and/or $\underset{\sim}{A}$ are coercive (or strongly elliptic) in the respective Sobolev spaces.

3.4 The general case of FEM-BEM Galerkin methods with finer FEM mesh

For the cases as formulated in Section 3.2 with A_c we are also able to prove stability and convergence for the coupled FEM-BEM method with (1.47) if the finite elements in the domain Ω_1 are refined faster than the boundary elements on the coupling boundary Γ_c . The refinement on the exterior boundary Γ can be performed independently.

Theorem 3.4: *For the coupled FEM-BEM with* (3.4) *let the equations* (1.47) *be uniquely solvable and let the family* $\mathcal{V}_h$ *satisfy* (3.5). *Furthermore, let*

$$h = o(\bar{h}) \quad for \quad \bar{h} \to 0 \tag{3.22}$$

and require the inverse assumption for the boundary elements $S_{\bar{r}}^{\bar{k},\bar{r}}(\Gamma_c)$ *on the coupling boundary. Then there exists* $h_o > 0$ *such that for all* $0 < \bar{h},\ \bar{\bar{h}} \le h_o$ *the equations* (3.4) *are uniquely solvable. Moreover, we have asymptotic convergence of optimal order, i.e.* (3.6) .

Proof: For the proof we consider the case $f = 0$ only, since by using some solution u^* of the inhomogeneous differential equation

$$Lu^* = f,$$

the general case can be reduced to $f = 0$. The FEM-BEM Galerkin equations read as

$$a_1(u_h - u,\ v_h) = \langle (\Lambda_{\bar{h}} - \Lambda), B_c\gamma v_h\rangle_{L_2(\Gamma_c)} \quad \text{for all} \quad v_h \in S_h^{k,r}(\Omega_1), \tag{3.23}$$

$$\langle \varphi_{\bar{h}}, A_c(\Lambda_{\bar{h}} - \Lambda)\rangle_{L_2(\Gamma_c)} = \langle \varphi_{\bar{h}}, (K_c - I)(B_c\gamma)(u_h - u) + C_{12}(w_{\bar{\bar{h}}} - w)\rangle_{L_2(\Gamma_c)}$$
$$\text{for all} \quad \varphi_{\bar{h}} \in S_{\bar{h}}^{\bar{k},\bar{r}}(\Gamma_c), \tag{3.24}$$

$$\langle \varphi_{\bar{\bar{h}}}, A(w_{\bar{\bar{h}}} - w)\rangle_{L_2(\Gamma)} = \langle \varphi_{\bar{\bar{h}}}, C_{20}(u_h - u) \quad + C_{21}(\Lambda_{\bar{h}} - \Lambda)\rangle_{L_2(\Gamma)} \tag{3.25}$$

$$\text{for all } \varphi_{\bar{\bar{h}}} \in S_{\bar{\bar{h}}}^{\bar{\bar{k}},\bar{\bar{r}}}(\Gamma) .$$

Since a_1 is $H^m(\Omega_1)$ -coercive, we find from (3.23) with Theorems 1.1 and 1.2 the estimate

$$\|u_h - (u + N^{-1}(\Lambda_{\bar{h}} - \Lambda))\|_{H^m(\Omega_1)} \quad = \|u_h - N^{-1}\Lambda_{\bar{h}}\|_{H^m(\Omega_1)} \le ch\|\Lambda_{\bar{h}}\|_{H^{\alpha_c}+1(\Gamma_c)},$$

where $N^{-1}\Lambda_{\bar{h}}$ denotes the solution of the boundary value problem

$$L(N^{-1}\Lambda_{\bar{h}}) = 0, \quad S_c\gamma N^{-1}\Lambda_{\bar{h}} = \Lambda_{\bar{h}} \quad \text{on} \quad \Gamma_c.$$

For the right hand side of the estimate we use the inverse assumption on Γ_c to obtain

$$\|u_h - N^{-1}\Lambda_{\bar{h}}\|_{H^m(\Omega_1)} \le o(1)\|\Lambda_{\bar{h}}\|_{H^{\alpha_c}(\Gamma_c)}. \tag{3.26}$$

For (3.24), (3.25) we apply Theorem 2.3 separately to get

$$\|\Lambda_{\tilde{h}} - \Lambda - A_c^{-1}\left\{B_c\gamma(u-u_h) - K_cB_c\gamma(u-u_h) - C_{12}(w-w_{\tilde{\tilde{h}}})\right\}\|_{H^{\alpha_c}(\Gamma_c)}$$

$$\leq c\|\Lambda + A_c^{-1}\left\{B_c\gamma(u-u_h) - K_cB_c\gamma(u-u_h) - C_{12}(w-w_{\tilde{\tilde{h}}})\right\}\|_{H^{\alpha_c}(\Gamma_c)}, \tag{3.27}$$

$$\|w_{\tilde{\tilde{h}}} - w - A^{-1}\{C_{20}(u-u_h) + C_{21}(\Lambda - \Lambda_{\tilde{h}})\}\|_{H^{\alpha}(\Gamma)}$$

$$\leq c\,\tilde{\tilde{h}}\left\{\|w\|_{H^{\alpha+1}(\Gamma)} + \|u-u_h\|_{H^m(\Omega_1)} + \|\Lambda - \Lambda_{\tilde{h}}\|_{H^{\alpha_c}(\Gamma_c)}\right\}.$$

As in the proof of Theorem 1.2 we now show stability. If (3.6) were not true for $\ell = \tilde{\ell} = \tilde{\tilde{\ell}} = s = \tilde{s} = \tilde{\tilde{s}} = 0$, then we could find a subsequence of approximate solutions $(u_{h'}, \Lambda_{\tilde{h}'}, w_{\tilde{\tilde{h}}'})$ whose norms tend to infinity. Then introduce the bounded sequence

$$(\mathring{u}_{h'}, \mathring{\Lambda}_{\tilde{h}'}, \mathring{w}_{\tilde{\tilde{h}}''}) := (u_{h'}, \Lambda_{\tilde{h}'}, w_{\tilde{\tilde{h}}'})/\{\|u_{h'}\|_{H^m(\Omega_1)} + \|\Lambda_{\tilde{h}'}\|_{H^{\alpha_c}(\Omega_c)} + \|w_{\tilde{\tilde{h}}'}\|_{H^{\alpha}(\Gamma)}\}$$

which contains a weakly convergent subsequence of norms 1,

$$(\mathring{u}_{h''}, \mathring{\Lambda}_{\tilde{h}''}, \mathring{w}_{\tilde{\tilde{h}}''}) \rightharpoonup (\mathring{u}, \mathring{\Lambda}, \mathring{w}) \text{ in } H^m(\Omega_1) \times H^{\alpha_c}(\Gamma_c) \times H^{\alpha}(\Gamma).$$

Dividing (3.23) - (3.25) by

$$\left\{\|u_{h''}\|_{H^m(\Omega_1)} + \|\Lambda_{\tilde{h}''}\|_{H^{\alpha_c}(\Gamma_c)} + \|w_{\tilde{\tilde{h}}''}\|_{H^{\alpha}(\Gamma)}\right\} \to \infty.$$

we find that the weak limit $(\mathring{u}, \mathring{\Lambda}, \mathring{w})$ will satisfy the same homogeneous system of equations as in the proof of Theorem 3.2. Hence,

$$\mathring{u} = 0, \ \mathring{\Lambda} = 0, \ \mathring{w} = 0.$$

Moreover, from (3.26), (3.27) we get

$$\|\mathring{u}_{h''} - N^{-1}\mathring{\Lambda}_{\tilde{h}''}\|_{H^m(\Omega_1)} \to 0, \ \|\mathring{w}_{\tilde{\tilde{h}}''}\|_{H^{\alpha}(\Gamma)} \to 0 \quad \text{for } \tilde{h}'', \tilde{\tilde{h}}'' \to 0, \tag{3.28}$$

$$\|\mathring{\Lambda}_{\tilde{h}''} + A_c^{-1}\{B_c\gamma\,\mathring{u}_{h''} - K_cB_c\gamma\,\mathring{u}_{h''}\}\|_{H^{\alpha_c}(\Gamma_c)} + c'\|\mathring{w}_{\tilde{\tilde{h}}''}\|_{H^{\alpha}(\Gamma)} \tag{3.29}$$

$$\leq c\|B_c\gamma\,\mathring{u}_{h''} - K_cB_c\gamma\,\mathring{u}_{h''}\|_{H^{-\alpha_c}(\Gamma_c)}$$

for $\tilde{h}'' \leq \tilde{h}_0$ with $\tilde{h}_0 > 0$ sufficiently small. The first limit in (3.28) can also be written as

$$\|S_c\gamma \overset{\circ}{u}_{h''} - \overset{\circ}{\Lambda}_{\tilde{h}''}\|_{H^{\alpha_c}(\Gamma_c)} \to 0 \quad \text{for} \quad \tilde{h}'' \to 0.$$

Hence, (3.23) yields

$$a_1(\overset{\circ}{u}_{h''}, v_{h''}) - \langle S_c\gamma \overset{\circ}{u}_{h''}, B_c\gamma v_{h''}\rangle_{L_2(\Gamma_c)} = o(1)$$

for $\tilde{h}'' \to 0$ uniformly for all $v_{h''}$ with $\|v_{h''}\|_{H^m(\Omega_1)} \leq 1$. With (1.41) for Ω_2 and the extensions $\overset{\circ}{\tilde{u}}_{h''} \in \tilde{V}_h$, $\tilde{v}_{h''} \in \tilde{V}_h$, this becomes

$$a_1(\overset{\circ}{\tilde{u}}_{h''}, \tilde{v}_{h''}) + a_2(\overset{\circ}{\tilde{u}}_{h''}, \tilde{v}_{h''}) = o(1)$$

and the choice of $\tilde{v}_{h''} = \overset{\circ}{\tilde{u}}_{h''} \to 0$ yields with the coerciveness of a_1, a_2 ,

$$\|\overset{\circ}{u}_{h''}\|_{H^m(\Omega_1)} \to 0 \quad \text{for} \quad \tilde{h}'' \to 0.$$

Hence, the inequality (3.29) gives

$$\|\overset{\circ}{\Lambda}_{h''}\|_{H^{\alpha_c}(\Gamma_c)} \to 0$$

in contradiction to

$$\|\overset{\circ}{u}_{h''}\|_{H^m(\Omega_1)} + \|\overset{\circ}{\Lambda}_{\tilde{h}''}\|_{H^{\alpha_c}(\Gamma_c)} + \|w_{\tilde{\tilde{h}}''}\|_{H^{\alpha}(\Gamma)} = 1.$$

As a consequence, with an appropriately chosen $h_0 > 0$, the stability inequality

$$\|u_h\|_{H^m(\Omega_1)} + \|\Lambda_{\tilde{h}}\|_{H^{\alpha_c}(\Gamma_c)} + \|w_{\tilde{\tilde{h}}}\|_{H^{\alpha}(\Gamma)}$$

$$\leq c\left\{\|u\|_{H^m(\Omega_1)} + \|\Lambda\|_{H^{\alpha_c}(\Gamma_c)} + \|w\|_{H^{\alpha}(\Gamma)}\right\}$$

must be valid for all $0 < \tilde{h}, \tilde{\tilde{h}} \leq h_0$. This stability estimate implies (1.4) and, hence, Theorem 1.1. For sufficiently smooth boundaries Γ_C and Γ , the system of equations corresponding to (3.23) - (3.25) is a system of "strongly elliptic pseudo-differential equations" on $\Omega_1 \times \Gamma_C \times \Gamma$ and, therefore, provides the shift theorem, i.e. regularity assumption (1.12) [73]. Hence, Theorem 1.3 is also applicable which implies the proposed estimate (3.6). □

3.5 FEM-BEM collocation for $n = 2$ with finer FEM mesh

The arguments of Section 3.4 can be modified for the coupling method (3.20) with boundary element collocation if the finite elements are refined faster than the boundary elements on the coupling boundary Γ_C. However, here we require quasiuniform meshes in Ω_1 as well as on Γ_C.

Theorem 3.5: *For the coupled FEM-BEM collocation (3.20) let (1.47) be uniquely solvable and let $m + \alpha_C \le r \le k$ for $S_h^{k,r}(\Omega_1)$ and the conditions (2.17) be satisfied. Furthermore, let the meshes in Ω_1 and on Γ_C be quasiuniform providing the inverse assumptions and let*

$$h = o(\bar{h}) \quad \text{for} \quad \bar{h} \to o.$$

Then there exists $h_o > 0$ such that for all $o < \bar{h}, \tilde{\tilde{h}} \le h_o$ the equations (3.20) are uniquely solvable. Moreover, we have asymptotic convergence of optimal order, i.e. (3.21).

Proof: The proof is very similar to the proofs of Theorems 3.3 and 3.4. However, here we use Theorem 2.5 and replace α_c by $2\alpha_c$, m by $m + \alpha_c$ and α by 2α. For quasiuniform FEM and with $m + \alpha_c \le r$ we find from Theorems 1.1 and 1.2 the estimate,

$$\begin{aligned} \|u_h - N^{-1}\Lambda_{\bar{h}}\|_{H^{m+\alpha_c}(\Omega_1)} &\le ch\|\Lambda_{\bar{h}}\|_{H^{2\alpha_c+1}(\Gamma_c)} \\ &\le o(1)\|\Lambda_{\bar{h}}\|_{H^{2\alpha_c}(\Gamma_c)} \qquad \text{for } \bar{h} \to 0. \end{aligned}$$

Instead of (3.27) we now have with Theorem 2.5,

$$\begin{aligned} &\|\Lambda_{\bar{h}} - \Lambda - A_c^{-1}\left\{B_c\gamma(u-u_h) - K_cB_c\gamma(u-u_h) - C_{12}(w - w_{\tilde{\tilde{h}}})\right\}\|_{H^{2\alpha_c}(\Gamma_c)} \\ &\quad \le c\|\Lambda + A_c^{-1}\left\{B_c\gamma(u-u_h) - K_cB_c\gamma(u-u_h) - C_{12}(w - w_{\tilde{\tilde{h}}})\right\}\|_{H^{2\alpha_c}(\Gamma_c)}, \end{aligned}$$

$$\begin{aligned} &\|w_{\tilde{\tilde{h}}} - w - A^{-1}\{C_{20}(u-u_h) + C_{21}(\Lambda - \Lambda_{\bar{h}})\|\}_{H^{2\alpha}(\Gamma)} \\ &\quad \le c\,\tilde{\tilde{h}}\left\{\|w\|_{H^{2\alpha+1}(\Gamma)} + \|u-u_h\|_{H^{m+\alpha_c}(\Gamma_1)} + \|\Lambda - \Lambda_{\bar{h}}\|_{H^{2\alpha_c}(\Gamma_c)}\right\}. \end{aligned} \tag{3.30}$$

If the coupled FEM-BEM collocation (3.20) were not stable we now could find a subsequence with norms one and

$$(\mathring{u}_{h''},\ \mathring{\Lambda}_{\tilde{h}''},\ \mathring{w}_{\tilde{\tilde{h}}''}) \to (0,0,0) \ \text{ in } \ H^{m+\alpha_c}(\Gamma_1)\times H^{2\alpha_c}(\Gamma_c)\times H^{2\alpha}(\Gamma).$$

As in the proof of Theorem 3.4, we could prove $\|\mathring{w}_{h''}\|_{H^{2\alpha}(\Gamma)} \to 0$ and

$$\|S_c\gamma\,\mathring{u}_{h''} - \mathring{\Lambda}_{\tilde{h}''}\|_{H^{2\alpha_c}(\Gamma_c)} \to 0$$

for $\tilde{h}'' \to 0$, $\tilde{\tilde{h}}'' \to 0$ which implies

$$a_1(\mathring{u}_{h''},\ v_{h''}) - \langle S_c\gamma\,\mathring{u}_{h''},\ B_c\gamma v_{h''}\rangle_{L_2(\Gamma_c)} = o(1)$$

uniformly for all $v_{h''}$ with $\|v_{h''}\|_{H^{m-\alpha_c}(\Omega_c)} \le 1$. Since the strongly coercive real part of $a_1 + a_2$ is equivalent to the $H^m(\Gamma_1 \overset{\bullet}{\cup} \Omega_2)$ scalar product and $H^{m-\alpha_c} \subset H^m \subset H^{m+\alpha_c}(\Omega_1)$ is a duality with respect to $H^m(\Gamma_1)$ we find

$$\|\mathring{u}_{h''}\|_{H^{m+\alpha_c}(\Omega_1)} \le c \sup_{\|v_{h''}\|_{H^{m-\alpha_c}(\Gamma_c)} \le 1} |a_1(\mathring{u}_{h''},\ v_{h''}) + a_2(\mathring{u}_{h''},\ \tilde{v}_{h''})| = o(1),$$

i.e. $\|\mathring{u}_{h''}\|_{H^{m+\alpha_c}(\Omega_1)} \to 0$.

The first of the inequalities (3.30) would finally yield

$$\|\mathring{\Lambda}_{\tilde{h}''}\|_{H^{2\alpha_c}(\Gamma_c)} \to 0 \ ,$$

in contradiction to

$$\|\mathring{u}_h\|_{H^{m+\alpha_c}(\Gamma_c)} + \|\mathring{\Lambda}_{\tilde{h}''}\|_{H^{2\alpha_c}(\Gamma_c)} + \|\mathring{w}_{\tilde{\tilde{h}}''}\|_{H^{2\alpha}(\Gamma)} = 1.$$

Hence, choosing $h_0 \ge 0$ appropriately, we find stability,

$$\|u_h\|_{H^{m+\alpha_c}(\Omega_1)} + \|\Lambda_{\tilde{h}}\|_{H^{2\alpha_c}(\Gamma_c)} + \|w_{\tilde{\tilde{h}}}\|_{H^{2\alpha}(\Gamma)}$$

$$\le c\left\{\|u\|_{H^{m+\alpha_c}(\Omega_1)} + \|\Lambda\|_{H^{2\alpha_c}(\Gamma_c)} + \|w\|_{H^{2\alpha}(\Gamma)}\right\}.$$

The remaining arguments are the same as in the proof of Theorem 3.4. □

4 Symmetric Galerkin coupling of BEM and FEM

As in Chapter 3 we use the FEM-BEM trial spaces $\mathcal{V}_h$ (3.2), (3.3). For the discretization, however, we now use the Galerkin method for (1.54), i. e.

Find $\mathcal{U}_h = \left(u_h,\ \Lambda_{\tilde{h}},\ w_{\tilde{\tilde{h}}}\right) \in \mathcal{V}_h = S_h^{k,r}(\Omega_1) \times S_{\tilde{h}}^{k,\tilde{r}}(\Gamma_C) \times S_{\tilde{\tilde{h}}}^{k,\tilde{\tilde{r}}}(\Gamma)$ *such that*

$$\begin{aligned}
\mathcal{A}\left(\mathcal{U}_h,\ \mathcal{Y}_h\right) &:= 2a_1\left(u_h,\ v_h\right) + \langle D_C B_C \gamma u_h,\ B_C \gamma v_h\rangle_{L_2(\Gamma_C)} - \langle A_C \Lambda_{\tilde{h}},\ \phi_{\tilde{h}}\rangle_{L_2(\Gamma_C)} \\
&\quad -\langle \Lambda_{\tilde{h}},\ B_C \gamma v_h\rangle_{L_2(\Gamma_C)} - \langle B_C \gamma u_h,\ \phi_{\tilde{h}}\rangle_{L_2(\Gamma_C)} \\
&\quad +\langle \Lambda_{\tilde{h}},\ K_C B_C \gamma v_h\rangle_{L_2(\Gamma_C)} + \langle K_C B_C \gamma u_h,\ \phi_{\tilde{h}}\rangle_{L_2(\Gamma_C)} \\
&\quad +\langle C_{12}' w_{\tilde{\tilde{h}}},\ B_C \gamma v_h\rangle_{L_2(\Gamma_C)} + \langle C_{12} w_{\tilde{\tilde{h}}},\ \phi_{\tilde{h}}\rangle_{L_2(\Gamma_C)} \\
&\quad +\langle A w_{\tilde{\tilde{h}}},\ z_{\tilde{\tilde{h}}}\rangle_{L_2(\Gamma)} + \langle C_{20} u_h,\ z_{\tilde{\tilde{h}}}\rangle_{L_2(\Gamma)} + \langle C_{21} \Lambda_{\tilde{h}},\ z_{\tilde{\tilde{h}}}\rangle_{L_2(\Gamma)} \\
&= \mathcal{L}(\mathcal{Y}_h)
\end{aligned}$$

$$\text{for all} \qquad \mathcal{Y}_h = \left(v_h,\ \phi_{\tilde{h}},\ z_{\tilde{\tilde{h}}}\right) \in \mathcal{V}_h\,. \tag{4.1}$$

This symmetric Galerkin method was already applied for electromagnetic field computations in plasma physics (see reference in [26]), in wave scattering by elastic objects in fluids [46] and in elasticity (Remark iii) below).

Theorem 4.1 [25]: *Let the assumptions of Theorem 1.4 be satisfied and let* (1.53), (1.54) *be uniquely solvable. Then under the restrictions* (3.5) *there exists* $h_o > 0$ *such that for all* $0 < h,\ \tilde{h},\ \tilde{\tilde{h}} \le h_o$ *the FEM-BEM Galerkin equations* (4.1) *are uniquely solvable and we have asymptotic convergence of optional order, i. e.* (3.6).

Proof: The FEM-BEM Galerkin equations (4.1) are equivalent to the FEM-BEM Galerkin approximation applied to (1.56). For $\mathcal{A}(\mathcal{U},\ \mathcal{Y})$ in (1.56), however, Theorem 1.5 provides the assumptions of Theorem 1.2. Moreover, for the equations (1.55) there holds the shift theorem since it holds separately for each of the equations in Ω_1 on Γ_c and on Γ if the coupling terms are skipped. Hence, Theorem 1.3 implies the proposed asymptotic error estimate (3.6). □

Remarks:

i) A different proof of Theorem 4.1 follows with arguments similar to those of the proof of Theorem 7 in [92] which provide the BBL-condition (1.4) for $\mathcal{V}_h$ if h is sufficiently small.

ii) Theorem 4.1 can also be obtained from Grannell's arguments in [40] since (4.1) is equivalent to a mixed formulation of a hybrid FEM method with global elements in Ω_2 generated by boundary potentials of the BEM basis on Γ_C .

iii) If a_1 becomes nonlinear in the interior of Ω_1 then Theorem 4.1 will probably remain valid. Corresponding coupling for nonlinear problems can be found in [62], [74], [75], [76].

iv) If the differential operators L in (1.23) is of second order, i. e. $m = 1$, then Theorem 4.1 remains valid for Lipschitz domains and substructuring as in Figure 3 due to Costabel [24], [27]. In (3.4), however, we need the restrictions $o \leq s,\ \ell,\ \tilde{s},\ \tilde{\ell} < \frac{1}{2}$.

5 Concluding Remarks

The above results are leaving open several questions. As we have seen, in the general case the convergence can only be justified for the symmetric Galerkin coupling FEM-BEM or for the more common unsymmetric formulations either if $h = o(\tilde{h})$ or $\tilde{h} = o(h)$ (see Figure 5). In practice however, most codes are using unsymmetric coupling with BEM-collocation and refinement according to $h/\tilde{h} \leq c\tilde{h}/h \leq \tilde{c}h/\tilde{h}$, the case that we cannot handle yet. It would be very interesting to see from numerical experiments whether the different refinements really improve the accuracy or whether these assumptions are only of technical nature and not significant.

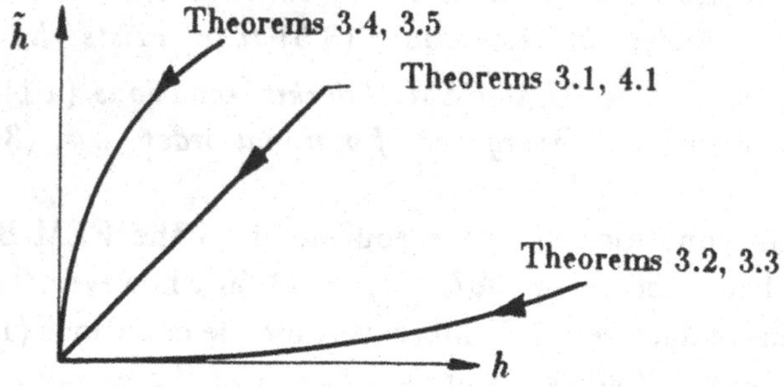

Figure 5: Justified convergence results

Also note that in practice often the FEM-BEM coupling is based on BEM collocation and on pointwise discrete coupling as, e. g. with

$$\Lambda_{\tilde{h}}(x_\ell) = (S_c \gamma u_h)(x_\ell) \ , \quad \ell = 1, \ldots , M_c$$

instead of the variational formulation in the first equation in (3.20) (see [9], [10], [11], [15], [20], [21], [34], [35], [53], [54], [56], [58], [63], [64], [65], [83], [88], [93]). For this case we have not any rigorous justification yet.

Another open question is the general justification of the FEM-BEM coupling methods when Γ_C and Γ have common endpoints, the case which is very important in substructuring and in domain-decomposition methods which play a decisive role in many applications.

References

1. Arnold D.N. and Wendland W.L. (1983), "On the asymptotic convergence of collocation methods", *Math. Comp.* 41, pp. 349-381.

2. Arnold D.N. and Wendland W.L. (1985), "The convergence of spline collocation for strongly elliptic equations on curves", *Numer. Math.* 47, pp. 317-341.

3. Atluri S.N. and Zhang J.-D. (1986), "A boundary/interior element method for quasi-static and transient response analyses of shallow shells", Preprint, Centre Adv. Comp. Mech., Civil Engrg., Georgia Tech., Atlanta GA 30332, USA.

4. Atluri S.N., Zhang J.-D. and O'Donoghue P.E. (1986), "Analysis and control of finite deformations of plates and shells: Formulations and interior boundary element algorithms", Preprint, Centre Adv. Comp. Mech., Civil Engrg., Georgia Tech., Atlanta GA 30322, USA.

5. Aubin J.P. (1972), *Elliptic Boundary Value Problems*, Wiley-Intersc. (New York).

6. Aziz A.K. and Kellog R.B. (1981), "Finite element analysis of a scattering problem", *Math. Comp.* 37, pp. 261-272.

7. Babuška I. (1970), "Error bounds for finite element method", *Numer. Math.*16, pp. 322-333.

8. Babuška I. and Aziz A.K. (1972), "Survey lectures on the mathematical foundations of the finite element method", in *The Mathematical Foundation of the*

Finite Element Method with Applications to Partial Differential Equations, Aziz A.K., ed., Academic Press (New York), pp. 3-359.

9. Beer G. and Meek J.L. (1981), "The coupling of boundary and finite element methods for infinite domain problems in elasto-plasticity", in *Boundary Element Methods III*, Brebbia C.A., ed., Springer-Verlag (Berlin, Heidelberg, New York), pp. 575-591.

10. Bettess P., Kelly D.W. and Zienkiewicz O.C. (1977), "The coupling of the finite element method and boundary solution procedures", *Int. J. Numer. Math. Engrg.* 11, pp. 355-375.

11. Bettess P., Kelly D.W. and Zienkiewicz O.C. (1979), "Marriage à la mode. The best of both worlds (Finite elements and boundary integrals)", in *Energy Methods in Finite Element Analysis*, Glowinski R., Rodin E.Y. and Zienkiewicz O.C., eds., J. Wiley (Chichester).

12. Bielak J. and MacCamy R. (1984), "Mixed variational FEM for interface problems", in *Unification of Finite Element Methods*, Kardestuncer H., ed., Elsevier North Holland (Amsterdam), pp. 149-165.

13. Brady B.H.G., Gulthard M.A. and Lemos J.V. (1984), "A hybrid distinct element-boundary element method for semi-infinite and infinite body problems", in *Computational Techniques and Applications CTAC-83*, Noye J. and Fletcher C., eds., Elsevier, North-Holland (Amsterdam), pp. 307-316.

14. Bramble J.H. and Pasciak J.E. (1987), "A boundary parametric approximation to the linearized scalar potential magnetostatic field problem", BNL 35476 Internal Report AMD 959, Cornell Univ. (Ithaca).

15. Brebbia C.A., Telles J.C.F. and Wrobel L.C. (1984), *Boundary Element Techniques*, Springer-Verlag (Berlin).

16. Brezzi F. and Johnson C. (1979), "On the coupling of boundary integral and finite element methods", *Calcolo* 16, pp. 189-201.

17. Brezzi F., Johnson C. and Nedelec J.C. (1978), "On the coupling of boundary and finite element methods", in *IVth LIBLICE Conf., Basic Problems of Numerical Analysis* (Pilsen).

18. Burnet-Fauchez M. (1987), "The use of boundary element-finite element coupling method in electrical engineering", in *Boundary Elements IX, Vol.* 3, Brebbia C.A., Wendland W.L., and Kuhn G., eds., Springer-Verlag (Berlin, Heidelberg, New York, Tokyo) pp. 611-624.

19. Céa J. (1964), "Approximation variationelle des Problèmes aux limites", *Ann. Inst. Fourier (Grenoble)*14, pp. 345-444.

20. Cen Z., Wang X. and Du Y. (1986), "A coupled finite element-boundary element method for three-dimensional elastoplasticity analysis", in *Boundary Elements*, Du Quighua, ed., Pergamon Press (Oxford) pp. 311-318.

21. Chen G.R. and Qiau J. (1986), "The coupling of BEM and FEM used in stress analysis of the structure on the half infinite foundation", in *Boundary Elements*, Du Quinghua, ed., Pergamon Press (Oxford) pp. 689-696.

22. Ciarlet P.G. (1978), *The Finite Element Method for Elliptic Problems*, North-Holland (Amsterdam, New York, Oxford).

23. Colton D. and Kress R. (1983), *Integral Equation Methods in Scattering Theory*, John Wiley & Son (New York).

24. Costabel M. (1984), "Starke Elliptiziät von Randintegraloperatoren erster Art", Habilitationschrift, Technische Hochschule Darmstadt, Germany.

25. Costabel M. (1987) "Symmetric methods for the coupling of finite elements and boundary elements", in *Boundary Elements IX, Vol.* 1, Brebbia C.A., Wendland W.L. and Kuhn G., eds., Springer-Verlag (Berlin, Heidelberg) pp. 411-420.

26. Costabel M. (1987) "Principles of boundary element methods", *Computer Physics Reports* 6, pp. 243-274.

27. Costabel M. (1985/88) "Boundary integral operators on Lipschitz domains: Elementary results", (Preprint 898 Math. TH Darmstadt), *SIAM J. Math. Anal.*, to appear.

28. Costabel M. and Stephan E.P. (1987), "A Galerkin method for a three-dimensional time-dependent interface problem in electromagnetics", in *Electromagnetisme- Magnetostatique, Problems non linéaires appliqués*, Lecture Notes, INRIA (Rocquencourt).

29. Costabel M. and Stephan E.P. (1988) "Coupling of finite elements and boundary elements for transmission problems of elastic waves in $\mathbf{R}^3$ ", in *Advanced Boundary Element Methods*, Cruse T.A., ed., Springer-Verlag (Berlin, Heidelberg, New York, Tokyo) pp. 117-124.

30. Costabel M. and Stephan E.P. (1988), "Coupling of finite elements and boundary elements for inhomogeneous transmission problems in $\mathbf{R}^3$ ", in: *The Mathematics of Finite Elements and Applications VI*, Whiteman J., ed., Academic Press (London), to appear.

31. Costabel M. and Stephan E.P. (1986/88), "Duality estimates for the numerical solution of integral equations", (Preprint 1027 TH Darmstadt), to appear.

32. Costabel M. and Wendland W.L. (1986), "Strong ellipticity of boundary integral operators", *J. Reine Angew. Math.* 372, pp. 34-62.

33. Dendrou B.A. and Dendrou S.A. (1981), "A finite element - boundary integral scheme to simulate rock-effects on the liner of an underground intersection", in *Boundary Element Methods*, Brebbia C.A., ed., Springer-Verlag (Berlin, Heidelberg, New York), pp. 592-608.

34. Du Y. and Cen Z.Z. (1986), "A hybrid or mixed approach for the combination of BEM and FEM technique", in *Proc. Int. China-American Workshop on Finite Element Methods*, (Chengdu, China).

35. Du Q., Yao Z. and Cen Z.Z. (1987), "On some coupled problems in mechanics by the coupling technique of boundary element and finite element", in *Boundary Elements IX*, Vol. 1, Brebbia C.A., Wendland W.L. and Kuhn G., eds., Springer-Verlag (Berlin, Heidelberg, New York, Tokyo) pp. 421-434.

36. Drapalik V. and Janovsky V. (1984), *On a Potential Problem with Incident Wave as a Field Source*, Tech. Report KNM-068/84, Dept. Numer. Anal. Charles University (Prague).

37. Feng K. (1983), "Finite element method and natural boundary reduction", in *Proc. Intern. Congress Math.*, August 16-24, (Warsaw), pp. 1439-1453.

38. Feng K., Yu De Hao (1983), "Canonical integral equations of elliptic boundary value problems and their numerical solutions", in *Proc. China-France Symp. On The Finite Element Method*, April 1982, Beijing, Gordon and Breach (New York), pp. 211-215.

39. Flassig U. (1984), *Das 2. Fundamentalproblem der ebenen Elastizität*, Diplom Thesis, Technical University Darmstadt (Darmstadt, West Germany).

40. Grannell J.J. (1987) "On simplified hybrid methods for coupling of finite elements and boundary elements", in *Boundary Elements IX*, Vol.1, Brebbia C.A., Wendland W.L., and Kuhn G., eds., Springer-Verlag (Berlin, Heidelberg, New York, Tokyo) pp. 447-460.

41. Han H. (1988), "A new class of variational formulations for the coupling of finite and boundary element methods", (Preprint Univ. Maryland, College Park, U. S. A.) .

42. Hebeker F.K. (1988), "A boundary element method for Stokes equations in 3-D exterior domains", in *The Mathematics of Finite Elements and Applications VI*, Whiteman J.R., ed., Academic Press (London), pp. 257-263.

43. Helfrich H.P. (1981), "Simultaneous approximation in negative norms of arbitrary order", *RAIRO Numer. Anal.* 15, pp. 231-235.

44. Herrera I. (1985), "Unified approach to numerical methods, Part II: Finite elements, boundary methods, and their coupling", *Numer. Methods Part. Diff. Equs.* 1, pp. 159-186.

45. Hildebrandt St. and Wienholtz E. (1964), "Constructive proofs of representation theorems in separable Hilbert space", *Comm. Pure Appl. Math.* 17, pp. 369-373.

46. Hsiao G.C., Kleinman R.E. and Shuetz L.S. (1988), "On variational formulations of boundary value problems for fluid-solid interactions" in: *Elastic Wave Propagation*, to appear.

47. Hsiao G.C. and Porter J.F. (1986), "A hybrid method for an exterior boundary value problem based on asymptotic expansion, boundary integral equation and finite element approximations", in *Innovative Numerical Methods in Engineering*, Shaw R.P. et al, eds., Springer-Verlag (Berlin, Heidelberg, New York, Tokyo), pp. 83-88.

48. Hsiao G.C. and Porter J.F. (1986), "The coupling of BEM and FEM for exterior boundary value problems", in *Boundary Elements*, Du Qinghua, ed., Pergamon Press (Oxford), pp. 77-86.

49. Hsiao G.C. and Porter J.F. (1988), "The coupling of BEM and FEM for the two-dimensional viscous flow problem", *Applicable Anal.* 27, pp. 79-108.

50. Hsiao G.C. and Wendland W.L. (1981), "The Aubin-Nitsche lemma for integral equations", *J. Integral Equations* 3, pp. 299-315.

51. Hsiao G.C. and Wendland W.L. (1985), "On a boundary integral method for some exterior problems in elasticity", *Procedings of Tbilisi University*, UDK 539. 3, *Math. Mech. Astron.* 257, pp. 31-60.

52. Johnson J. and Nedelec J.C. (1980), "On coupling of boundary integral and finite element methods ", *Math. Comp.* 35, pp. 1063-1079.

53. Kobayashi S. and Mori K. (1986), "Three-dimensional dynamic analysis of soil-structure interactions by boundary integral equations - finite element combined method", in *Innovative Numerical Methods in Engineering*, Shaw R.P. et al, eds., Springer Verlag (Berlin, Heidelberg, New York, Tokyo), pp. 613-618.

54. Kobayashi S., Nishimura N. and Mori K. (1986), "Applications of a boundary element-finite element combined method to three- dimensional viscoelastodynamic problems", in *Boundary Elements*, Du Qinghua, ed., Pergamon Press (Oxford), pp. 67-76.

55. Lee C.S. and Yoo Y.M. (1985), "Investigation of the boundary element method for engineering application", in *Boundary Elements VII*, Brebbia C. and Maier G., eds., Springer-Verlag (Berlin, Heidelberg, New York, Tokyo), pp. 14.93-14.107.

56. Li Hong-Bao, Han Guo-Ming, Mang H.A. and Torzicky P.A. (1986), "A new method for the coupling of finite element and boundary element discretized subdomains of elastic bodies", *Comput. Methods Appl. Mech. Engrg.* 54, pp. 161-185.

57. Lions J.L. and Magenes E. (1972), *Non-Homogeneous Boundary Value Problems and Applications Vol. I*, Springer-Verlag (Berlin, Heidelberg, New York).

58. Liu G. and Wang G. (1986), "A boundary element-finite element method of analysis for viscoelastic stress in tunnels", in *Boundary Elements*, Du Qinghua, ed., Pergamon Press (Oxford) pp. 593-600.

59. MacCamy R.C. and Marin S.P. (1980), "A finite element method for exterior interface problems", *Internat. J. Math and Math. Sci.* 3, pp. 311-350.

60. MacCamy R.C. and Suri M. (1987), "A time dependent interface problem for two-dimensional eddy currents", *Quart. Appl. Math.* 44, pp. 675-690.

61. Mackerle J. and Andersson T. (1984), "Boundary element software in engineering", *Advances in Eng. Software* 6, pp. 66-102.

62. Maier G. and Polizzotto C. (1987), "A Galerkin approach to boundary element elastoplastic analysis", *Comput. Methods Appl. Mech. Engrg.* 60, pp. 175-194.

63. Mang H.A., Chen Z.Y. and Torzicky P. (1988), "On the symmetricability of coupling matrices for BEM-FEM discretizations of solids", in *Advanced Boundary Element Methods*, Cruse T.A., ed., Springer-Verlag (Berlin, Heidelberg) pp. 241-248.

64. Mansur W.J. and Marques E. (1988), "Coupling of boundary and finite element methods: application to potential problems", *Engrg. Analysis*, to appear.

65. Margulies M. (1981), "Combination of the boundary element and finite element methods", in *Progress in Boundary Element Methods* 1 , Brebbia C.A., ed., Pentech Press (London, Plymouth), pp. 258-288.

66. Mikhlin S.G. and Prössdorf S. (1986), *Singular Integral Operators*, Springer-Verlag (Berlin, Heidelberg, New York, Tokyo).

67. Mikhlin S.G. (1965), *The Problem of the Minimum of a Quadratic Functional*, Holden Day Inc. (San Fransisco)(Orig. 1952, Leningrad Moscow).

68. Nedelec J.C. (1976), Curved finite element methods for the solution of singular integral equations on surfaces in $\mathbf{R}^3$ ", *Comput. Methods Appl. Mech. Engrg.* 8, pp. 61-80.

69. Nedelec J.C. (1977), *Approximation des Equations Intégrales en Mécanique et en Physique*, Lecture Notes, Centre de Mathématiques Appliquées, Ecole Polytechnique, Palaiseau, France.

70. Nitsche J.A. (1968), "Ein Kriterium für die Quasi-Optimalität des Ritzschen Verfahrens", *Numer. Math.* 11, pp. 346-348.

71. Nitsche J.A. (1968), "Zur Konvergenz des Ritzschen Verfahrens und der Fehlerquadratmethode I", *Internat. Schriftenreihe Numer. Math.* 8, pp. 97-103.

72. Ohtsu M. (1985), "BEM formulation based on the variational principle and coupling analysis with FEM", in *Boundary Elements VII*, Brebbia C. and Maier G., eds., Springer-Verlag (Berlin, Heidelberg, New York, Tokyo), pp. 11.13.-11.22.

73. Petersen B.E. (1983), *Introduction to the Fourier Transform and Pseudo-Differential Operators*, Pitman (Boston, London, Melbourne).

74. Polizzotto C. (1987), "A symmetric-definite BEM formulation for the elasto-plastic rate problem", in *Boundary Elements IX*, Vol 2, Brebbia C.A., Wendland W.L. and Kuhn G., eds., Springer-Verlag (Berlin, Heidelberg, New York, Tokyo) pp. 315-334.

75. Polizzotto, C. (1988), "A consistent formulation of the BEM within elastoplasticity" in *Advanced Boundary Element Methods*, Cruse T.A., ed., Springer-Verlag (Berlin, Heidelberg, New York, Tokyo), pp. 315-324.

76. Polizzotto C. (1988), "An energy approach to the boundary element method. Part I: Elastic solids; Part II: Elastic-plastic solids. To appear.

77. Protopsaltis B. (1985), *Rand-Integralgleichungen für dünne elastische Platten und ihre Kopplung mit finiten Elementen*, Doctoral Thesis, Technical University Munich (München, West Germany).

78. Riesz F. and Sz.- Nagy B. (1956), *Vorlesungen über Funktionalanalysis*, Dt. Verlag d. Wiss. (Berlin).

79. Schmidt G. (1984), "The convergence of Galerkin and collocation methods with splines for pseudodifferential equations on closed curves", *Z. Anal. Anwendungen* 3, pp. 371-384.

80. Schnack E. (1985), "Stress analysis with a combination of HSM and BEM", *The Mathematics of Finite Elements and Applications V*, Whiteman J., ed., Academic Press (London), pp. 273-282.

81. Sequeira A. (1983), "The coupling of boundary integral and finite element methods for the bidimensional exterior steady Stokes problem", *Math. Methods Appl. Sci.* 5, pp. 356-375.

82. Sequeira A. (1986), "On the computer implementation of a coupled boundary and finite element method for the bidimensional exterior steady Stokes problem", *Math. Methods Appl. Sci.* 8, pp. 117-133.

83. Shao K. (1985), "Axi-symmetric eddy current problems by a coupled FE-BE method", *Boundary Elements VII*, Brebbia C. and Maier G., eds., Springer-Verlag (Berlin, Heidelberg, New York, Tokyo), pp. 2.65-2.76.

84. Shaw R. and Falby W. (1978), "FEBIE - a combined finite element - boundary integral equation approach", *Comp. Fluids* 6, pp. 153-160.

85. Stephan E. and Wendland W.L. (1976), "Remarks to Galerkin and least squares methods with finite elements for general elliptic problems", *Manuscripta Geodaetica* 1, pp. 93-123.

86. Stummel F. (1969), *Rand- und Eigenwertaufgaben in Sobolevschen Räumen*, Lecture Notes in Math. 102, Springer-Verlag (Berlin, Heidelberg, New York).

87. Stummel F. (1971), "Diskrete Konvergenz linearer Operatoren I und II", *Math. Zeitschr.* 120, pp. 231-264.

88. Swoboda G.A., Mertz W.G. and Beer G. (1986), "Application of coupled FEM-BEM analysis for three-dimensional tunnel analysis", in *Boundary Elements*, du Qinghua, ed., Pergamon Press (Oxford), pp. 537-550.

89. Taylor M.E. (1981)*Pseudodifferential Operators*, Princeton Univ. Press (Princeton, N.J.).

90. Wendland W.L. (1983), "Boundary element methods and their asymptotic convergence", in *Theoretical Acoustics and Numerical Treatment*, Filippi P., ed., CISM Courses and Lectures No. 277, Springer-Verlag (Wien, New York). pp. 135-216.

91. Wendland W.L. (1986), "On asymptotic error estimates for the combined boundary and finite element method", in *Innovative Numerical Methods in Engineering*, Shaw R. et al., eds., Springer-Verlag (Berlin, Heidelberg, New York, Tokyo), pp. 55-70.

92. Wendland W.L. (1987), "Strongly elliptic boundary integral equations", in *The State of the Art in Numerical Analysis*, Iserles A. and Powell M., eds., Clarendon Press (Oxford), pp. 511-561.

93. Wu J.C. (1984), "Fundamental solutions and numerical methods in fluids", *Int. J. Numer. Methods Fluids* 4, pp. 185-201.

94. Yu D.-H. (1983), "Coupling canonical boundary element method with FEM to solve a harmonic problem over a cracked domain", *J. Comp. Math.* 1, pp. 195-202.

95. Yu D.-H. (1984), "Canonical boundary element method for plane elasticity problems", *J. Comp. Math.* 2, pp. 180-189.

96. Yu D.-H. (1985), "A system of plane elasticity canonical integral equations and its application", *J. Comp. Math.* 3, pp. 219 - 227.

97. Zi-Cai Li and Guo-Ping Liang (1983), "On the simplified hybrid-combined method", *Math. Comp.* 41, pp. 13-25.

Printed in the United States
By Bookmasters